Fishes and Crustaceans: Biology and Ecology in a Changing Marine Environment

Fishes and Crustaceans: Biology and Ecology in a Changing Marine Environment

Guest Editors

Sabrina Colella
Giorgia Gioacchini

MDPI

Basel • Beijing • Wuhan • Barcelona • Belgrade • Novi Sad • Cluj • Manchester

Guest Editors

Sabrina Colella
Institute for Marine Biological Resources and Biotechnology (IRBIM)
National Research Council (CNR)
Ancona
Italy

Giorgia Gioacchini
Department of Life and Environmental Sciences
Università Politecnica delle Marche
Ancona
Italy

Editorial Office
MDPI AG
Grosspeteranlage 5
4052 Basel, Switzerland

This is a reprint of the Special Issue, published open access by the journal *Animals* (ISSN 2076-2615), freely accessible at: https://www.mdpi.com/journal/animals/special_issues/246151LHJ5.

For citation purposes, cite each article independently as indicated on the article page online and as indicated below:

Lastname, A.A.; Lastname, B.B. Article Title. *Journal Name* **Year**, *Volume Number*, Page Range.

ISBN 978-3-7258-6049-4 (Hbk)
ISBN 978-3-7258-6050-0 (PDF)
https://doi.org/10.3390/books978-3-7258-6050-0

Cover image courtesy of Martina Scanu and Giulia Chemello

Contents

About the Editors vii

Preface ix

Ninglu Zhang, Rui Yang, Zhengyi Fu, Gang Yu and Zhenhua Ma
Mechanisms of Digestive Enzyme Response to Acute Salinity Stress in Juvenile Yellowfin Tuna (*Thunnus albacares*)
Reprinted from: *Animals* **2023**, *13*, 3454, https://doi.org/10.3390/ani13223454 1

Alexander G. Dvoretsky, Fatima A. Bichkaeva, Nina F. Baranova and Vladimir G. Dvoretsky
Fatty Acids in the Eggs of Red King Crabs from the Barents Sea
Reprinted from: *Animals* **2024**, *14*, 348, https://doi.org/10.3390/ani14020348 17

Martina Scanu, Carlo Froglia, Fabio Grati and Luca Bolognini
Estimate of Growth Parameters of *Penaeus kerathurus* (Forskäl, 1775) (Crustacea, Penaeidae) in the Northern Adriatic Sea
Reprinted from: *Animals* **2024**, *14*, 1068, https://doi.org/10.3390/ani14071068 30

Laura Ciaralli, Tommaso Valente, Eleonora Monfardini, Giovanni Libralato, Loredana Manfra, Daniela Berto, et al.
Rose or Red, but Still under Threat: Comparing Microplastics Ingestion between Two Sympatric Marine Crustacean Species (*Aristaeomorpha foliacea* and *Parapenaeus longirostris*)
Reprinted from: *Animals* **2024**, *14*, 2212, https://doi.org/10.3390/ani14152212 44

Qiang Ma, Renxiao Zhang, Yuliang Wei, Mengqing Liang and Houguo Xu
Effects of Intermittent and Chronic Hypoxia on Fish Size and Nutrient Metabolism in Tiger Puffer (*Takifugu rubripes*)
Reprinted from: *Animals* **2024**, *14*, 2470, https://doi.org/10.3390/ani14172470 65

Luca Caracausi, Zaira Da Ros, Alice Premici, Enrico Gennari and Emanuela Fanelli
Trophic Ecology of the Pyjama Shark *Poroderma africanum* (Gmelin, 1789) Elucidated by Stable Isotopes
Reprinted from: *Animals* **2024**, *14*, 2559, https://doi.org/10.3390/ani14172559 80

Domen Trkov, Danijel Ivajnšič, Marcelo Kovačić and Lovrenc Lipej
Diet of Three Cryptobenthic Clingfish Species and the Factors Influencing It
Reprinted from: *Animals* **2024**, *14*, 2835, https://doi.org/10.3390/ani14192835 92

Monica Panfili, Stefano Guicciardi o Guizzardi, Emanuela Frapiccini, Cristina Truzzi, Federico Girolametti, Mauro Marini, et al.
Influence of Contaminants Mercury and PAHs on Somatic Indexes of the European Hake (*Merluccius merluccius*, L. 1758)
Reprinted from: *Animals* **2024**, *14*, 2938, https://doi.org/10.3390/ani14202938 114

Alexander G. Dvoretsky and Vladimir G. Dvoretsky
Stock Dynamics of Female Red King Crab in a Small Bay of the Barents Sea in Relation to Environmental Factors
Reprinted from: *Animals* **2025**, *15*, 99, https://doi.org/10.3390/ani15010099 136

About the Editors

Sabrina Colella

Sabrina Colella studied Biological Sciences at the University of Urbino. Her collaboration with the National Research Council (CNR) began during the 1990s, and she is currently a Researcher at the Institute for Marine Biological Resources and Biotechnology (IRBIM), Ancona Office, Italy. The research activity carried out at the CNR, allowed her to gain experience on ecology, biology, and population dynamics of fishery resources.

In particular, her research is mainly focused on following topics: growth via age determination and estimations of growth rates by otolith macro-increment counts in adult specimens of demersal species; the reproductive biology of fishes and crustaceans through multidisciplinary approaches; pollutants and lipids in relation to several aspects of reproductive cycles.

She was member of several working groups of the FAO-ADRIAMED Project, International Council for the Exploration of the Sea (ICES), and General Commission for Mediterranean Fisheries (GFCM), examining topics related to age determination and growth, recreational fishing, and reproductive studies. Colella has also been involved in several research projects funded by the FAO, the European Union, Italian Ministries and Local Administrations. Since 2002 she has been involved in the EU Data Collection Framework (DCF-PLNRDA) in relation to the sampling of biological data aimed to stock assessments of small pelagics and demersal species, (GSAs 17 and18). Currently she is the supervisor of biological data collection for demersal species and the contact manager for GSA 17 fishing companies within the PLNRDA EU Data Collection Framework for Italian fisheries.

Giorgia Gioacchini

Giorgia Gioacchini is an Associate Professor at the Polytechnic University of Marche (Ancona, Italy), where she teaches Cell Biology and Reproductive Biology of Marine Vertebrates. Her research focuses on the reproduction, development, and welfare of marine organisms, with a particular emphasis on the impacts of environmental change and pollutants, including microplastics, on the reproductive biology, embryonic development, and health status of key Mediterranean species such as bluefin tuna, swordfish, sardine, sharks, *Sepia officinalis*, and the loggerhead turtle *Caretta caretta*. She applies a multidisciplinary approach that integrates histology, immunohistochemistry, molecular biology, genomics, transcriptomics, enzymatic assays, and spectroscopic analyses. She has authored over 100 peer-reviewed papers and several book chapters. She has coordinated and participated in numerous national and international research projects, including ICCAT programs on swordfish biology and genetics, and initiatives on aquaculture sustainability and microplastic biomonitoring. She is actively engaged in academic teaching, supervision, and public outreach and serves on national and international scientific societies such as GEI-SIBSC, ISFE, ESCE, and ISFRP.

Preface

This Special Issue brings together a collection of original research articles dedicated to the biology and ecology of fishes and crustaceans in the context of rapidly changing marine environments. The contributions explore a variety of themes, including reproductive biology, growth dynamics, physiological and hematological responses, and the effects of environmental stressors and human activities on marine and coastal species. By focusing on species of both ecological and economic importance, these studies provide updated insights into how aquatic organisms respond and adapt to dynamic conditions, with implications for aquaculture development, fishery management, and biodiversity conservation. Collectively, this Special Issue highlights the role of early life stages, physiological markers, and ecological indicators as powerful tools to assess population health, resilience, and sustainability. The evidence presented here offers valuable perspectives for researchers, practitioners, and policymakers seeking to balance resource use with the protection of marine ecosystems.

Sabrina Colella and Giorgia Gioacchini
Guest Editors

animals

MDPI

Article

Mechanisms of Digestive Enzyme Response to Acute Salinity Stress in Juvenile Yellowfin Tuna (*Thunnus albacares*)

Ninglu Zhang [1,2,†], Rui Yang [1,2,†], Zhengyi Fu [1,2,3], Gang Yu [1,2] and Zhenhua Ma [1,2,3,*]

1 Key Laboratory of Efficient Utilization and Processing of Marine Fishery Resources of Hainan Province, Sanya Tropical Fisheries Research Institute, Sanya 572018, China; ningluz@163.com (N.Z.); janeyhn4321@yeah.net (R.Y.); zhengyifu@163.com (Z.F.); gyu0928@163.com (G.Y.)
2 South China Sea Fisheries Research Institute, Chinese Academy of Fishery Sciences, Guangzhou 510300, China
3 College of Science and Engineering, Flinders University, Adelaide 5001, Australia
* Correspondence: zhenhua.ma@scsfri.ac.cn
† These authors contributed equally to this work.

Simple Summary: Regarding a better understanding of the process of changes in the digestive physiological state of yellowfin tuna (*Thunnus albacares*) and the distribution of digestive enzymes, it will provide data to support common problems during yellowfin tuna culture. For yellowfin tuna, the digestive state affects the measure of yellowfin tuna's physical health and plays a crucial relationship to its ontogeny in terms of nutrition and immune regulation. However, there are fewer studies on the digestive physiological state of farmed yellowfin tuna in China. In the present study, a control salinity of 32‰ and an experimental salinity of 29‰ in natural seawater were treated for 48 h under abrupt salinity change to identify the digestive enzyme activities in different tissues (stomach, foregut, and pyloric cecum) at different times (0 h, 12 h, 24 h, 48 h). The results of the study will provide data to support the aquaculture process of juvenile yellowfin tuna.

Abstract: This study investigates the effect of a sudden change in salinity for 48 h on the digestive enzyme activity of juvenile yellowfin tuna. The treatment included a control salinity of 32‰ in natural seawater and an experimental salinity of 29‰. Acute stress experiments were carried out on 72 juvenile yellowfin tuna (646.52 ± 66.32 g) for 48 h to determine changes in digestive enzyme activity in different intestinal sections over time (0 h, 12 h, 24 h, 48 h). The activities of pepsin, trypsin, α-amylase, lipase, and chymotrypsin in the digestive organs (stomach, foregut, and pyloric ceca) of juvenile yellowfin tuna were measured. Pepsin and pancreatic protease in the experimental group were significantly lower than in the control group ($p < 0.05$). α-amylase showed a fluctuating trend of decreasing and then increasing, and its activity trend was pyloric ceca > foregut > stomach. The lipase activity of gastric tissues decreased at the beginning and then increased, reaching a minimum at 24 h (2.74 ± 1.99 U·g protein^{-1}). The change of lipase in the pyloric ceca and foregut was increasing and then decreasing. The lipase activity trend was pyloric ceca > foregut > stomach. The chymotrypsin showed a decreasing and increasing trend and then stabilized at 48 h with a pattern of pyloric ceca > foregut > stomach. Similarly, the gut villi morphology was not significantly altered in the acutely salinity-stressed compared to the non-salinity-stressed. This study suggests that salinity may change the digestive function of juvenile yellowfin tuna, thereby affecting fish feeding, growth, and development. On the contrary, yellowfin tuna is highly adapted to 29‰ salinity. However, excessive stress may negatively affect digestive enzyme activity and reduce fish digestibility. This study may provide a scientific basis for a coastal aquaculture water environment for yellowfin tuna farming, which may guide the development and cultivation of aquaculture.

Keywords: fish; gut; acute hyposalinity stress; digestibility; intestinal morphology

1. Introduction

Thunnus albacares belongs to the order Perciformes, Mackerelidae, Tuna, commonly to referred as yellowfin tuna [1]. Yellowfin tuna is a high-speed swimming fish with pelagic migratory behavior, it is found mainly in the tropical and subtropical waters of the Pacific, Indian, and Atlantic Oceans, and in China is found in the South China Sea, East China Sea and off the coast of Taiwan [2,3]. As an economically valued tuna species, the global annual catch of yellowfin tuna exceeds 1.4 million tons until 2022, making it the second most harvested tuna species in the world, after skipjack tuna (International Seafood Sustainability Foundation 2022). Yellowfin tuna is a fast-growing species of tuna with high flesh quality [4]. It is the species of choice for offshore aquaculture [1,5,6]. Currently, research on yellowfin tuna is focused on biology [7,8], stock and fisheries research [9,10], food nutrient analysis [11,12] evaluation [13,14], and growth and culture. Artificial farming of yellowfin tuna in China is still in its infancy. The Deep Seawater Aquaculture Technology and Species Development Innovation Team of the Chinese Academy of Fisheries Sciences has realized indoor recirculating water and offshore deep-water net tank culture of yellowfin tuna in Lingshui Li Autonomous County, Hainan Province. The team has long been dedicated to research on the culture biology and disease control of yellowfin tuna and has made important progress in indoor recirculating water and offshore deep-water net tank culture [15,16].

Salinity is an important environmental condition for fish life and is one of the most critical factors affecting the physiological activity of fish [17]. Salinity affects the structure, digestive enzyme activity, and physiological function of the corresponding tissues and organs in the fish digestive tract, which in turn affects the digestion and absorption of food, ultimately affecting the growth and development of fish and disrupting their normal physiological and behavioral activities [18–20]. Different fish have different abilities and ranges of adaptation to salinity. Generally speaking, fish have the lowest metabolic rates and highest growth rates in environments close to the appropriate salinity for their long-term life. When salinities are too high or too low, fish use more energy for osmotic pressure regulation, resulting in reduced activity of their digestive enzymes, reduced muscle quality, and compromised growth and survival rates. Fluctuations in salinity are also evident in the coastal areas of China. Due to the low-lying topography of China's coastal areas, they are susceptible to weather extremes, such as typhoons and hurricanes. In turn, these weather extremes may trigger floods and tides, leading to fluctuations in salinity. In addition, river injection may also be a cause of salinity fluctuations which poses a threat to cage and land-based farming that rely on naturally filtered seawater. A low salinity of 28.5‰ has been observed off the coast of Hainan, China [20], but the optimal salinity for yellowfin tuna growth is 31.2 to 33.3‰ [21]. Changes in fish intestinal viability and histomorphology are receiving increasing attention for the long-term purpose of sustainable aquaculture production systems [22]. Limited at this stage to the development of aquaculture-built ecosystem states, fish intestinal enzyme activity and gut status can be inferred from aspects of fish feeding, digestion, energy balance, and health [23], but links to tuna are still lacking. Gut viability and morphology may vary between fish populations [24], so specific information is needed, particularly during changes in the culture environment and fish rearing.

Pepsin is a digestive protease secreted by the master cell of the gastric mucosa in the stomach. Its function is to break down proteins in food into small peptide fragments. Pepsin is not produced directly by the cell. The master cell secretes pepsinogen, which is stimulated by gastric acid or pepsin to form pepsin [25]. This is a protective mechanism that prevents pepsin from digesting its own proteins within the cells. Trypsin is a serine protein hydrolase and is one of the most widely used proteases available [26,27]. Derived from the hepatopancreas, intestine, and blind pyloric sac of fish, trypsin is the main endogenous protein hydrolase of the fish gut [28], which selectively hydrolyses peptide bonds in proteins made up of the carboxyl groups of lysine or arginine. In vertebrates, it functions as a digestive enzyme [29], which breaks down proteins in food and promotes nutrient absorption. In the pancreas, trypsin is synthesized as precursor trypsinogen and then secreted as a component of pancreatic juice into the duodenum, where it is activated by enterokinase or self-catalysis. Amylase, another indicator of an animal's digestive capacity, can be divided into α-amylase and β-amylase depending on the configuration of the hydrolysis product. α-amylase can exist in various extreme environments and is often found in the digestive system of aquatic animals [30,31]. Lipases, also known as triacylglycerol acyl hydrolases, catalyze the hydrolysis of triglycerides at the oil-water interface to produce the corresponding fatty acids and glycerol [32,33]. Chymotrypsin is an endoprotease secreted by the pyloric caeca that specifically hydrolyses peptide bonds formed by carboxyl groups of aromatic amino acids or amino acids with large non-polar side chains [34].

Temporary environmental fluctuations may pose an unknown challenge to yellowfin tuna larvae in artificial breeding experiments. We aim to study the changes in digestive enzyme activity with salinity during the growth of juvenile tuna and explore the adaptability of the digestive enzyme system of juvenile tuna to salinity, to accumulate experimental data and research materials for the breeding, ecophysiology, and domestication of yellowfin tuna due to low salinity water.

2. Materials and Methods

2.1. Experimental Methods and Design

The Lingshui Research Station provided the juvenile yellowfin tuna, Tropical Aquatic Research and Center, South China Sea Fisheries Research Institute, Chinese Academy of Fisheries Sciences, China. The body length is 28.97 ± 2.17 cm, and the body weight is 646.52 ± 66.32 g. Seventy-two juvenile yellowfin tuna were randomly placed in six 5000 L fiberglass tanks equipped with a recirculating filtered seawater system for a 7-day period of domestication. Fresh miscellaneous fish pieces (4 cm × 2 cm) were fed daily from 08:30 to 09:00 at 5–8% of body weight per day. No feeding was provided the day before and during the experiment. The experimental installation was divided into a fiberglass water tank, a supply tank, a filtration basin, a circulating water control system, and an oxygen pump. To maintain the water temperature, salinity, and light control, automatic water changes were carried out through the electrical control system. During the experiment, each fiberglass tank was supplied with filtered seawater at a water exchange rate of 300% of the daily tank capacity. Each experiment was replicated three times with 32‰ as the control group and 29‰ as the pressure group (Figure 1).

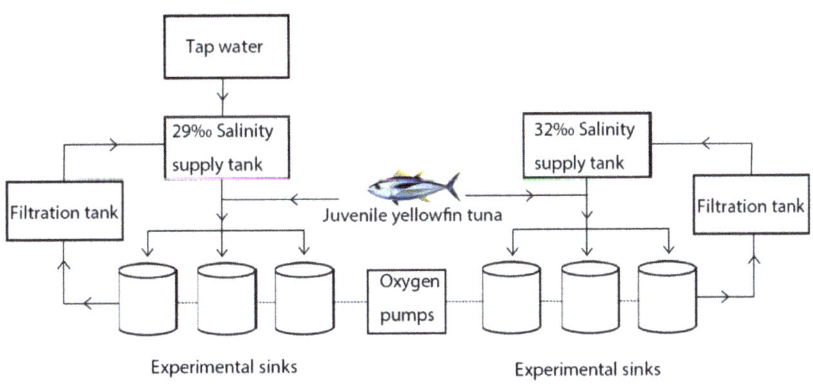

Figure 1. Experimental design of acute hyposalinity on digestive enzyme activity and intestinal tissue sectioning in juvenile yellowfin tuna (*Thunnus albacares*).

The salinity of the stress group was adjusted gradually, at a rate of 1‰ to prevent instability of salinity caused by a rapid rate therefore, naturally filtered seawater was added to tap water that had been aerated for 24 h. The experiment was initiated when the salinity of the water in the stress group reached 29‰. The experiment was carried out over a period of 48 h. Illumination time was maintained at 14:10 h (light:dark). Salinity, water temperature, dissolved oxygen (DO), and pH were monitored using the HQ40d multi-parameter instrument (HQ40d18, Hach, Loveland, CO, USA), and $NaNO_2$ and NH_3-N were monitored using the Chinese biotechnology (Zhecheng Biotechnology, Beijing, China). During the experiment, the temperature of the water is maintained at 29.5 ± 0.5 °C, DO > 7.50 mg·L^{-1}, pH 7.93 ± 0.12, $NaNO_2$ < 0.1 mg·L^{-1} and NH_3-N < 0.05 mg·L^{-1}.

2.2. Analytical Method

Samples were collected from each experimental group at 0 h, 6 h, 24 h, and 48 h of the experiment. Three juvenile yellowfin tuna were randomly collected from the experimental and control groups at each time point for digestive enzyme activity assay and histological analysis. Fish were anesthetized using 200 mg·L^{-1} MS-222 and dissected on ice trays. The stomach, pyloric ceca, and foregut samples were quickly removed, rinsed with pre-cooled saline, blotted on filter paper, placed in frozen centrifuge tubes, and stored at −80 °C. Samples were partially removed prior to the assay, thawed in a refrigerator at 0–4 °C, 0.1–0.2 g of the tissue sample were weighed, then pre-chilled saline was added at 9 times the volume and homogenized using a glass homogenizer on ice with 0.2 M NaCl and centrifuged (0–4 °C, 2500 r·min^{-1}, 10 min). Incubate supernatant in enzyme substrate and the digested enzyme activity was read on a spectrophotometer (Synergy H1, BioTek Instruments, Winooski, VT, USA). The triplicate method is used for each data item. Juvenile yellowfin tuna foregut was collected and fixed in 4% paraformaldehyde (500 ML, BL 539A, Biosharp, Hefei, China). The fixed tissue was embedded in paraffin blocks using a Leica RM 2016 rotary slicer (Shanghai Leica Instrument Co., Ltd., Shanghai, China) and cut into a series of cross sections (4 μm thick). A histological analysis was carried out using hematoxylin-eosin (HE) staining. Each slide with tissue sections was permanently fixed with a neutral ball. Sections were observed, photographed, and preserved using an inverted biomicroscope (DMI8, Leica, Wetzlar, Germany).

A protein quantification kit (catalog No: A045-4, built in Nanjing, China) was used to determine the protein content by the Thomas Brilliant Blue method using bovine serum protein as a standard. The protein concentration by the microplate colorimetric method was incubated at 37 °C for 30 min at 562 nm. The Pepsin Assay Kit (Catalogue No: A080-1-1, built in Nanjing, China) was used to determine the amount of pepsin in gastric tissue. Pepsin hydrolyzes protein to produce phenol-containing amino acids, and the phenol-containing amino acids reduce the phenol reagent to a blue substance by colorimetric incubation at 37 °C for 20 min at 660 nm and the absorbance is measured to calculate its activity. An activity unit is defined as 1 μg of tyrosine per mg of histone at 37 °C, enzymatic minute of proteolysis produces 1 μg of tyrosine which is equivalent to 1 enzyme activity unit (1 enzyme activity unit = 1 μg of tyrosine·min·mg^{-1} of histone). Trypsin activity in tissues was determined with a kit (Catalogue No: A080-2, built in Nanjing, China). Trypsin catalyzes the hydrolysis of the ester chain of the substrate ethyl arginate, causing an increase in its absorbance value at 253 nm, and the enzyme activity can be calculated from the change in absorbance. By colorimetric method, the absorbance values were measured, and the activity was calculated by adjusting distilled water to zero using a UV spectrophotometer at 253 nm using a 0.5 cm quartz cuvette. A unit of enzyme activity is defined as a change in absorbance of 0.003 per minute of trypsin per mg of protein at pH 8.0, 37 °C. The α-amylase Assay Kit (Catalogue No: C016-1-1, built in Nanjing, China) was used to determine the activity of α-amylase in fish visceral tissues. α-amylase hydrolyses starch to produce glucose, maltose, and dextrin. The amount of starch hydrolyzed can be deduced from the shade of blue, and the activity of α-amylase can be calculated by adding iodine solution to the unhydrolyzed starch to produce a blue complex if the substrate concentration is known and in excess. The absorbance values were measured, and the activity was calculated by a colorimetric method using a 1 cm optical diameter colorimetric cup adjusted to zero with distilled water at 660 nm. The unit of activity was 1 unit of amylase activity equal to 1 mg of protein in the tissue hydrolyzing 10 mg of starch at 37 °C for 30 min with the substrate. The Lipase Assay Kit (Catalogue No: A054-1-1, built in Nanjing, China) was used to determine the lipase activity in fish visceral tissues. The emulsion made from triglycerides and water has an emulsified nature due to the absorption and scattering of incident light by its micelles. The triglycerides in the micelles are hydrolyzed by the action of lipase, causing the micelles to split and the scattered light or turbidity to be reduced. As a result, the rate of reduction is related to the lipase activity. This was determined by a colorimetric method in which a spectrophotometer is metered at 420 nm, a 1 cm optical diameter glass cuvette zeroed with the Tris buffer allowed the uptake of the tissue to be measured and its viability was calculated. The unit of activity is one unit of enzyme activity for each 1 µmol of substrate consumed per gram of histone that reacted with the substrate in this reaction system for 1 min at 37 °C. The Chymotrypsin Assay Kit (Catalogue No: A080-3-1, built in Nanjing, China) was used for the determination of chymotrypsin activity in animal tissues. The chymotrypsin assay uses casein as a substrate. Chymotrypsin hydrolyses the protein to produce phenol-containing amino acids, and the phenol reagent is reduced to a blue substance by the phenol-containing amino acids. The chymotrypsin activity can be determined using colorimetry. Incubate for 20 min at 37 °C, 660 nm, with a 1 cm optical diameter cuvette, zeroed with distilled water, colorimetric, and calculate the activity. Activity units are equivalent to 1 enzyme activity unit per mg of histone protein per minute of protein breakdown at 37 °C to produce 1 µg of amino acids.

2.3. Statistical Analysis

SPSS 26.0 statistical software was used for statistical analysis of all data, and differences in enzyme activity between groups were compared between control and stress groups using a two-way ANOVA. Differences were considered significant when $p < 0.05$ and not significant when $p > 0.05$. Plots were made using Origin software (2019 edition). Data results were expressed as mean ± standard deviation (mean ± SD).

3. Results

3.1. Effect of Salinity Changes on Pepsin and Trypsin Activity

Under stress, pepsin activity increased with time from 0 h to 48 h (Figure 2a). There was no significant difference between 0 h and 48 h ($p > 0.05$), and the maximum value occurred at 24 h (14.208 ± 2.774 U·mg protein^{-1}), followed by a decrease to a minimum value (10.887 ± 1.440 U·mg protein^{-1}) at the 48 h. Pepsin activity in fish from the stress group was significantly lower than that of the control group at 6 and 48 h ($p < 0.05$).

In the stress group, there was a slight fluctuation in trypsin activity in the pylorus of juvenile yellowfin tuna between 0 h to 48 h as the duration of stress increased (Figure 2b), showing a trend of increasing (35.8%) and then decreasing (−24.3%), reaching a maximum value (2357.97 ± 602.61 U·mg protein^{-1}) at 24 h. There was no significant difference within groups between sampling times and within the same sampling time. In addition, there was no significant difference between different sampling times, versus within the same sampling period, between the control and stress groups ($p > 0.05$).

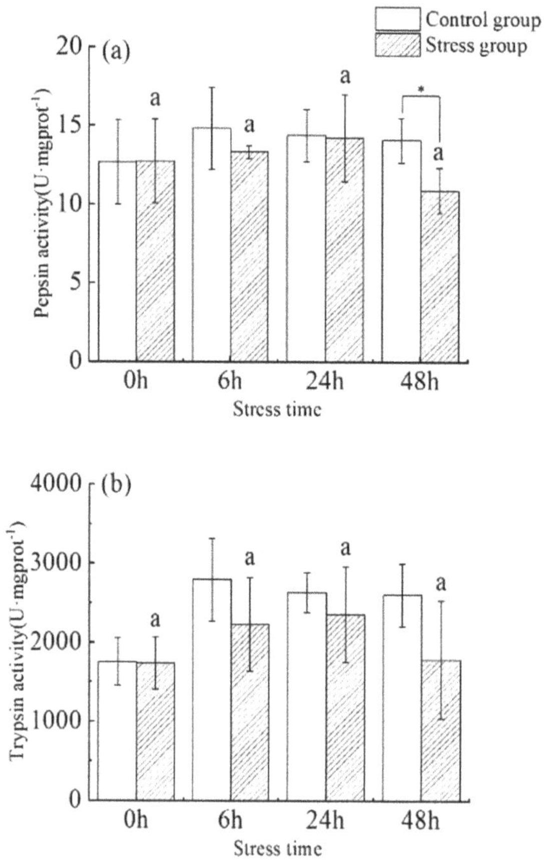

Figure 2. Effect of acute hyposalinity stress on pepsin and trypsin activity in juvenile yellowfin tuna (*Thunnus albacares*), ($n = 9$). (**a**) Pepsin activity, (**b**) trypsin activity. At the same salinity, different letters at different time points indicate a significant difference ($p < 0.05$). Different letters indicate differences in experimental groups at different times, and * indicates differences between the experimental or control groups.

3.2. *Effect of Salinity Changes on α-Amylase Activity*

As shown in Figure 3, in the stomach tissue (Figure 3a) of juvenile yellowfin tuna, there was a slight change in α−amylase activity in the stress group from 0 h to 48 h with increasing stress time, but there was no significant difference between the control and stress groups at different sampling times and the same sampling time ($p > 0.05$). In the foregut tissue (Figure 3b) of juvenile yellowfin tuna, α−amylase activity was significantly higher in the stress group from 6 h to 48 h than in the control group ($p < 0.05$) and was not significantly different ($p > 0.05$), reaching a maximum at 24 h (1.25 ± 0.22 U·mg protein^{-1}). In the pyloric ceca (Figure 3c) tissue of juvenile yellowfin tuna, the α-amylase activity of the stress group was significantly higher than that of the control group from 24 h to 48 h ($p < 0.05$), and the α-amylase activity of the stress group showed a trend of decreasing and then increasing compared with that of the control group. After 48 h of acute salinity stress, the α-amylase activity of each digestive organ was ranked as follows: pyloric ceca > foregut > stomach.

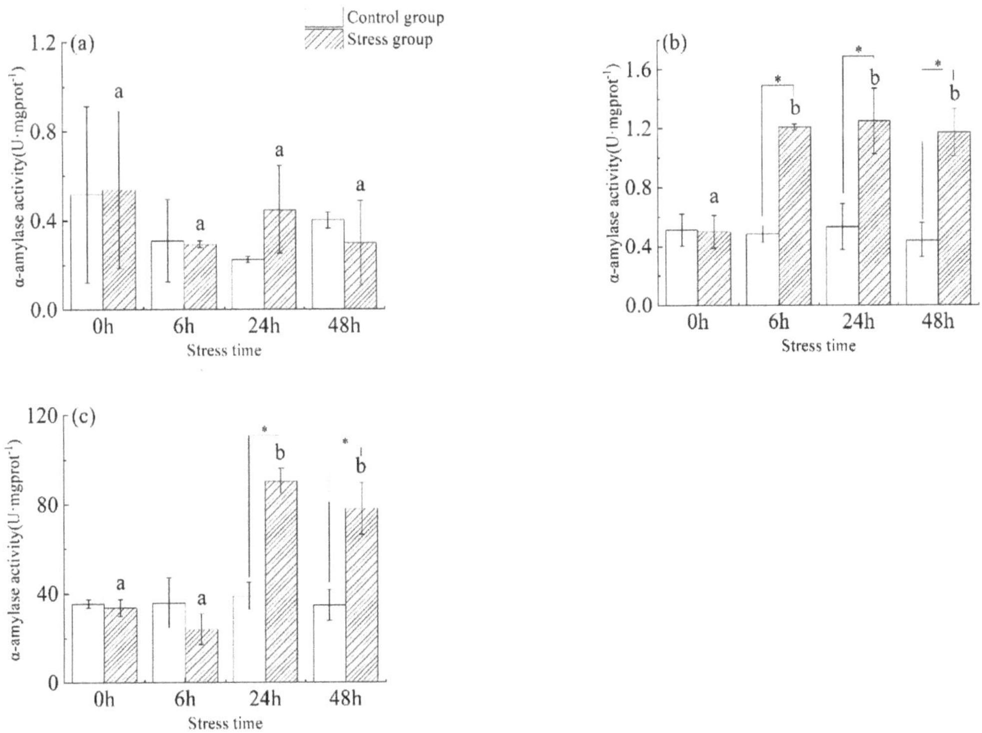

Figure 3. Effect of acute hyposalinity stress on α-amylase activity in juvenile yellowfin tuna (*Thunnus albacares*), (n = 9). α-Amylase activity values of yellowfin tuna stomach tissue (**a**), foregut tissue (**b**), and pyloric ceca tissue (**c**). At the same salinity, different letters at different time points indicate a significant difference ($p < 0.05$). Different letters indicate differences in experimental groups at different times, and * indicates differences between the experimental or control groups.

3.3. Effect of Salinity Changes on Lipase Activity

As shown in Figure 4, the lipase activity in the stomach (Figure 4a) of juvenile yellowfin tuna in the stress group was significantly lower than that of the control group from 6 h to 48 h ($p < 0.05$). Gastric tissue lipase activity in the stressed group showed a trend of decreasing (−94.02%) and then increasing (112.47%), reaching a minimum value (2.74 ± 1.99 U·g protein^{-1}) at 24 h. In the stress group, lipase activity in the foregut (Figure 4b) first increased by (31.69%) and then decreased (−38.45%), reaching a maximum value (36.70 ± 11.22 U·g protein^{-1}) at 24 h and a minimum value (22.59 ± 2.51 U·g protein^{-1}) at 48 h. In the pyloric ceca (Figure 4c) of juvenile yellowfin tuna from the stress group, lipase activity was significantly higher ($p < 0.05$) than in the control group from 6 h to 48 h and reached its highest activity at 24 h (25.08 ± 1.76 U·g protein^{-1}). After 48 h of acute salinity stress, the lipase activity of each digestive organ was ranked as follows: pyloric ceca > foregut > stomach.

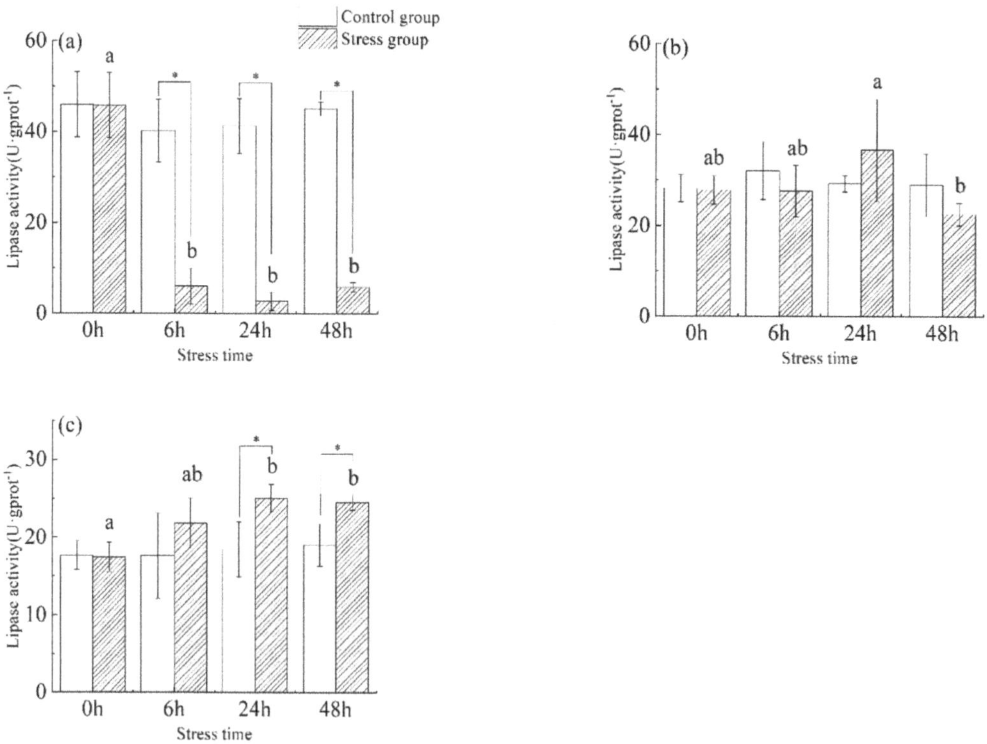

Figure 4. Effect of acute hyposalinity stress on lipase activity in juvenile yellowfin tuna (*Thunnus albacares*), (n = 9). Lipase activity values of yellowfin tuna stomach tissue (**a**), foregut tissue (**b**), and pyloric ceca tissue (**c**). At the same salinity, different letters at different time points indicate a significant difference ($p < 0.05$). Different letters indicate differences in experimental groups at different times, and * indicates differences between the experimental or control groups.

3.4. Effect of Salinity Changes on Chymotrypsin Activity

As shown in Figure 5, the chymotrypsin activity in the stomach (Figure 5a) of juvenile yellowfin tuna showed a trend of increasing (34.3%) and then decreasing (−38.1%), and the chymotrypsin activity in the stomach was significantly higher in the stress group compared to the control group before 24 h ($p < 0.05$) and stabilized at 48 h. The chymotrypsin activity in the stomach reached a maximum at 24 h (16.22 ± 0.81 U·mg protein^{-1}) and was significantly higher than in the control group ($p < 0.05$). Chymotrypsin activity in the foregut (Figure 5b) of juvenile yellowfin tuna in the stress group fluctuated slightly from 0 h to 48 h with increasing stress time, with a significant difference between 6 h and 24 h ($p < 0.05$). The chymotrypsin activity in the pyloric ceca (Figure 5c) of juvenile yellowfin tuna experienced a trend of decreasing (−22.4%) and then increasing (39.7%), reaching a minimum value (12.05 ± 0.55 U·mg protein^{-1}) at 6 h. The differences in chymotrypsin activity at 24 and 48 h were statistically significant ($p < 0.05$) compared to the control group, the chymotrypsin activity of each digestive organ was ranked as follows: pyloric ceca > foregut > stomach.

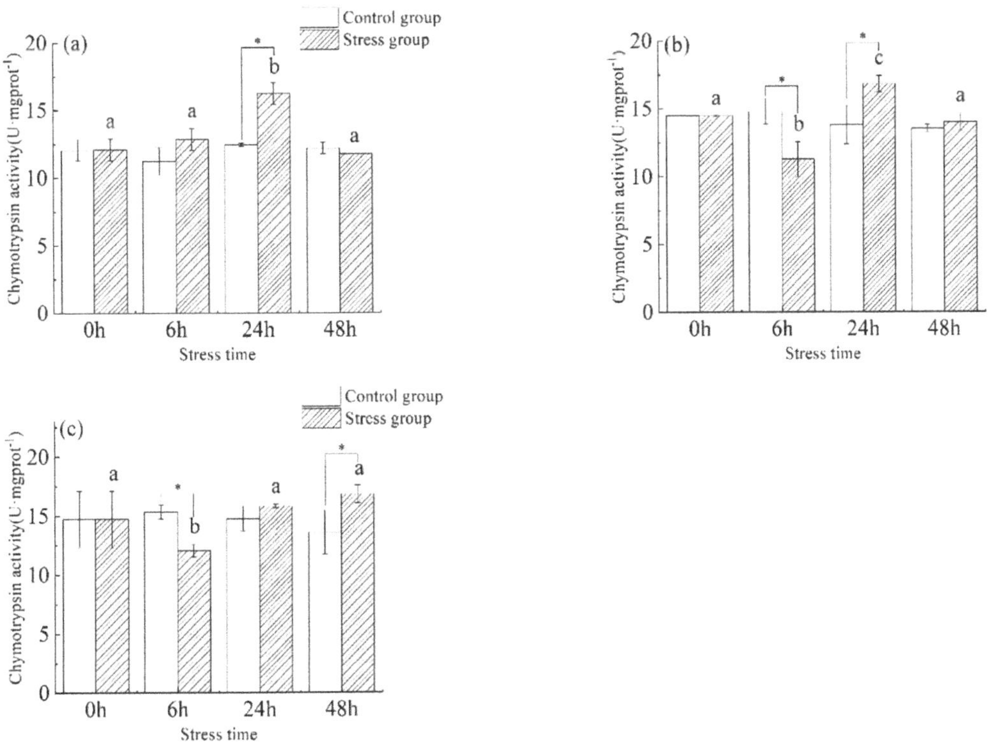

Figure 5. Effect of acute hyposalinity stress on chymotrypsin activity in juvenile yellowfin tuna (*Thunnus albacares*), ($n = 9$). Chymotrypsin activity values of yellowfin tuna stomach tissue (**a**), foregut tissue (**b**), and pyloric ceca tissue (**c**). At the same salinity, different letters at different time points indicate a significant difference ($p < 0.05$). Different letters indicate differences in experimental groups at different times, and * indicates differences between the experimental or control groups.

3.5. *Effect of Salinity Changes on Foregut Tissue*

Tissue sections of the foregut of each experimental and stress group are shown in Figure 6. The foregut villi were well developed from 6 h–48 h in all experimental groups, and there was no significant difference in villi density. In Figure 6A, the villi appeared sparse, and the intestinal folds were slightly shorter than those in groups B, C, and D, but they were all tightly arranged and morphologically intact. The foregut in the stress group was similar to that of the experimental group, with a few villi having blurred margins and being more closely arranged.

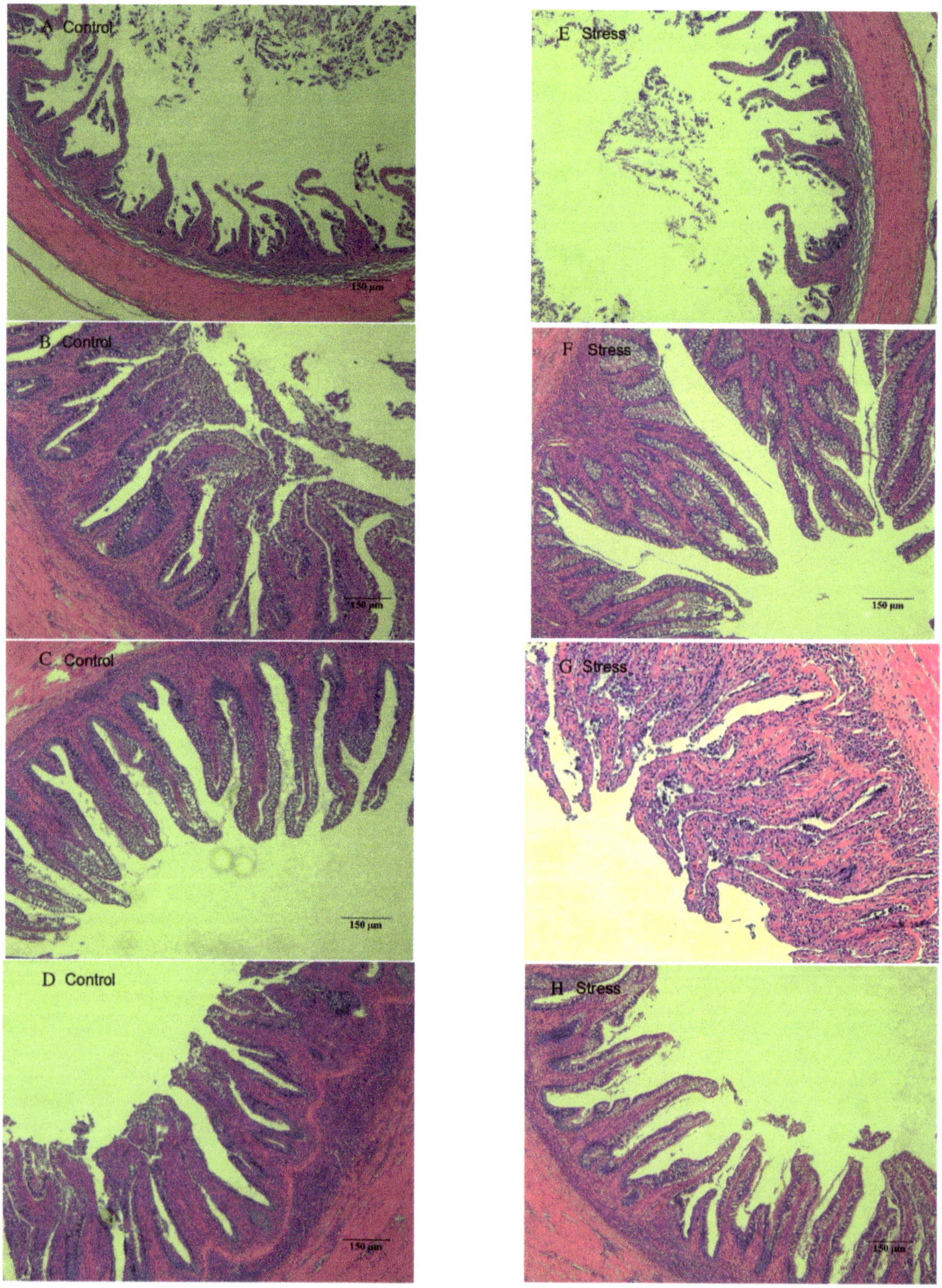

Figure 6. Foregut tissue section of juvenile yellowfin tuna (*Thunnus albacares*) (**A**), Control 0 h; (**B**), Control 6 h; (**C**), Control 24 h; (**D**), Control 48 h; (**E**), Stress 0 h; (**F**), Stress 6 h; (**G**), Stress 24 h; (**H**), Stress 48 h.

4. Discussion

Digestive enzymes influence nutrient absorption and transformation, and salinity has an important effect on their activity and efficiency [35]. The effect of salinity on the activity of digestive enzymes in fish depends on the species, life stage, salinity range, and exposure time of the fish [36]. In general, low salinity generally inhibits the activity of proteases, amylases, and lipases in the digestive tract of fish, affecting the digestion, absorption, and growth of food [37,38]. However, there are some broad-saline or salt-tolerant fish species whose digestive enzyme activity increases with decreasing salinity within a certain low salinity range, showing a strong ability to adapt [39]. It has been shown that changes in salinity within a certain range lead to changes in the activity of digestive enzymes in the digestive tract of fish, which can be broadly classified into three categories: firstly, activation of digestive enzymes [40]. Secondly, the inhibition of digestive enzymes [41], and thirdly, no significant effect on digestive enzymes [42]. In this experiment, with the decrease of salinity on the digestive enzyme activity of the digestive organs of juvenile yellowfin tuna, pepsin, and trypsin showed a decreasing trend, α-amylase, lipase, and chymotrypsin showed an increasing trend except for the stomach tissue which showed a decreasing trend. This proved that the decrease in salinity had an inhibitory effect on the digestive enzyme activity of juvenile yellowfin tuna. Analysis shows that during domestication, when juvenile yellowfin tuna swallow low-salinity seawater, the cells absorb large amounts of water, and the stomach acid becomes overly diluted, resulting in reduced digestibility [43–45]. At the same time, the ingestion of large amounts of seawater leads to an increase in inorganic ions in the digestive tract of juvenile yellowfin tuna, many of which are activators or inhibitors of digestive enzymes that can have a direct effect on the enzymes, thus affecting changes in digestive enzyme activity [46–48].

For example, Li Xuejiao et al. [49] studied the effects of short-time salinity changes on the blood biochemical parameters and digestive enzyme activity of black sea bream (*Acanthopagrus schlegelii*). Mozanzadeh et al. [50] studied the intestinal protease, amylase, and lipase activity of yellow mackerel (*Acanthopagrus latus*) and Asian sea bass (*Lates calcarifer*) at different salinity gradients for 24 h. Kawai et al. [51] showed some variation in digestive enzyme activity in different digestive organs of fish, which can exhibit some tissue-organ specificity. Our study showed that among the digestive enzymes of juvenile yellowfin tuna, trypsin activity was the highest in α-amylase, lipase, and chymotrypsin activity in pancreatic tissues under the same salinity environment, with obvious tissue organ characteristics. The results of the salmon (*Oncorhynchus keta*) [52] study showed that an appropriate increase in salinity could promote pepsin activity, presumably due to the chloride ion activating the protease. In contrast, in the present study, when juvenile yellowfin tuna were under acute hypersaline stress conditions, the pepsin activity of the stress group was lower than that of the control group by 24 h, and the peptidase activity decreased significantly with time. It is assumed that the activation of pepsin in the stomach of juvenile yellowfin tuna was reduced due to the increased concentration of Na^+, K^+, and Cl^- plasma. Hieu et al. [37] reported that the trypsin activity of striped catfish (*Pangasianodon hypophthalmus*) did not vary with salinity, which is consistent with the results of this experiment. This indicates that salinity did not affect trypsin in pancreatic tissues.

Studies in flower eels (*Anguilla marmorata*) and Pacific bicolor eels (*Anguilla bicolor pacifica*) showed a decrease in gastric and intestinal amylase activity with increasing salinity [53]. Contrary to our results some studies have shown that lower concentrations of chloride ions can activate α-amylase activity, while higher concentrations inhibit α-amylase activity [54,55]. Studies on cannabis salmon (*Oncorhynchus keta*) have shown that the direct effect of pH and inorganic ions on enzymes is the main reason salinity affects digestive enzyme activity [56,57]. In our experiment, the amylase activity in the stomach and pyloric ceca of juvenile yellowfin tuna showed a decrease and then an increase.

The results of the Zhang Longgang et al. [58] study on *Scortum barcoo* show that the inhibition of lipase activity decreases with increasing salinity from 0 to 13, and increases with salinity above 13, with the activation of lipase activity. However, in our study, the stomach tissue of juvenile yellowfin tuna was inhibited by the decrease in salinity and the lipase activity of its foregut and pancreatic tissues showed an increase, which may be due to the different changes in the lipase activity of the digestive system caused by the absence of feeding during the experiment. In the present study, juvenile yellowfin tuna foregut and pancreatic tissues decreased lipase activity with increasing treatment time after 24 h, presumably due to longer treatment times and adaptation to their changing environment. The results of a study on American shad (*Alosa sapidissima*) [38] showed that after treatment with salinities of 0, 7, 14, 21, and 28 $g \cdot L^{-1}$, the specific activity of chymotrypsin was highest in the 21 $g \cdot L^{-1}$ treatment group and did not change significantly in the other treatment groups (7, 14 and 28 $g \cdot L^{-1}$).

There are not many studies on the effects of chymotrypsin digestion, and from this study, it appears that acute low salt has an agonistic effect on various tissues of juvenile yellowfin tuna [59]. Therefore, a suitable salinity level will increase the activity of digestive enzymes in fish to avoid metabolic disorders in the gastrointestinal tract caused by oxidative stress [38,60]. At the same time, too high or too low salinity has a negative effect on the activity of digestive enzymes and will lead to a reduction in the digestive capacity of fish. Salinity directly affects the enzyme activity of the fish, and enzyme activity directly affects the body functions of the fish [61]. In summary, the pyloric blind sac is more sensitive to acute low-salt stress, followed by the foregut and stomach. Under acute hypersaline conditions, there was a significant effect on digestive enzyme activity in all digestive organs of juvenile yellowfin tuna between time points, but digestive enzyme activity basically stabilized at 48 h in all salinity groups, indicating that juvenile yellowfin tuna have slowly adapted to their environment and have a strong ability to regulate.

The intestinal villi increase the digestive and absorption area of the intestine, making it easier for the nutrients in food to be broken down and absorbed by the digestive juices. The intestinal villi contain cells that secrete various digestive enzymes such as amylase, lipase, and protease. These enzymes play an important role in the absorption of nutrients and in improving growth performance. These enzymes help to break down large molecules such as carbohydrates, fats, and proteins. The presence of intestinal villi are important for the growth and development of the animal, as they improve nutrient utilization and bio-efficiency [62]. The height and width of the intestinal villi are usually regarded as the main indicators to evaluate the functional state of the intestine. The larger the area of the intestinal villi, the larger the absorption area of the intestine and the better the digestion and absorption capacity of the organism [63]. In addition, the intestinal environment influences intestinal villus development, and oxidative stress damage is one of the important factors affecting its development [64]. Salinity directly affects the enzyme activity of the fish, and enzyme activity directly affects the body functions of the fish [65]. This study showed that the morphological structure of the foregut was altered between the control group 0 h and the stress group 0 h, possibly due to the impaired quality of their fillets. Acute hyposalinity had no significant effect on the foregut villi of juvenile yellowfin tuna.

The aim of this paper is to characterize how salinity affects digestive enzyme activity in fish tissues and to focus on the activation and inhibition. There was no significant effect of changes in salinity on digestive enzyme activity based on its previous studies, therefore we worked backward to the fact that digestive enzyme activity affects changes in the digestive organs of the fish body, and thus affects fish feeding, growth, and development. Thus, the focus on aquaculture water bodies is also a focus on their aquaculture industry. Therefore, the moderate reduction of seawater salinity in the culture sea area has basically no effect on its growth and development, which provides basic information for perfecting the domestication and culture of yellowfin tuna under seawater culture conditions.

5. Conclusions

In this study, we investigated the effects of acute low salinity on the digestive enzyme activities and intestinal status of juvenile yellowfin tuna (Tetraodontidae), whose digestive capacity could be adapted to a salinity level of 29‰ within 48 h. Under salinity stress, digestive enzyme activity in the fish stomach, foregut, and pyloric ceca resumed to average levels after 48 h. In addition, the tissue quality of the fish foregut was more consistent, indicating that digestive enzyme activity in fish can function normally within this salinity variation to avoid metabolic disturbance in the gastrointestinal tract caused by oxidative stress. In summary, the pyloric ceca are more sensitive to acute low-salt stress, followed by the foregut and stomach. This study contributes to an accurate understanding of the mechanisms affecting digestive physiology under salinity stress conditions. In response to the net-pen culture of yellowfin tuna in harsh environments in the coastal areas of China, the growth and development in an orderly manner improve fish survival and production, thus guiding the development and cultivation of the aquaculture industry.

Author Contributions: Conceptualization, Z.F. and N.Z.; Methodology, R.Y.; Software, N.Z.; Validation, Z.F.; Formal Analysis, R.Y.; Investigation, G.Y. and R.Y.; Resources, G.Y.; Data Curation, R.Y.; Writing–Original Draft Preparation, N.Z.; Writing—Review & Editing, Z.F. and Z.M.; Visualization, Z.F.; Supervision, G.Y. and Z.M.; Project Administration, Z.M., R.Y. and G.Y. All authors have read and agreed to the published version of the manuscript.

Funding: This work was supported by Hainan Major Science and Technology Project (Grant number [ZDKJ2021011]); Central Public-interest Scientific Institution Basal Research Fund, CAFS (Grant number [2023TD58]); Central Public-Interest Scientific Institution Basal Research Fund South China Sea Fisheries Research Institute, CAFS (Grant number [2021SD09]); the Project of Sanya Yazhou Bay Science and Technology City (Grant number [SKJC-2022-PTDX-015]).

Institutional Review Board Statement: The animal study protocol was approved by the Institutional Review Board (or Ethics Committee) of Animal Care and Use Committee of the South China Sea Fisheries Research Institute, Chinese Academy of Fishery Sciences (BIOL5346, 9 May 2022).

Data Availability Statement: The original contributions presented in the study are included in the article. Further inquiries can be directed to the corresponding authors.

Conflicts of Interest: The authors have no relevant financial or non-financial interest to disclose.

References

1. Surhone, L.M.; Tennoe, M.T.; Henssonow, S.F. *Yellowfin Tuna*; Betascript Publishing: New York, NY, USA, 2010.
2. Sun, R.-X.; Sun, Y.; Xie, X.-D.; Yang, B.-Z.; Cao, L.-Y.; Luo, S.; Wang, Y.-Y.; Mai, B.-X. Bioaccumulation and human health risk assessment of DDT and its metabolites (DDTs) in yellowfin tuna (*Thunnus albacares*) and their prey from the South China Sea. *Mar. Pollut. Bull.* **2020**, *158*, 111396. [CrossRef]
3. FAO (Fisheries and Aquaculture Department). *The State of World Fisheries and Aquaculture*; FAO Fisheries Report; FAO: Rome, Italy, 2020; Volume 2020, pp. 6–8.
4. Murua, H.; Rodriguez-Marin, E.; Neilson, J.D.; Farley, J.H.; Juan-Jordá, M.J. Fast versus slow growing tuna species: Age, growth, and implications for population dynamics and fisheries management. *Rev. Fish Biol. Fish.* **2017**, *27*, 733–773. [CrossRef]
5. Tian, Z.P.; Wang, F.; Tian, S.Q.; Ma, Q.Y. Stock assessment for Atlantic yellowfin tuna based on extended surplus production model considering life history. *Acta Oceanol. Sin.* **2022**, *41*, 41–51. [CrossRef]
6. Zhou, W.F.; Hu, H.J.; Wei, F.; Jin, S.F. Impact of Abnormal Climatic Events on the CPUE of Yellowfin Tuna Fishing in the Central and Western Pacific. *Sustainability* **2022**, *14*, 1217. [CrossRef]
7. Pacicco, A.E.; Brown-Peterson, N.J.; Murie, D.J.; Allman, R.J.; Snodgrass, D.; Franks, J.S. Reproductive biology of yellowfin tuna (*Thunnus albacares*) in the northcentral U.S. Gulf of Mexico. *Fish. Res.* **2023**, *261*, 106620. [CrossRef]
8. Schaefer, K.M.; Fuller, D.W. Spatiotemporal variability in the reproductive biology of yellowfin tuna (*Thunnus albacares*) in the eastern Pacific Ocean. *Fish. Res.* **2022**, *248*, 106225. [CrossRef]
9. Hoyle, S.D.; Williams, A.J.; Minte-Vera, C.V.; Maunder, M.N. Approaches for estimating natural mortality in tuna stock assessments: Application to global yellowfin tuna stocks. *Fish. Res.* **2023**, *257*, 106498. [CrossRef]
10. Filous, A.; Friedlander, A.M.; Griffin, L.; Lennox, R.J.; Danylchuk, A.J.; Mereb, G.; Golbuu, Y. Movements of juvenile yellowfin tuna (*Thunnus albacares*) within the coastal FAD network adjacent to the Palau National Marine Sanctuary: Implications for local fisheries development. *Fish. Res.* **2020**, *230*, 105688. [CrossRef]
11. Sánchez-Parra, M.; Lopez, A.; Muñoz-Redondo, J.M.; Montenegro-Gómez, J.C.; Pérez-Aparicio, J.; Pereira-Caro, G.; Rodríguez-Solana, R.; Moreno-Rojas, J.M.; Ordóñez-Díaz, J.L. Study of the influence of the fishing season and the storage temperature in the fishing vessel on the biogenic amine and volatile profiles in fresh yellowfin tuna (*Thunnus albacares*) and dry-cured mojama. *J. Food Compos. Anal.* **2022**, *114*, 104845. [CrossRef]

12. Alexi, N.; Hvam, J.; Lund, B.W.; Nsubuga, L.; Hansen, R.M.d.O.; Thamsborg, K.; Lofink, F.; Byrne, D.V.; Leisner, J.J. Potential of novel cadaverine biosensor technology to predict shelf life of chilled yellowfin tuna (*Thunnus albacares*). *Food Control.* **2020**, *119*, 107458. [CrossRef]
13. Huang, C.-H.; Hsieh, C.-Y.; Lee, Y.-C.; Ou, T.-Y.; Chang, T.-H.; Lee, S.-H.; Tseng, C.-H.; Tsai, Y.-H. Inhibitory Effects of High-Hydrostatic-Pressure Processing on Growth and Histamine Formation of Histamine-Forming Bacteria in Yellowfin Tuna Meat during Storage. *Biology* **2022**, *11*, 702. [CrossRef] [PubMed]
14. Rahmah, A.; Mardhatillah, I.; Damora, A.; Muhammad, M.; Nurfadillah, N. Application of Surplus Production Model to the Yellowfin Tuna *Thunnus albacares* in the northern and western parts of Aceh waters. *IOP Conf. Ser. Earth Environ. Sci.* **2021**, *869*, 012072. [CrossRef]
15. Nguyen, K.Q.; Phan, H.T.; Tran, P.D.; Van Nguyen, B.; Van Do, T.; Nguyen, L.T.; Van To, P.; Vu, N.K. Length-length, Length–weight, and Weight-weight Relationships of Yellowfin (*Thunnus albacares*) and Bigeye (*Thunnus obesus*) Tuna Collected From the Commercial Handlines Fisheries in the South China Sea. *Int. J. Mar. Sci.* **2022**, *38*, 911–917. [CrossRef]
16. Liu, H.Y.; Fu, Z.Y.; Zhou, S.J.; Hu, J.; Yang, R.; Yu, G.; Ma, Z.H. The Complete Mitochondrial Genome of Pennella sp. Parasitizing *Thunnus albacares*. *Front. Cells Infect. Microbiol.* **2022**, *12*, 945152. [CrossRef]
17. Sloman, K.A.; Wilson, R.W.; Balshine, S. Fish physiology. In *Behaviour and Physiology of Fish*; Elsevier: Amsterdam, The Netherlands, 2005; Volume 24, pp. 1–480.
18. You, H.Z.; Sun, Z.J.; Zhang, Q.; Zhang, Z.K. Effects of salinity on feeding growth and body composition of juvenile leopard gill perch. *J. Dalian Ocean. Univ.* **2013**, *28*, 89–93.
19. Liu, W.; Zhi, B.-J.; Zhan, P.-R.; Guan, H.-H.; Qin, D.-L. Effects of salinity on haematological biochemistrical indices and liver tissue in juvenile *Oncorhynchus keta*. *J. Appl. Ecol.* **2010**, *21*, 2411–2417.
20. Jiang, Z.; Huang, X.; Zhang, J. Dynamics of nonstructural carbohydrates in seagrass *Thalassia hemprichii* and its response to shading. *Acta Oceanol. Sin.* **2013**, *32*, 61–67. [CrossRef]
21. Artetxe-Arratea, I.; Fraile, I.; Marsac, F.; Farley, J.H.; Murua, H. A review of the fisheries, life history and stock structure of tropical tuna (skipjack Katsuwonus pelamis, yellowfin *Thunnus albacares* and bigeye *Thunnus obesus*) in the Indian Ocean. *Adv. Mar. Biol.* **2020**, *88*, 39–89.
22. Ashouri, G.; Soofiani, N.M.; Hoseinifar, S.H.; Jalali, S.A.H.; Morshedi, V.; Valinassab, T.; Bagheri, D.; Van Doan, H.; Mozanzadeh, M.T.; Carnevali, O. Influence of dietary sodium alginate and *Pediococcus acidilactici* on liver antioxidant status, intestinal lysozyme gene expression, histomorphology, microbiota, and digestive enzymes activity, in Asian sea bass (*Lates calcarifer*) juveniles. *Aquaculture* **2020**, *518*, 734638. [CrossRef]
23. Saleh, N.E.; Helal, M.; Ali, N.G.; Abbas, E.; Abdel-Tawwab, M. Effects of using vital wheat gluten in practical diets on growth, intestinal histopathology, proinflammation-related gene expression, and resistance of white seabream (*Diplodus sargus*) to Staphylococcus epidermidis infection. *Aquaculture* **2021**, *537*, 736508. [CrossRef]
24. Khoa, T.N.D.; Hayasaka, O.; Matsui, H.; Waqalevu, V.; Honda, A.; Nakajima, K.; Yamashita, H.; Ishikawa, M.; Shiozaki, K.; Kotani, T. Changes in early digestive tract morphology, enzyme expression and activity of Kawakawa tuna (*Euthynnus affinis*). *Aquaculture* **2021**, *530*, 735935. [CrossRef]
25. Fruton, J.S. A History of Pepsin and Related Enzymes. *Q. Rev. Biol.* **2002**, *77*, 127–147. [CrossRef] [PubMed]
26. Zvereva, E.A.; Zaichik, B.T.; Eremin, S.A.; Zherdev, A.V.; Dzantiev, B.B. Enzyme immunoassay for detection of Sudan I dye and its application to the control of foodstuffs. *J. Anal. Chem.* **2016**, *71*, 944–948. [CrossRef]
27. Psochiou, E.; Sarropoulou, E.; Mamuris, Z.; Moutou, K.A. Sequence analysis and tissue expression pattern of *Sparus aurata* chymotrypsinogens and trypsinogen. *Comp. Biochem. Physiol. Part B Biochem. Mol. Biol.* **2007**, *147*, 367–377. [CrossRef] [PubMed]
28. Darias, M.J.; Murray, H.M.; Gallant, J.W.; Douglas, S.E.; Yúfera, M.; Martínez-Rodríguez, G. The spatiotemporal expression pattern of trypsinogen and bile salt-activated lipase during the larval development of red porgy (*Pagrus pagrus*, Pisces, Sparidae). *Mar. Biol.* **2007**, *152*, 109–118. [CrossRef]
29. Ohshima, Y.; Suzuki, Y.; Nakatani, A.; Nohara, D. Refolding of Fully Reduced Bovine Pancreatic Trypsin. *J. Biosci. Bioeng.* **2008**, *106*, 345–349. [CrossRef]
30. Hidalgo, M.C.; Urea, E.; Sanz, A. Comparative study of digestive enzymes in fish with different nutritional habits. Proteolytic and amylase activities. *Aquaculture* **1999**, *170*, 267–283. [CrossRef]
31. Kumar, V. *Enzymes in Human and Animal Nutrition Principles and Perspectives*; Academic Press: Cambridge, MA, USA, 2018.
32. Anthonsen, H.W.; Baptista, A.; Drabls, F.; Martel, P.; Petersen, S.B.; Sebastio, M.; Vaz, L. Lipases and esterases: A review of their sequences, structure and evolution—Sciencedirect. *Biotechnol. Annu. Rev.* **1995**, *1*, 315–371. [CrossRef]
33. Sandoval, G. Chemoen-zymatic Synthesis of Nitrogen Polymers with Biomedical Applications Catalyzed by Lipases: An overview. In *Lipases and Phospholipases (Methods and Protocols)*; Methods in Molecular Biology; Humana Press: New York, NY, USA, 2018; Volume 1835, pp. 359–376. [CrossRef]
34. Guo, J.P.; Wang, J.; Li, H.J.; Ma, T.; Qi, Z.G.; Hao, Z.L.; Chen, J.Y. Effect of different weaning ages on the main digestive enzyme activities in the small intestinal contents of lamb. *Chin. Anim. Husb. Vet. Med. Intest. Contents Lamb* **2017**, *44*, 2603–2612.
35. Gheisvandi, N.; Hajimoradloo, A.; Ghorbani, R.; Hoseinifar, S.H. The effects of gradual or abrupt changes in salinity on digestive enzymes activity of Caspian kutum, *Rutilus kutum* (Kamensky, 1901) larvae. *J. Appl. Ichthyol.* **2015**, *31*, 1107–1112. [CrossRef]
36. Navarro-Guillén, C.; Yúfera, M.; Perera, E. Biochemical features and modulation of digestive enzymes by environmental temperature in the greater amberjack, *Seriola dumerili*. *Front. Mar. Sci.* **2022**, *9*, 960746. [CrossRef]

37. Hieu, D.Q.; Hang, B.T.B.; Huong, D.T.T.; El Kertaoui, N.; Farnir, F.; Phuong, N.T.; Kestemont, P. Salinity affects growth performance, physiology, immune responses and temperature resistance in striped catfish (*Pangasianodon hypophthalmus*) during its early life stages. *Fish Physiol. Biochem.* **2021**, *47*, 1995–2013. [CrossRef] [PubMed]
38. Liu, Z.-F.; Gao, X.-Q.; Yu, J.-X.; Qian, X.-M.; Xue, G.-P.; Zhang, Q.-Y.; Liu, B.-L.; Hong, L. Effects of different salinities on growth performance, survival, digestive enzyme activity, immune response, and muscle fatty acid composition in juvenile American shad (*Alosa sapidissima*). *Fish Physiol. Biochem.* **2016**, *43*, 761–773. [CrossRef]
39. Hamed, S.S.; Jiddawi, N.S.; Poj, B. Effect of salinity levels on growth, feed utilization, body composition and digestive enzymes activities of juvenile silver pompano *Trachinotus blochii*. *AkiNik Publ.* **2016**, *4*, 279–283.
40. Squires, E.J.; Haard, N.F.; Feltham, L.A.W. Gastric proteases of the Greenland cod *Gadus ogac*. I. Isolation and kinetic properties. *Biochem. Cell Biol.* **1986**, *64*, 205–214. [CrossRef] [PubMed]
41. Lin, H. *Fish Physiology*; Guangdong Higher Education Press: Guangzhou, China, 2011; pp. 213–219.
42. Lee-Shing, F.; Shu-Fen, C. Effect of salinity on the activities of digestive proteases from the tilapia fish, *Oreochromis niloticus* in different culture environments. *Comp. Biochem. Physiol. Part A Physiol.* **1989**, *93*, 439–443. [CrossRef]
43. Noda, M.; Murakami, K. Studies on proteinases from the digestive organs of sardine. II. Purification and characterization of two acid proteinases from the stomach. *Biochim. Biophys. Acta (BBA) Enzym.* **1981**, *658*, 27–34. [CrossRef]
44. Buentello, J.A.; Pohlenz, C.; Margulies, D.; Scholey, V.P.; Wexler, J.B.; Tovar-Ramírez, D.; Neill, W.H.; Hinojosa-Baltazar, P.; Gatlin, D.M. A preliminary study of digestive enzyme activities and amino acid composition of early juvenile yellowfin tuna (*Thunnus albacares*). *Aquaculture* **2011**, *312*, 205–211. [CrossRef]
45. Kihara, M. Pepsin-like protease activity and the gastric digestion within ex vivo Pacific bluefin tuna *Thunnus orientalis* stomachs, as a gastric digestion model. *Anim. Feed. Sci. Technol.* **2015**, *206*, 87–99. [CrossRef]
46. Chakrabarti, R.; Sharma, J.G. Digestive physiology of fish larvae during ontogenic development: A brief overview. *Indian J. Anim. Sci.* **2005**, *75*, 1337–1347. [CrossRef]
47. Kolkovski, S. Digestive enzymes in fish larvae and juveniles—Implications and applications to formulated diets. *Aquaculture* **2001**, *200*, 181–201. [CrossRef]
48. Zhou, S.J.; Zhang, N.L.; Fu, Z.Y.; Yu, G.; Ma, Z.H.; Zhao, L. Impact of Salinity Changes on the Antioxidation of Juvenile Yellowfin Tuna (*Thunnus albacares*). *J. Mar. Sci. Eng.* **2023**, *11*, 132. [CrossRef]
49. Li, X.J.; Shen, Y.D.; Bao, Y.G.; Wu, Z.X.; Yang, B.Q.; Jiao, L.F.; Zhang, C.D.; Tocher, D.R.; Zhou, Q.C.; Jin, M. Physiological responses and adaptive strategies to acute low-salinity environmental stress of the euryhaline marine fish black seabream (*Acanthopagrus schlegelii*). *Aquaculture* **2022**, *554*, 738117. [CrossRef]
50. Mozanzadeh, M.T.; Safari, O.; Oosooli, R.; Mehrjooyan, S.; Najafabadi, M.Z.; Hoseini, S.J.; Saghavi, H.; Monem, J. The effect of salinity on growth performance, digestive and antioxidant enzymes, humoral immunity and stress indices in two euryhaline fish species: Yellowfin seabream (*Acanthopagrus latus*) and Asian seabass (*Lates calcarifer*). *Aquaculture* **2020**, *534*, 736329. [CrossRef]
51. Kawai, S.-I.; Ikeda, S. Studies on digestive enzymes of fishes II. Effect of dietary change on activities of digestive enzymes in carp intestine. *Bull. Jpn. Soc. Sci. Fish.* **1972**, *38*, 265–270. [CrossRef]
52. Sánchez-Chiang, L.; Cisternas, E.; Ponce, O. Partial purification of pepsins from adult and juvenile salmon fish *Oncorhynchus keta*. Effect of NaCl on proteolytic activities. *Comp. Biochem. Physiol. Part B Comp. Biochem.* **1987**, *87*, 793–797. [CrossRef]
53. Luo, M.Z.; Guan, R.Z.; Jin, H. Effects of the salinity on the growth performance and digestive enzyme activities of *Anguilla marmorata* elver and *A. bicolor* pacifica elver. *Acta Hydrobiol. Sin.* **2015**, *39*, 653–660.
54. Chiu, Y.N.; Benitez, L.V. Studies on the carbohydrases in the digestive tract of the milkfish chanos chanos. *Mar. Biol.* **1981**, *61*, 247–254. [CrossRef]
55. Soltan, N.M.; Soaudy, M.R.; Abdella, M.M.; Hassaan, M.S. Partial dietary fishmeal replacement with mixture of plant protein sources supplemented with exogenous enzymes modify growth performance, digestibility, intestinal morphology, haemato-biochemical and immune responses for Nile tilapia, *Oreochromis niloticus*. *Anim. Feed. Sci. Technol.* **2023**, *299*, 115642. [CrossRef]
56. Zhi, B.J.; Liu, W.; Zhao, C.G.; Duan, Y.Y. Effects of salinity on digestive enzyme and alkaline phosphatase activity of young chum salmon (*Oncorhynchus keta* Walbaum). *J. Shanghai Ocean. Univ.* **2009**, *31*, 627–632.
57. Yang, J.W.; Zhou, Y.G.; Huang, M.; Xiong, Y.H.; Wang, F.; Gao, Q.F.; Sun, D.J.; Dong, S.L. Comparative studies on the digestive and antioxidant enzyme activities between juvenile rainbow (*Oncorhynchus mykiss*) and steelhead troutof rainbow trout (*O. mykiss*). *J. Ocean. Univ. China* **2019**, *49*, 119–128.
58. Zhang, L.G.; An, L.; Sun, D.; Fu, P.S. Effects of salinity on digestive enzyme activities of juvenile Jade Perch *Scortum barcoo*. *Chin. J. Fish.* **2021**, *24*, 21–24.
59. Rivera-Ingraham, G.A.; Lignot, J.-H. Osmoregulation, bioenergetics and oxidative stress in coastal marine invertebrates: Raising the questions for future research. *J. Exp. Biol.* **2017**, *220*, 1749–1760. [CrossRef] [PubMed]
60. Pimentel, M.S.; Faleiro, F.; Diniz, M.; Machado, J.; Pousão-Ferreira, P.; Peck, M.A.; Pörtner, H.O.; Rosa, R. Oxidative Stress and Digestive Enzyme Activity of Flatfish Larvae in a Changing Ocean. *PLoS ONE* **2015**, *10*, e0134082. [CrossRef]
61. Pujante, I.M.; Moyano, F.J.; Martos-Sitcha, J.A.; Mancera, J.M.; Martínez-Rodríguez, G. Effect of different salinities on gene expression and activity of digestive enzymes in the thick-lipped grey mullet (*Chelon labrosus*). *Fish Physiol. Biochem.* **2017**, *44*, 349–373. [CrossRef]

62. Yu, L.J.; Wen, H.; Jiang, M.; Wu, F.; Tian, J.; Lu, X.; Xiao, J.R.; Liu, W. Effects of ferulic acid on intestinal enzyme activities, morphology, microbiome composition of genetically improved farmed tilapia (*Oreochromis niloticus*) fed oxidized fish oil. *Aquaculture* **2020**, *528*, 735543. [CrossRef]
63. Huang, B.; Zhang, S.; Dong, X.H.; Chi, S.Y.; Yang, Q.H.; Liu, H.Y.; Tan, B.P.; Xie, S.W. Effects of fishmeal replacement by black soldier fly on growth performance, digestive enzyme activity, intestine morphology, intestinal flora and immune response of pearl gentian grouper (*Epinephelus fuscoguttatus* ♀× *Epinephelus lanceolatus* ♂). *Fish Shellfish. Immunol.* **2022**, *120*, 497–506. [CrossRef]
64. Pe Tan, P.; Zhu, W.L.; Zhang, P.; Wang, L.G.; Chen, R.Y.; Xu, D.D. Dietary soybean lecithin inclusion promotes growth, development, and intestinal morphology of yellow drum (*Nibea albiflora*) larvae. *Aquaculture* **2022**, *559*, 738446. [CrossRef]
65. Chen, J.; Xu, P.; Wen, H.; Xue, M.; Wang, Q.; He, J.; He, C.; Su, S.; Li, J.; Yu, F.; et al. Hypothermia-mediated oxidative stress induces immunosuppression, morphological impairment and cell fate disorder in the intestine of freshwater drum, *Aplodinotus grunniens*. *Aquaculture* **2023**, *575*, 739805. [CrossRef]

animals

MDPI

Communication

Fatty Acids in the Eggs of Red King Crabs from the Barents Sea

Alexander G. Dvoretsky [1,*], Fatima A. Bichkaeva [2], Nina F. Baranova [2] and Vladimir G. Dvoretsky [1,*]

1 Murmansk Marine Biological Institute of the Russian Academy of Sciences (MMBI RAS), 183038 Murmansk, Russia
2 N. Laverov Federal Center for Integrated Arctic Research of the Ural Branch of the Russian Academy of Sciences (FECIAR UrB RAS), 163000 Arkhangelsk, Russia
* Correspondence: ag-dvoretsky@yandex.ru (A.G.D.); v-dvoretsky@yandex.ru (V.G.D.)

Simple Summary: Seafood by-products contain a variety of valuable components, including bioactive peptides and essential fatty acids. The red king crab is a large, commercially important crustacean supporting profitable fisheries in the Barents Sea. In Norway, female red king crabs are included in the fishery. Large adult females carry egg masses, but red king crab eggs have not yet been studied for fatty acid content. In this paper, we provide information regarding the fatty acid profiles of red king crab eggs. We found a higher proportion of polyunsaturated fatty acids in comparison to saturated and monounsaturated fatty acids. Total pools of fatty acid did not differ significantly in terms of the stage of embryo development, female size, limb injury status, and habitat conditions. Individual comparisons, however, indicated significant differences for some fatty acids, providing evidence that they may play a role in physiological processes. Red king crab eggs may be considered a product with high nutritional value and are recommended for wide use in the food, pharmaceutical, and biomedical industries.

Abstract: The red king crab, *Paralithodes camtschaticus*, was introduced into the Barents Sea where, after a period of 30 years of adaptation, it has established a new population. This population has been commercially exploited over the past two decades, supporting profitable fisheries in both Russia and Norway. Biochemical studies aimed at assessing fatty acid profiles have been conducted, focusing primarily on the edible parts of red king crabs. Only recently have by-products been included in this research. Capture of female red king crabs is prohibited in Russia but is allowed in Norway. The fatty acids of the egg masses carried by these females have not yet been studied. To fill this knowledge gap, we assayed the fatty acid composition of eggs using gas–liquid chromatography. Our results showed a predominance of polyunsaturated fatty acids, while the concentrations of saturated and monounsaturated fatty acids were similar. Multivariate comparisons showed no significant differences in fatty acid profiles in terms of egg developmental stage (nauplius vs. metanauplius), habitat conditions (soft vs. hard bottoms), female size class, or number of autotomized limbs. However, individual comparisons showed some differences in fatty acids, the most important being the lower content of docosahexaenoic acid in eggs at the metanauplius stage compared to eggs at the nauplius stage, which is likely due to its essential role in the development of red king crab embryos. The total fatty acid content (53.94 mg g^{-1}) was 2–87 times higher in eggs than in other red king crab tissues, confirming the critical role that fatty acids play in maintaining physiological processes during vitellogenesis. The high content of essential fatty acids and an optimal omega-3-to-omega-6 ratio (4.9) suggest that red king crab eggs are a good product for a healthy diet and a valuable source for extracting essential fatty acids.

Keywords: red king crab; *Paralithodes camtschaticus*; fatty acids; eggs; nauplius; metanauplius; Barents Sea

1. Introduction

The red king crab, *Paralithodes camtschaticus* (Tilesius, 1815), is a large crustacean species with a wide distribution range from British Columbia (Canada) to the North Pacific, southwest to Korea at 140° E, and north at approximately 170° E through the Bering Sea [1]. There are three distinct populations of this species occupying seabed locations from the intertidal region to the continental slope around the North Pacific [1]. Besides these native populations, a non-native population was intentionally introduced into the Barents Sea in the 1960s by Russian scientists to establish a new profitable resource for the local fisheries in that region where no commercial crab stocks existed at that time [2]. After a 30-year period of adaptation and range expansion, this species had successfully formed a new self-sustaining population in the Barents Sea [2]. Further population growth and range extension both eastward and westward along the coastline of the Barents Sea supported a significant increase in the abundance and biomass of this species. By the late 2000s, the abundance had become sufficient to develop a local king crab industry, leading to the initiation of commercial fishing for red king crab in 2002 in Norway and 2004 in Russia [3]. In Russian waters, the population has demonstrated significant biomass fluctuations, mainly due to inadequate fishing pressure over the first decade and changes in the climate regime over the second decade of exploration [3]. However, the annual catch rates of red king crabs have increased from 10,820 t in 2020 to 11,629 t in 2021 and to 12,529 t in 2022 [3].

Red king crabs have been extensively researched, focusing on their distribution patterns, population dynamics [4], growth and reproduction [2,3], behavior and migrations [5], feeding and competition with native fauna [6–8], and symbiotic relationships [9,10], as well as fishery, management, and conservation aspects [3,4,11]. Biochemical assays have been conducted to describe their hormonal profiles in relation to life-history traits [12,13] and the nutritional quality and chemical composition of their edible parts [14]. Important by-products, such as the hepatopancreas [15] and shell [16], are used for the production of chitin, chitosan, α-glucosidase inhibitors, and the anticancer agent prodigiosin, as well as proteolytic enzymes and non-protein components such as essential fatty acids [17]. Recently, the fatty acid composition of other by-products, such as hemolymph and heart muscles, which are currently discarded at sea after standard onboard processing, has also been investigated [18,19].

Eggs are another red king crab product that is not typically processed, largely due to a lack of information on their biochemical composition and poor adaptation of laboratory protocols to assess conditions at capture [17]. In Russia, the fishing of female red king crabs is prohibited, which includes the harvesting of their eggs [3]. However, Norwegian authorities have established a regime that involves the harvesting of female red king crabs, with a quota of 100–120 t per year in a quota-regulated area. Furthermore, for maintaining a sustainable red king crab population, the harvest of female red king crabs is unrestricted within an open-access fishing area established in Norwegian waters west of 26° E, to prevent further westward and southward expansion of the red king crab population [20]. In Russia, red king crab roe is also consumed by tourists and local fishermen in Russia as a delicacy and exotic product. Female red king crabs are considered a potential for aquaculture, for conservation and scientific purposes, and, to some degree, for the food industry.

Lipids and fatty acids are important components in the development of crustaceans and are considered essential to their growth and success in molting [21]. Previous studies have shown that fatty acids in crustacean eggs can provide crucial energy stores for later developmental stages, and can provide valuable information on environmental factors affecting natural populations [22,23]. Furthermore, these findings can assist in the development of reliable nutritional protocols for cultivated species in hatchery settings [24,25].

This study aims to explore the fatty acid profiles of red king crab eggs in the Barents Sea region, with the objective of improving our understanding of the nutritional quality of red king crab by-products. The hypothesis is that the fatty acid profiles of the eggs will differ in relation to developmental stage, female size, number of injured limbs, and benthic habitats.

2. Materials and Methods

Our study was undertaken in Dalnezelenetskaya Bay, a small gulf with a total area of 2.23 km^2 located on the Eastern Murman coast of the Kola Peninsula [26]. A more detailed description of our study site is available in our previous publications [27,28]. In July 2016, a total of 37 egg-bearing female red king crabs were collected at depths ranging from 8 to 32 m by scientific SCUBA divers familiar with the nearshore habitats of the coastal Barents Sea. Previous investigations have established that sample sizes of 3 to 15 crustaceans are adequate to characterize patterns in fatty acid compositions across different developmental stages of eggs [23,25,29,30].

When capturing the crabs, divers registered temperature, depth, and the type of benthic habitat where each female was collected. In accordance with information obtained during diving surveys, the seabed locations in the study area were divided into two categories: hard bottoms—habitats composed mainly of rock, boulders, outcroppings, and algal kelps—and soft bottoms—habitats where sand alone or in different combinations with pebbles and shells were present. Female red king crabs were delivered to the coastal laboratory for biological analysis [31], which included visual inspection for the egg category, shell condition, and presence/absence of injured legs. Each female was measured for carapace length (CL, the greatest straight-line distance from the posterior margin of the right eye orbit to the medial-posterior margin of the carapace) using a vernier caliper to the nearest 0.1 mm, and weighed on an electronic balance to the nearest 1 g. All crabs collected had light coxa, spines, dactyls, and ventral surface of exoskeleton, and their legs were full of muscle tissue, i.e., characteristics indicative of new shells (2–12 months post ecdysis) [31].

After biological analysis, two portions of eggs were removed from each egg clutch and analyzed. One subsample (10–15 g) was frozen and used for fatty acid analyses, while the other (1–2 g) was fixed in Bouin solution and examined under a stereo-microscope to determine the stage of embryo development. Eggs were separated with diluted bleach, placed between 2 transparent sheets and photographed under a digital camera. Egg diameter was measured using ImageView 4.11 software [32], after adjustments and calibrations, with 15 eggs randomly selected from each subsample for photographic documentation. The development of red king crab embryos includes five stages: cleavage (I), gastrula (II), nauplius (III), metanauplius (IV), and late-stage zoeal egg (V) [33].

In the laboratory of the Federal Center for Integrated Arctic Research (Arkhangelsk), fatty acids were extracted, and fatty acid methyl esters were prepared according to the method of Folch et al. [34] with modifications [14]. The homogenized sample was dissolved into 10 mL of an extracting chloroform–methanol mixture and a solution of nonadecanoic acid in chloroform. The resulting solution was mixed for 30 min and then held in a thermostat for 10–12 h at 25 °C for lipid extraction. The solution was then filtered and mixed with a chloroform–methanol (2:1) solution to achieve the final sample volume of 15 mL. A 0.74% water solution of $CaCl_2$ (3 mL) was added to the sample and then stored in a refrigerator for 12 h. After stratification, the top layer containing water-soluble impurities was removed, while the lower layer was mixed with methanol and then evaporated to dryness using a vacuum evaporator Multivapor P12 (pressure 318 mbar, temperature 50 °C). The evaporated extract was added to 0.2 mL of the chloroform–methanol mixture, mixed for 5 min, and dissolved in a 1.5% solution of H_2SO_4 in methanol (2 mL). The sample was incubated in a water bath for 30 min at 90 °C. The sample was then placed in 0.8 mL of distilled water and incubated at ambient temperature for 2–4 h. The top fraction was pipetted into a 2 mL vial and evaporated again. The solution of fatty acid methyl esters (200 μL) was injected into an Agilent 7890A gas chromatograph (Agilent Technologies Inc., Wilmington, DE, USA) equipped with a flame ionization detector. The gas–liquid chromatography conditions were as follows: capillary column: Agilent DB-23 (60 m × 0.25 mm × 0.15 μm); carrier gas: nitrogen (injection volume 1 mm^3) with a 1 mL min^{-1} flow rate; injection port temperature: 270 °C; flame ionization detector temperature: 280 °C; the column was programmed from an initial temperature of 130 °C (0.5 min hold), rising to 170 °C at 8.5 °C min^{-1}, 206 °C at 2 °C min^{-1}, 220 °C at 0.7 °C min^{-1}, and 230 °C at 6 °C min^{-1}. Fatty acid methyl esters were identified by comparing their retention times with those of Nu Chek Prep Inc. (Elysian, MN, USA) 569 standards in the Agilent Chem Station B.04.03 software.

Gas–liquid chromatography is an established technique utilized for analyzing the types and quantities of lipids in foods. This method is regarded as highly sensitive, generating results with high accuracy and reproducibility. Although gas–liquid chromatography is time-consuming, labor-intensive, and necessitates experienced and adequately trained personnel to perform, it remains extensively employed to identify and quantify fatty acids in crustacean eggs [23,25,29,30,35–37].

A principal component analysis (PCA) was used to identify underlying patterns in egg fatty acid composition and to simplify data presentation and interpretation. The Bray–Curtis dissimilarity index was used to perform PERMANOVA with 9999 permutations on the raw data to investigate differences in fatty acid composition between two egg stages (violet and brown), three female size classes (121–133 mm, 134–146 mm, and >146 mm CL), three groups of females with different numbers of injured limbs (0, 1, and >1), and two different types of benthic communities (soft and hard). Differences in individual fatty acid concentrations were assessed using one-way analysis of variance (ANOVA) followed by Tukey's multiple comparison tests for normal data and the Kruskal–Wallis test (KWT) for non-normal data followed by Bonferroni multiple comparison tests. The Shapiro–Wilk test and modified Levene's test were applied to examine normal distribution and homoscedasticity of data. When necessary, the data were transformed to satisfy normality and homogeneity assumptions. ANOVA was used to evaluate differences in egg diameter between the two developmental stages, and a chi-squared test was used to compare the percentage occurrence of females with violet and brown eggs. Pearson correlation coefficients were calculated for log-transformed data to identify potential associations between fatty acid content in the eggs and ovaries of red king crabs. Ovary fatty acid content data for the same females used in the egg fatty acid analysis were extracted from our previous publication [19]. Statistical analyses were carried out using NCSS PASS 2004 and PAST 3.26.

3. Results

The female red king crabs used for biochemical assays ranged from 121.5 to 162.4 mm in CL and from 934 to 2548 g in body weight (Table 1).

Table 1. Carapace length and weight of female red king crabs collected for egg masses in Dalnezelenetskaya Bay, July 2016.

Egg Stage	N	X	SD	Min	Max
			Carapace length, mm		
Nauplius	28	140.2	9.4	121.5	162.4
Metanauplius	9	147.0	6.5	137.0	157.0
Combined	37	139.4	8.8	121.5	162.4
			Weight, g		
Nauplius	28	1913	329	1317	2548
Metanauplius	9	1773	401	934	2098
Combined	37	1879	347	934	2548

Note. N—sample size, X—mean, SD—standard deviation, Min—minimum, Max—maximum.

Visual observations indicated that 33 females had uneyed violet eggs (Figure 1a), while the remaining 4 individuals had uneyed brown eggs (Figure 1b).

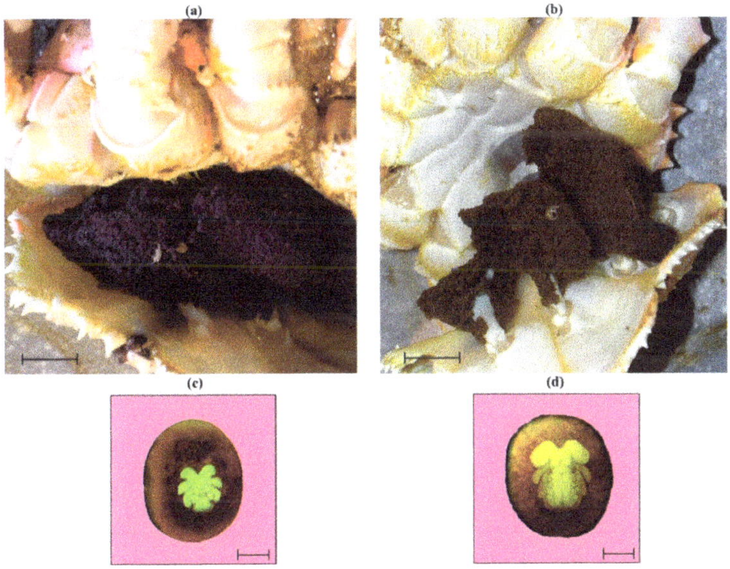

Figure 1. Eggs and developmental stages of red king crabs. (**a**)—violet eggs, scale bar 20 mm; (**b**)—brown eggs, scale bar 20 mm; (**c**)—nauplius stage, scale bar 300 μm; (**d**)—metanauplius stage, scale bar 300 μm.

However, stereo-microscope observations revealed that 28 of the 33 violet-egg clutches were at the nauplius stage (Figure 1c). The eggs from the remaining 5 violet-egg clutches and from all brown-egg clutches were at the metanauplius stage (Figure 1d) as defined and described by Nakanishi [33], Stevens [38], and Matyushkin [39]. Thus, the proportions of females with eggs at the nauplius and metanauplius stages were 75.7% and 24.3%, respectively. These percentages were significantly different from each other ($\chi^2 = 8.53$, $p = 0.004$). Additionally, egg diameter of nauplius-staged eggs (range, 850–1012 μm; mean ± SD, 939 ± 42 μm) was significantly lower than that of metanauplius-staged eggs (947–1049 μm; mean ± SD, 998 ± 26 μm) (ANOVA, F = 237.52, $p < 0.001$). Females with different-staged eggs had similar CL (ANOVA, F = 1.06, $p = 0.310$) and mean body weight (ANOVA, F = 1.12, $p = 0.298$).

Biochemical analysis detected 43 fatty acids in the eggs of red king crabs (Table 2).

Saturated fatty acids (SFAs) were mainly composed of palmitic (C16:0) and stearic (C18:0) acids. The mean levels of C16:0 were 8119 μg g^{-1} (15.2%) in nauplius-staged eggs and 7571 μg mL^{-1} (15.2%) in metanauplius-staged eggs, while for C18:0, these values accounted for 2505 μg g^{-1} (4.7%) and 2488 μg g^{-1} (5.0%), respectively. Oleic acid (C18:1n9C) was the predominant monounsaturated fatty acid (MUFA) with mean levels of 6987 μg g^{-1} (13.0%) in nauplius-staged eggs and 6685 μg g^{-1} (13.3%) in metanauplius-staged eggs, followed by palmitoleic acid (C16:1C) with mean levels of 3723 μg g^{-1} (6.9%) in nauplius-staged eggs and 3618 μg g^{-1} (7.2%) in metanauplius-staged eggs. Polyunsaturated fatty acids (PUFAs) were the dominant fatty acid types in the egg profiles of red king crabs. Arachidonic acid (C20:4n6) was the major n-6 PUFA, accounting for 2717 μg g^{-1} (5.0%) in nauplius-staged eggs and 2513 μg g^{-1} (5.0%) in metanauplius-staged eggs.

The most prevalent n-3 polyunsaturated fatty acids (PUFAs) were eicosapentaenoic acid (EPA, C20:5n3) and docosahexaenoic acid (DHA, C22:6n3). EPA constituted 13,568 μg g^{-1} (25.1%) and 12,954 μg g^{-1} (25.8%) in nauplius-staged and metanauplius-staged eggs, respectively. The mean levels of DHA were 8537 μg g^{-1} (13.8%) in nauplius-staged eggs and 7007 μg g^{-1} (14.0%) in metanauplius-staged eggs. The percentage of PUFA content was significantly higher than that of saturated fatty acids (SFAs) and monounsaturated fatty acids (MUFAs) in both nauplius- (ANOVA, F = 4309.06, $p < 0.001$) and metanauplius-staged eggs (KWT, H = 20.22, $p < 0.001$). In contrast, pairwise comparisons showed insignificant differences between the proportions of SFAs and MUFAs ($p > 0.05$ in both cases).

Table 2. Fatty acid composition of different stages of red king crab eggs from Dalnezelenetskaya Bay, July 2016.

Fatty Acid	Level, µg g^{-1}				Proportion, %			
	Nauplius		Metanauplius		Nauplius		Metanauplius	
	X ± SE	Min–Max	X ± SE	Min–Max	X ± SE	Min–Max	X ± SE	Min–Max
C6:0	1.8 ± 0.1	1.3–3.3	2.4 ± 0.4	1.1–4.2	0.002 ± 0	0–0.005	0.005 ± 0.001	0.002–0.008
C8:0	3.7 ± 0.2	2.1–5.6	3.4 ± 0.4	1.9–5.1	0.007 ± 0	0.004–0.011	0.007 ± 0.001	0.003–0.011
C9:0	3.3 ± 0.2	1.7–6.7	4.2 ± 0.4	1.8–5.8	0.006 ± 0.001	0.003–0.012	0.009 ± 0.001	0.003–0.013
C10:0	7.2 ± 0.4	4.5–11	7.6 ± 0.4	6.1–9.7	0.013 ± 0.001	0.009–0.022	0.015 ± 0.001	0.013–0.02
C11:0	3.7 ± 0.3	1.5–9.1	4.8 ± 0.6	2.7–8	0.007 ± 0.001	0.003–0.017	0.01 ± 0.001	0.005–0.016
C12:0	146 ± 6	84–207	147 ± 12	116–234	0.27 ± 0.01	0.19–0.36	0.29 ± 0.01	0.26–0.35
C13:0	21.2 ± 0.9	9.3–31.3	19.7 ± 1.4	13.2–25.8	0.039 ± 0.001	0.029–0.055	0.039 ± 0.002	0.029–0.049
C14:0	925 ± 37	480–1335	854 ± 60	534–1025	1.72 ± 0.04	1.45–2.31	1.71 ± 0.1	1.11–1.98
C15:0	491 ± 17	278–687	426 ± 13	364–499	0.92 ± 0.02	0.74–1.12	0.86 ± 0.02	0.74–0.94
C16:0	8119 ± 289	5453–12,450	7571 ± 338	6447–9969	15.2 ± 0.2	13.4–17.3	15.2 ± 0.4	14.1–17.1
C17:0	367 ± 12	242–478	329 ± 10	266–380	0.69 ± 0.01	0.55–0.84	0.66 ± 0.02	0.56–0.75
C18:0	2505 ± 92	1572–3468	2488 ± 91	2215–3081	4.7 ± 0.1	3.8–5.1	5 ± 0.1	4.5–5.3
C20:0	131 ± 7	51–199	141 ± 14	87–187	0.24 ± 0.01	0.16–0.35	0.28 ± 0.02	0.19–0.34
C21:0	36.8 ± 4.6	7.1–74.8	26.4 ± 7.1	10.7–73	0.065 ± 0.007	0.02–0.138	0.051 ± 0.012	0.024–0.112
C22:0	26.6 ± 2.5	5.8–59.5	33.6 ± 3.8	18.7–44.5	0.049 ± 0.004	0.012–0.104	0.067 ± 0.007	0.039–0.091
C23:0	102 ± 8	43–252	153 ± 41	53–448	0.19 ± 0.01	0.09–0.47	0.29 ± 0.06	0.11–0.66
C24:0	306 ± 31	91–697	185 ± 18	116–250	0.55 ± 0.05	0.23–1.03	0.37 ± 0.04	0.26–0.57
C14:1t	15.2 ± 0	15.2–15.2	4.1 ± 0	4.1–4.1	0.001 ± 0.001	0–0.033	0.001 ± 0.001	0–0.008
C14:1C	11.5 ± 0.9	3.2–28	11.1 ± 1.2	6–15.1	0.021 ± 0.002	0.006–0.06	0.022 ± 0.002	0.013–0.031
C15:1	5.2 ± 1.2	0.8–13.5	3.4 ± 0.1	3.1–3.6	0.004 ± 0.001	0–0.024	0.002 ± 0.001	0–0.007
C16:1t	95 ± 5	22–139	102 ± 6	82–133	0.17 ± 0.01	0–0.28	0.21 ± 0.01	0.17–0.28
C16:1C	3723 ± 155	2113–5142	3618 ± 237	2973–5350	6.9 ± 0.1	6.1–7.8	7.2 ± 0.1	6.6–7.9
C17:1	6.2 ± 0.7	2.9–14.6	7.9 ± 1.1	4.7–12.3	0.009 ± 0.001	0–0.021	0.014 ± 0.003	0–0.025
C18:1n9t	288 ± 11	191–413	264 ± 18	138–344	0.541 ± 0.016	0.383–0.709	0.529 ± 0.033	0.286–0.615
C18:1n9C	6987 ± 303	4266–11,360	6685 ± 385	5726–9485	13 ± 0.2	11.2–15.2	13.3 ± 0.3	12.5–15.5
C20:1	1114 ± 57	562–1897	1107 ± 82	784–1508	2.1 ± 0.1	1.4–3.4	2.2 ± 0.1	1.8–2.9
C22:1	105 ± 6	59–200	131 ± 15	83–205	0.2 ± 0.01	0.12–0.32	0.26 ± 0.03	0.18–0.42
C24:1	49.4 ± 2.8	24.9–83.2	49 ± 6	25.8–87	0.092 ± 0.005	0.049–0.152	0.096 ± 0.008	0.053–0.128
C18:2n6t	153 ± 25	13–424	106 ± 49	10–366	0.27 ± 0.04	0–0.87	0.23 ± 0.11	0.02–0.8
C18:2n6C	664 ± 29	336–995	574 ± 33	467–778	1.2 ± 0	1–1.5	1.1 ± 0	1–1.3
C18:3n3	344 ± 22	152–587	325 ± 39	215–605	0.63 ± 0.03	0.41–0.96	0.63 ± 0.04	0.48–0.89
C18:3n6	203 ± 14	86–341	172 ± 13	101–228	0.38 ± 0.02	0.18–0.62	0.35 ± 0.03	0.15–0.44
C20:2n6	620 ± 26	364–913	608 ± 37	472–747	1.2 ± 0	0.9–1.4	1.2 ± 0.1	1–1.5
C20:3n6	102 ± 6	56–216	108 ± 9	83–160	0.19 ± 0.01	0.13–0.34	0.22 ± 0.01	0.17–0.27
C20:4n6	2717 ± 129	1183–4251	2513 ± 210	1672–3349	5.1 ± 0.2	3.6–6.7	5 ± 0.4	3.7–6.8
C22:2n6	7 ± 0.9	0.9–22.6	5.9 ± 1.4	1.6–14	0.014 ± 0.002	0.002–0.042	0.012 ± 0.003	0.003–0.029
C20:5n3	13,568 ± 592	7530–18,394	12,954 ± 790	10,126–17,596	25.1 ± 0.4	21.2–28.7	25.8 ± 0.8	21–28
C22:6n3	8537 ± 372	5007–12,065	7007 ± 456	5715–9962	15.8 ± 0.3	13.1–19.1	14 ± 0.6	11.9–17
C20:3n3	178 ± 13	60–404	182 ± 22	98–270	0.33 ± 0.02	0.19–0.64	0.36 ± 0.04	0.22–0.49
C22:4n6	208 ± 11	136–415	177 ± 9	136–221	0.4 ± 0.02	0.24–0.86	0.36 ± 0.02	0.31–0.44
C22:3n3	6.3 ± 1	0.9–15.7	3.4 ± 1.2	1–11.1	0.011 ± 0.002	0–0.029	0.007 ± 0.003	0.002–0.025
C22:5n6	146 ± 6	83–215	122 ± 6	93–157	0.27 ± 0.01	0.2–0.4	0.24 ± 0.01	0.21–0.27
C22:5n3	849 ± 33	568–1237	831 ± 69	566–1162	1.6 ± 0.1	0.9–2.6	1.7 ± 0.2	1.3–2.5
∑SFA	13,194 ± 456	8369–18,916	12,730 ± 538	10,919–16,217	24.6 ± 0.2	23.1–26.6	25.5 ± 0.7	23.8–30.1
∑MUFA	12,377 ± 496	7304–18,425	11,976 ± 688	10,296–17,069	23 ± 0.3	20.5–25.7	23.9 ± 0.4	23–26.1
∑PUFA	28,296 ± 1092	15,831–37,468	25,684 ± 1252	22,454–34,507	52.4 ± 0.3	49.4–55.4	51.3 ± 0.5	48.2–52.7
Total	53,867 ± 2006	31,504–74,809	50,060 ± 2392	43,831–67,793	100 ± 0	100–100	100 ± 0	100–100
∑n-3	23,482 ± 936	13,534–32,105	21,301 ± 1096	18,908–29,395	–	–	–	–
∑n-6	4814 ± 198	2297–7057	4386 ± 245	3546–5316	–	–	–	–
∑n-9	8544 ± 350	5117–13,143	8236 ± 464	6998–11,567	–	–	–	–
∑n-7	3819 ± 157	2182–5268	3727 ± 241	3074–5487	–	–	–	–
n-3/n-6	4.9 ± 0.1	3.6–6.3	4.9 ± 0.2	3.9–5.8	–	–	–	–

Note. X—mean, SE—standard error, SFA—saturated fatty acids, MUFA—monounsaturated fatty acids, PUFA—polyunsaturated fatty acids.

A PCA biplot of the fatty acid data for red king crab eggs showed that principal components 1 and 2 (PC1 and PC2) cumulatively explained a majority of the variance in the data (94.9%) (Figure 2).

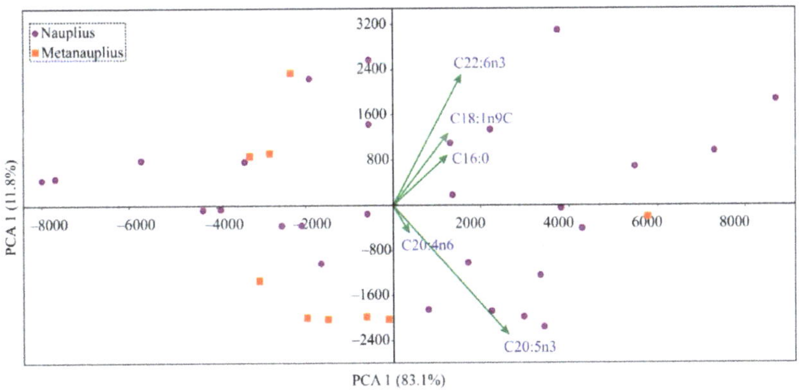

Figure 2. Principal component analysis (PCA) of the fatty acid composition of red king crab eggs in Dalnezelenetskaya Bay, July 2016.

The first axis separated females with eggs containing higher (positive scores) and lower (negative scores) concentrations of DHA and EPA. Almost all females with metanauplius-staged eggs were positioned on the left side of the biplot. However, many points representing females with nauplius-staged eggs also had negative PC1 scores, resulting in overlap between the two groups and insignificant differences in the fatty acid profiles between the different-staged eggs (PERMANOVA, F = 0.53, p = 0.533). Comparisons conducted for individual fatty acids confirmed the results of the PCA; a significantly lower value of DHA was found in metanauplius-staged eggs compared to nauplius-staged eggs (ANOVA, F = 4.66, p = 0.039). Additionally, significant differences were found for C9:0 and C:15 (ANOVA, $p < 0.05$), but these fatty acids had very low contributions to the total content.

A multivariate comparison revealed that the fatty acid profiles of females of different sizes were similar (PERMANOVA, F = 0.72, p = 0.517), except for three cases (C18:2n6t, C22:2, and C22:4n6), where significant differences were recorded (ANOVA or KWT, $p < 0.05$) with lower values found in the intermediate size class (134–146 mm CL) (Figure S1). The number of autotomized limbs did not affect the fatty acid profiles in the eggs (Table S1, PERMANOVA, F = 0.72, p = 0.517), but some less important fatty acids demonstrated significant variations. For C8:0, intact females had a higher concentration than females with injured legs (ANOVA, F = 4.98, p = 0.013), whereas the highest concentrations of C9:0, C13:0, C15:0, C17:0, and C22:5n6 were found in the eggs of females with one autotomized leg (ANOVA, $p < 0.05$) (Figure S2). Fatty acid profiles in the eggs of females captured on soft and hard bottoms did not differ significantly (PERMANOVA, F = 0.38, p = 0.642), and individual comparisons also showed insignificant results (Table S2, ANOVA or KWT, $p < 0.05$). Correlation coefficients indicated no significant difference between the concentration of SFAs, MUFAs, PUFAs, and total fatty acids in ovaries [19] and corresponding levels in the eggs (k = –0.014 . . . –0.059, p = 0.728–0.935).

4. Discussion

Fatty acids are crucial components of lipid-containing molecules such as triacylglycerols, fats, and waxes, and due to their high caloric content they serve as important chemical feedstocks in basic metabolic processes [40]. In crustacean invertebrates, the importance of fatty acids has been shown to increase significantly in response to high energy requirements during key physiological processes such as molting, limb regeneration, and especially mating and spawning [21,41,42].

Red king crab embryos develop inside eggs and have no external food sources. Therefore, the total fatty acid concentration in eggs (combined data for violet and brown eggs, 53,940 $\mu g\ g^{-1}$) is higher than in hemolymph (617 $\mu g\ g^{-1}$), leg muscles (2930 $\mu g\ g^{-1}$), cardiac muscles (5240 $\mu g\ g^{-1}$), ovaries (21,990 $\mu g\ g^{-1}$), and even hepatopancreases of females (25,590 μg g^{-1}) [14,15,18,19]. It is known that in decapods, dietary lipids accumulate in the hepatopancreas and are transferred to the ovary, and then to hatching eggs during the annual reproductive cycle [43,44]. Embryo development in red king crabs lasts 11–11.5 months [3]. Thus, these high fatty acid concentrations in the eggs are necessary to ensure the developmental processes during this period. Similar variations in the fatty acid content in different tissues with higher concentrations in eggs were reported for the blue swimmer crab *Portunus pelagicus* [23,45].

The fatty acid profiles in the red king crab eggs were dominated by PUFAs, while the SFA and MUFA contents were similar. A predominance of PUFAs has also been reported for the eggs of the crabs *Maja brachydactyla* [46], *Chionoecetes opilio* [29], *Cancer setosus* [47], and *Charybdis japonica* [37]. Similar proportions of PUFAs and MUFAs have been found in embryos of the lobster *Nephrops norvegicus* [30]. In contrast, MUFAs were relatively more prevalent than PUFAs in the eggs of the crab species *Uca rapax* [48], *Armases cinereum* [49], *Eriocheir sinensis* [25], and *Shinkaia crosnieri* [22], and the shrimp species *Chorismus antarcticus, Notocrangon antarcticus,* and *Nematocarcinus lanceopes* [50]. SFA-dominated profiles have been observed for the eggs of the crabs *Uca annulipes* [51], *Uca inversa, Uca urvillei, Uca chlorophthalmus, Uca vocans* [35], and *Portunus pelagicus* [52,53]. A balanced SFA/MUFA/PUFA ratio has been observed in the blue crab *Callinectes sapidus* [36].

According to Donaldson and Byesrdorfes [31], stages I, II, and III are considered "uneyed violet eggs", stage IV represents "uneyed brown eggs", and stage V is "eyed brown" or "orange eggs". However, results from previous studies [39] and our data indicate that this classification is not accurate. In the Barents Sea in July, the proportions of females with eggs at stages III and IV were reported to be 63% and 37%, respectively [39]. Our ratio of 76%:24% is slightly different from these values, likely due to differences in environmental conditions in our sampling location. It is known that after fertilization and deposition of eggs on the female's abdomen, the development of the eggs to hatching depends on water temperature and requires a specific number of heat units (degree-days) [38,54]. The study by Matyushkin [39] was conducted in Ura Bay, where the water temperature is higher than in Dalnezelenetskaya Bay [27,55]. Red king females in that location may rear eggs under more suitable conditions, yielding a faster egg maturation rate. Mating of red king crabs in the Barents Sea can occur any time from late January to early June, with a peak during March–April [56,57]. Therefore, we can assume that females with eggs at stage III spawned 1–2 months earlier than females with eggs at stage II.

Despite the developmental events that occur in red king crab embryos from the nauplius stage to the metanauplius stage, which are accompanied by a significant increase in egg diameter, the fatty acid content in the eggs remained consistent across different stages (Figure 2). This pattern may be explained by the low contribution of fatty acids to bioenergetic investments during these stages of development. During the nauplius stage, a white blastodisc is formed, and the optic lobes and rudiments of the mandibles, antennules, antennae, labrum, and pleopods appear [38]. During the metanauplius stage, there is further development of all the cephalic appendages mentioned above, plus two pairs of maxillipeds and three pairs of maxillae, and protuberances from the thorax begin to appear [38]. These processes may have specific requirements, which could explain the significantly lower content of two minor and one major fatty acid (DHA) in the eggs. Further development of red king crab embryos leads to significant transformations, including elongation of the unsegmented telson, its stretching over the head, and segmentation. The embryo gradually outgrows the yolk, and by the eighth month, all major body parts have appeared [38]. The embryo then increases in size until hatching [33] when the egg lipid content is completely consumed [58].

Similar patterns of fatty acid concentrations by egg maturation have been reported for many crustacean species. In general, the fatty acid content in early-staged eggs (stages I and II) or intermediate-staged eggs (stages II and III) did not differ significantly [36,37,46,48,49,59]. However, as vitellogenesis progressed, the amounts of almost all important fatty acids decreased considerably [23,24,35,37,51,53]. In other species, there was a continuous decrease in the fatty acid contents throughout embryo development [29,47,60]. It should be noted that the egg-staging systems used for these species do not directly correspond to the classification proposed for red king crabs, but they do reflect significant changes in the embryo's morphology and/or physiology.

We observed that the total pool of fatty acids in eggs was not significantly affected by female size or limb injury status. Therefore, the significant variations found for certain fatty acids may be related to their role in the autotomy process, which can significantly affect the physiology of red king crabs [41]. This phenomenon has been shown for the Chinese mitten crab, *Eriocheir sinensis*, for which changes in the ratio of specific fatty acids were observed in relation to autotomy [61]. Alternatively, the observed variations may also be attributed to changes in the physiological status of different-sized red king crab specimens, which could be associated with their different migration activity [5], or to differences in food habits among female red king crabs [58]. In our previous study, we found that the fatty acid profiles in hemolymph differed significantly between red king crabs captured on soft and hard sediments, indicating the role of feeding in determining the total pool of fatty acids [18]. Similar effects have been observed in other crab species under both natural [62] and laboratory conditions [63,64]. Our study also revealed insignificant dissimilarity between the fatty acid signatures of red king crab females from different habitats. This result is not surprising, as the females had already extruded their clutches of eggs well before being collected and were likely outside the study area at the time. Similarly, we found no significant correlations between the fatty acid contents in the ovaries and eggs. This can be explained by the fact that the current ovaries had no direct links to the previously extruded eggs.

In the Barents Sea, the nauplius stage of red king crabs occurs from May to September, peaking in July at 43%. The metanauplius stage occurs in almost equal proportions (37–40%) from July through September [39]. Given the tendency of fatty acid content to decrease during embryo development, as observed in other crab species, we can conclude that the harvest of red king crab females for high-quality eggs with high concentrations of essential fatty acids should be completed before October, when eggs of most females are at the late zoea stage. In Norway, the annual female quota for the 2023 fishery season is 120 t. Based on the mean weight of a female of 1550 g, the mean egg weight of 0.61 mg, and the mean weight of an egg clutch of 131.8 g, or 8.5% of the total female weight [57,65], we can estimate that Norwegian fishermen can receive 10.2 t of raw red king crab eggs.

It is well established that EPA and DHA are highly valuable compounds that can be obtained through consumption or extraction from seafood. These fatty acids have gained significant market value due to their associated beneficial health effects [66]. The consumption of seafood products and by-products containing long-chain n-3 polyunsaturated fatty acids (PUFAs) has been shown to have positive effects on human health, with demonstrated anticancer, antidepressant, anti-diabetic, antihyperglycemic, antihypertensive, anti-inflammatory, antimicrobial, antioxidant, anti-rheumatic, and immunomodulatory properties [67–71].

In accordance with the Joint Food and Agriculture Organization of the United Nations/World Health Organization Expert Consultation on Fats and Fatty Acids in Human Nutrition recommendations, the total daily intake of n-3 fatty acids can range between 0.5% and 2% of energy intake, with a minimum dietary requirement of alpha-linolenic acid (ALA, C18:3n3) of more than 0.5% of energy for adults to prevent deficiency-related symptoms. The higher value of 2% of energy intake for ALA in combination with EPA and DHA, i.e., 0.25–2.0 g per day, can be considered indicative of a healthy diet. For adult males and non-pregnant/non-lactating adult females, the recommended daily intake of EPA + DHA is 0.25 g. However, for adult pregnant and lactating females, the minimum daily intake for optimal adult health and fetal and infant development is 0.3 g EPA + DHA, of which at least 0.2 g should be DHA [72].

Due to their higher concentrations of fatty acids, particularly n-3 polyunsaturated fatty acids, and a well-balanced n-3/n-6 ratio in comparison to other tissues such as muscles, red king crab eggs can be recommended for direct human consumption.

5. Conclusions

Our study provides novel insights into the fatty acid composition of different-staged eggs of the commercially important red king crab (*Paralithodes camtschaticus*) inhabiting the Barents Sea. We detected high concentrations of fatty acids reaching 50–53 mg g^{-1}. These results indicate that lipids play a critical role in sustaining embryonic development in the red king crab. We observed that the fatty acid profiles of eggs at different developmental stages were largely similar, with the exception of docosahexaenoic acid (DHA), whose levels were significantly reduced in eggs at the metanauplius stage. This suggests DHA has an important nutritional function for embryos undergoing the transition from the nauplius to metanauplius phase. Notably, the fatty acid profiles of red king crabs of varying sizes, with different numbers of autotomized appendages, and from either hard or soft benthic habitats, were mostly analogous, apart from minor variations in select fatty acids. Considering the demand for alternative sources of essential omega-3 fatty acids, our results provide valuable information demonstrating that red king crab eggs are a nutrient-dense source of essential fatty acids and have an optimal ratio of omega-3 to omega-6 PUFAs, rendering them an exceptional food product for human consumption and use in pharmaceutical and food industries.

Supplementary Materials: The following supporting information can be downloaded at: https://www.mdpi.com/article/10.3390/ani14020348/s1, Figure S1. Significant variation in the content of fatty acids in the eggs of female red king crabs of different sizes from the Dalnezelenetskaya Bay, July 2016. Vertical bars show standard errors. Different letters show significant differences between groups; Figure S2. Significant variations in fatty acid content in eggs of female red king crabs with different numbers of injured legs in Dalnezelenetskaya Bay, July 2016. Vertical bars show standard errors. Different letters show significant differences between groups; Table S1. Fatty acid composition (μg g^{-1}) of red king crab eggs in crabs with different numbers of injured legs (0, 1, and >1) in Dalnezelenetskaya Bay, July 2016; Table S2. Fatty acid composition (μg g^{-1}) of red king crab eggs in crabs from hard- and soft-bottom habitats in Dalnezelenetskaya Bay, July 2016.

Author Contributions: Conceptualization, F.A.B. and A.G.D.; data curation, F.A.B., A.G.D., N.F.B. and V.G.D.; validation, F.A.B., N.F.B., A.G.D. and V.G.D.; formal analysis, F.A.B., N.F.B., A.G.D. and V.G.D.; methodology, F.A.B. and A.G.D.; software, A.G.D. and V.G.D.; visualization, A.G.D. and V.G.D., project administration, V.G.D.; writing—original draft, A.G.D., V.G.D. and F.A.B.; writing—review and editing, V.G.D., A.G.D. and F.A.B. All authors have read and agreed to the published version of the manuscript.

Funding: This article is based upon work that was supported by the Ministry of Science and Higher Education of the Russian Federation under contracts with MMBI RAS and FECIAR UrB RAS.

Institutional Review Board Statement: This study was conducted in accordance with Russian legislation guidelines (Federal Act 52-F3) and approved by the Institutional Review of Murmansk Marine Biological Institute RAS (No. 188-1252/14, 19 December 2023).

Informed Consent Statement: Not applicable.

Data Availability Statement: Data are contained within the article and Supplementary Materials.

Acknowledgments: The English was greatly improved thanks to careful editing and valuable remarks by J. Muehlbauer. We thank the four reviewers for their constructive comments.

Conflicts of Interest: The authors declare no conflicts of interest.

References

1. Stevens, B.G.; Lovrich, G.A. King crabs of the World: Species and distributions. In *King Crabs of the World: Biology and Fisheries Management*; Stevens, B.G., Ed.; CRC Press: Boca Raton, FL, USA; Taylor and Francis Group: Abingdon, UK, 2014; pp. 1–29.
2. Kuzmin, S.A.; Gudimova, E.N. *Introduction of the Kamchatka (Red King) Crab in the Barents Sea: Peculiarities of Biology, Perspectives of Fishery*; KSC RAS Press: Apatity, Russia, 2002. (In Russian)
3. Dvoretsky, A.G.; Dvoretsky, V.G. Ecology and distribution of red king crab larvae in the Barents Sea: A review. *Water* **2022**, *14*, 2328. [CrossRef]

4. Stesko, A.V. Distribution and status of the king crab stock in the Russian territorial waters of the Barents Sea. *Probl. Fish.* **2015**, *16*, 175–192. (In Russian) [CrossRef]
5. Talberg, N.B. Comparative characteristics in the red king crab migration pattern in the shallows of the Barents and Okhotsk Sea. *Coast. Hydrobiol. Investig. VNIRO Proc.* **2005**, *142*, 91–101. (In Russian)
6. Pavlova, L.V. Effect of juvenile red king crabs on zoobenthos in Kola Bay (Barents Sea). *Dokl. Biol. Sci.* **2008**, *422*, 312–315. [CrossRef] [PubMed]
7. Pavlova, L.V. Ration of the red king crab on coastal shoals of the Barents Sea. *Dokl. Biol. Sci.* **2015**, *463*, 200–204. [CrossRef] [PubMed]
8. Pavlova, L.V. The red king crab *Paralithodes camchaticus* (Tilesius, 1815) (Decapoda: Anomura): The use of species equality indicators to assess the influence on the benthos of the Barents Sea. *Russ. J. Mar. Biol.* **2021**, *47*, 508–514. [CrossRef]
9. Dvoretsky, A.G.; Dvoretsky, V.G. Distribution of amphipods *Ischyrocerus* on the red king crab, *Paralithodes camtschaticus:* Possible interactions with the host in the Barents Sea. *Estuar. Coast. Shelf Sci.* **2009**, *82*, 390–396. [CrossRef]
10. Dvoretsky, A.G.; Dvoretsky, V.G. Epibiotic communities of common crab species in the coastal Barents Sea: Biodiversity and infestation patterns. *Diversity* **2022**, *14*, 6. [CrossRef]
11. Pereladov, M.V.; Stesko, A.V. Features of distribution and biology of juvenile red king crab in the Barents Sea. In *The Red King Crab in the Barents Sea*, 3rd ed.; Bizikov, V.A., Stesko, A.V., Alekseev, D.O., Buyanovsky, A.I., Dolgov, A.V., Novikov, M.A., Pereladov, M.V., Sentyabov, E.V., Sokolov, K.M., Eds.; VNIRO: Moscow, Russia, 2021; pp. 240–262. (In Russian)
12. Dvoretsky, A.G.; Dvoretsky, V.G. Hemolymph molting hormone concentrations in red king crabs from the Barents Sea. *Polar Biol.* **2010**, *33*, 1293–1298. [CrossRef]
13. Dvoretsky, A.G.; Tipisova, E.V.; Elfimova, A.E.; Alikina, V.A.; Dvoretsky, V.G. Sex hormones in hemolymph of red king crabs from the Barents Sea. *Animals* **2021**, *11*, 2149. [CrossRef]
14. Dvoretsky, A.G.; Bichkaeva, F.A.; Baranova, N.F.; Dvoretsky, V.G. Fatty acid composition of the Barents Sea red king crab (*Paralithodes camtschaticus*) leg meat. *J. Food Compos. Anal.* **2021**, *98*, 103826. [CrossRef]
15. Dvoretsky, A.G.; Bichkaeva, F.A.; Baranova, N.F.; Dvoretsky, V.G. Fatty acid composition in the hepatopancreas of the Barents Sea red king crab. *Biol. Bull.* **2020**, *47*, 332–338. [CrossRef]
16. Yakovlev, I.A.; Lysøe, E.; Heldal, I.; Steen, H.; Hagen, S.B.; Clarke, J.L. Transcriptome profiling and in silico detection of the antimicrobial peptides of red king crab *Paralithodes camtschaticus. Sci. Rep.* **2020**, *10*, 12679. [CrossRef] [PubMed]
17. Ponomareva, T.; Timchenko, M.; Filippov, M.; Lapaev, S.; Sogorin, E. Prospects of red king crab hepatopancreas processing: Fundamental and applied biochemistry. *Recycling* **2021**, *6*, 3. [CrossRef]
18. Dvoretsky, A.G.; Bichkaeva, F.A.; Baranova, N.F.; Dvoretsky, V.G. Fatty acids in the circulatory system of an invasive king crab from the Barents Sea. *J. Food Compos. Anal.* **2022**, *110*, 104528. [CrossRef]
19. Dvoretsky, A.G.; Bichkaeva, F.A.; Baranova, N.F.; Dvoretsky, V.G. Fatty acid profiles in the gonads of red king crab (*Paralithodes camtschaticus*) from the Barents Sea. *Animals* **2023**, *13*, 336. [CrossRef]
20. Sundet, J.H.; Hoel, A.H. The Norwegian management of an introduced species: The Arctic red king crab fishery. *Mar. Policy* **2016**, *72*, 278–284. [CrossRef]
21. Kumar, V.; Sinha, A.K.; Romano, N.; Allen, K.M.; Bowman, B.A.; Thompson, K.R.; Tidwell, J.H. Metabolism and nutritive role of cholesterol in the growth, gonadal development, and reproduction of crustaceans. Reviews in Fisheries Science. *Aquaculture* **2018**, *26*, 254–273.
22. Saito, H. Characteristics of fatty acid composition of the deep-sea vent crab, *Shinkaia crosnieri* Baba and Williams. *Lipids* **2011**, *46*, 723–740. [CrossRef]
23. Romano, N.; Safee, M.A.; Ebrahimi, M.; Arshad, A. Fatty acid compositional changes during the embryonic development of the swimming crab, *Portunus pelagicus* (Portunidae: Decapoda). *Invert. Reprod. Develop.* **2016**, *60*, 112–117. [CrossRef]
24. Alava, V.R.; Quinitio, E.T.; De Pedro, J.B.; Priolo, F.M.P.; Orozco, Z.G.A.; Wille, M. Lipids and fatty acids in wild and pond-reared mud crab *Scylla serrata* (Forsskål) during ovarian maturation and spawning. *Aquac. Res.* **2007**, *38*, 1468–1477. [CrossRef]
25. Chang, G.; Wu, X.; Cheng, Y.; Zeng, C.; Yu, Z. Reproductive performance, offspring quality, proximate and fatty acid composition of normal and precocious Chinese mitten crab *Eriocheir sinensis. Aquaculture* **2017**, *469*, 137–143. [CrossRef]
26. Il'in, G.V.; Moiseev, D.V.; Shirokolobov, D.V.; Deryabin, A.A.; Pavlova, L.G. Long-term dynamics of hydrological conditions of the Zelenetskaya Bay, East Murman. *Vestn. MSTU* **2016**, *19*, 268–277. (In Russian)
27. Dvoretsky, A.G.; Dvoretsky, V.G. Aquaculture of green sea urchin in the Barents Sea: A brief review of Russian studies. *Rev. Aquac.* **2020**, *12*, 1280–1290. [CrossRef]
28. Dvoretsky, A.G.; Dvoretsky, V.G. New echinoderm-crab epibiotic associations from the coastal Barents Sea. *Animals* **2021**, *11*, 917. [CrossRef] [PubMed]
29. Reppond, K.; Rugolo, L.; de Oliveira, A.C. Change in biochemical composition during development of snow crab, *Chionoecetes opilio*, embryos. *J. Crustac. Biol.* **2008**, *28*, 519–527. [CrossRef]
30. Pochelon, P.N.; Da Silva, T.L.; Reis, A.; Dos Santos, A.; Queiroga, H.; Calado, R. Inter-individual and within-brood variability in the fatty acid profiles of Norway lobster, *Nephrops norvegicus* (L.) embryos. *Mar. Biol.* **2011**, *158*, 2825–2833. [CrossRef]
31. Donaldson, W.E.; Byersdorfer, S.E. *Biological Field Techniques for Lithodid Crabs*; Alaska Sea Grant College Program, University of Alaska Fairbanks: Fairbanks, AK, USA, 2005.
32. Abramoff, M.D.; Magalhães, P.J.; Ram, S.J. Image processing with Image. *J. Biophotonics Int.* **2004**, *11*, 36–41.

33. Nakanishi, T. Rearing condition of eggs, larvae and post-larvae of king crab. *Bull. Jpn. Sea Reg. Fish. Lab.* **1987**, *37*, 57–161.
34. Folch, J.; Less, M.; Sloane-Stanley, G.H. A simple method for the isolation and purification of total lipids from animal tissues. *J. Biol. Chem.* **1957**, *226*, 497–509. [CrossRef]
35. Torres, P.; Penha-Lopes, G.; Narciso, L.; Macia, A.; Paula, J. Fatty acids dynamics during embryonic development in genus *Uca* (Brachyura: Ocypodidae), from the mangroves of Inhaca Island, Mozambique. *Estuar. Coast. Shelf Sci.* **2008**, *80*, 307–313. [CrossRef]
36. Li, S.; Cheng, Y.; Zhou, B.; Hines, A.H. Changes in biochemical composition of newly spawned eggs, prehatching embryos and newly hatched larvae of the blue crab *Callinectes sapidus*. *J. Shellfish Res.* **2012**, *31*, 941–946. [CrossRef]
37. Xu, X.; Liu, X.; Tao, J. Changes in biochemical composition and digestive enzyme activity during the embryonic development of the marine crab, *Charybdis japonica* (Crustadea: Decapoda). *Zool. Sci.* **2013**, *30*, 160–166. [CrossRef] [PubMed]
38. Stevens, B.G. (Ed.) Embryo development and hatching of king crabs. In *King Crabs of the World: Biology and Fisheries Management*; CRC Press: Boca Raton, FL, USA; Taylor and Francis Group: Abingdon, UK, 2014; pp. 211–231.
39. Matyushkin, V.B. Determination of spawning periods by stages of embryo development in the red king crab *Paralithodes camtschaticus* female deposits. *Commer. Species Their Biol.* **2016**, *161*, 27–37. (In Russian)
40. de Mello, P.H.; Araujo, B.C.; Marques, V.H.; Branco, G.S.; Honji, R.M.; Moreira, R.G.; Rombenso, A.N.; Portella, M.C. Long-Chain polyunsaturated fatty acids n−3 (n−3 LC-PUFA) as phospholipids or triglycerides influence on *Epinephelus marginatus* juvenile fatty acid profile and liver morphophysiology. *Animals* **2022**, *12*, 951. [CrossRef] [PubMed]
41. Morado, F.J.; Shavey, C.A.; Ryazanova, T.; White, V.C. Diseases of king crabs and other anomalies. In *King Crabs of the World: Biology and Fisheries Management*; Stevens, B.G., Ed.; CRC Press: Boca Raton, FL, USA; Taylor and Francis Group: Abingdon, UK, 2014; pp. 139–210.
42. Subramoniam, T. *Sexual Biology and Reproduction in Crustaceans*; Academic Press: Cambridge, MA, USA; Elsevier: Amsterdam, The Netherlands, 2017.
43. Sun, P.; Jin, M.; Jiao, L.; Monroig, Ó.; Navarro, J.C.; Tocher, D.R.; Betancor, M.B.; Wang, X.; Yuan, Y.; Zhou, Q. Effects of dietary lipid level on growth, fatty acid profiles, antioxidant capacity and expression of genes involved in lipid metabolism in juvenile swimming crab, *Portunus trituberculatus*. *Brit. J. Nutr.* **2020**, *123*, 149–160. [CrossRef] [PubMed]
44. Dai, Y.J.; Jiang, G.Z.; Liu, W.B.; Abasubong, K.P.; Zhang, D.D.; Li, X.F.; Chi, C. Evaluation of dietary linoleic acid on growth as well as hepatopancreatic index, lipid accumulation oxidative stress and inflammation in Chinese mitten crabs (*Eriocheir sinensis*). *Aquac. Rep.* **2022**, *22*, 100983. [CrossRef]
45. Sullivan, M.; Su, X.Q.; Li, D. Distribution on n−3 polyunsaturated fatty acids in different edible portions of the blue swimmer crab (*Portunus pelagicus*). *Asia Pacific J. Clin. Nutr.* **2001**, *10*, S42.
46. Figueiredo, J.; Narciso, L. Egg volume, energy content and fatty acid profile of *Maja brachydactyla* (Crustacea: Brachyura: Majidae) during embryogenesis. *J. Mar. Biol. Assoc.* **2008**, *88*, 1401–1405. [CrossRef]
47. Fischer, S.; Thatje, S.; Graeve, M.; Paschke, K.; Kattner, G. Bioenergetics of early life-history stages of the brachyuran crab *Cancer setosus* in response to changes in temperature. *J. Exp. Mar. Biol. Ecol.* **2009**, *374*, 160–166. [CrossRef]
48. Figueiredo, J.; Penha-Lopes, G.; Anto, J.; Narciso, L.; Lin, J. Potential fertility and egg development (volume, water, lipid, and fatty acid content) through embryogenesis of *Uca rapax* (Decapoda: Brachyura: Ocypodidae). *J. Crustac. Biol.* **2008**, *28*, 528–533. [CrossRef]
49. Figueiredo, J.; Penha-Lopes, G.; Anto, J.; Narciso, L.; Lin, J. Fecundity, brood loss and egg development through embryogenesis of *Armases cinereum* (Decapoda: Grapsidae). *Mar. Biol.* **2008**, *154*, 287–294. [CrossRef]
50. Graeve, M.; Wehrtmann, I. Lipid and fatty acid composition of Antarctic shrimp eggs (Decapoda: Caridea). *Polar Biol.* **2003**, *26*, 55–61. [CrossRef]
51. Penha-Lopes, G.; Torres, P.; Narciso, L.; Cannicci, S.; Paula, J. Comparison of fecundity, embryo loss and fatty acid composition of mangrove crab species in sewage contaminated and pristine mangrove habitats in Mozambique. *J. Exp. Mar. Biol. Ecol.* **2009**, *381*, 25–32. [CrossRef]
52. Soundarapandian, P.; Singh, R.K. Biochemical composition of the eggs of commercially important crab *Portunus pelagicus* (Linnaeus). *Int. J. Zool. Res.* **2008**, *4*, 53–58. [CrossRef]
53. Hamid, A.; Wardiatno, Y.; Batu, D.T.L.; Riani, E. Changes in proximate and fatty acids of the eggs during embryo development in the blue swimming crab, *Portunus pelagicus* (Linnaeus 1758) at Lasongko bay, Southeast Sulawesi, Indonesia. *Indian J. Sci. Technol.* **2015**, *8*, 501–509.
54. Stevens, B.G. (Ed.) Development and biology of king crab larvae. In *King Crabs of the World: Biology and Fisheries Management*; CRC Press: Boca Raton, FL, USA; Taylor and Francis Group: Abingdon, UK, 2014; pp. 233–260.
55. Shatsky, A.V. Sea Urchins in the Genus Strongylocentrotus in the Murmansk Coast of the Barents Sea: Biology, Distribution, Harvesting Perspectives. Ph.D. Thesis, VNIRO, Moscow, Russia, 2012. (In Russian).
56. Matyushkin, V.B. Peculiarities of reproduction of the red king crab in fjord waters of the western Murman. In *The Red King Crab in the Barents Sea*; Berenboim, B.I., Ed.; PINRO Press: Murmansk, Russia, 2003; pp. 88–100. (In Russian)
57. Matyushkin, V.B. Reproductive parameters of female red king crab (*Paralithodes camtschaticus*, Tilesius) in the Ura inlet of the Barents Sea. *Tr. VNIRO* **2005**, *144*, 212–221. (In Russian)

58. Reppond, K.D. *Biochemistry of Red King Crab (Paralithodes camtschaticus) from Different Locations in Alaskan Waters. NMFS-NWFSC-102;* U.S. Department of Commerce, National Oceanic and Atmospheric Administration, National Marine Fisheries Service: Washington, DC, USA, 2009.
59. Spaziani, E.P.; Hinsch, G.W. Variation in selected unsaturated fatty acids during vitellogenesis in the Florida freshwater crayfish *Procambarus paeninsulanus*. *Invert. Reprod. Develop.* **1997**, *32*, 21–25. [CrossRef]
60. Wehrtmann, I.S.; Kattner, G. Changes in volume, biomass, and fatty acids of developing eggs in *Nauticaris magellanica* (Decapoda: Caridea): A latitudinal comparison. *J. Crustac. Biol.* **1998**, *18*, 413–422. [CrossRef]
61. Zhang, C.; Song, X.Z.; Zhang, Q.; Pang, Y.Y.; Lv, J.H.; Tang, B.P.; Cheng, Y.X.; Yang, X.Z. Changes in bud morphology, growth-related genes and nutritional status during cheliped regeneration in the Chinese mitten crab, *Eriocheir sinensis*. *PLoS ONE* **2018**, *13*, e0209617. [CrossRef]
62. Guo, F.; Lee, S.Y.; Kainz, M.J.; Brett, M.T. Fatty acids as dietary biomarkers in mangrove ecosystems: Current status and future perspective. *Sci. Total Environ.* **2020**, *739*, 139907. [CrossRef] [PubMed]
63. Tu, L.; Wu, X.; Wang, X.; Shi, W. Effects of fish oil replacement by blending vegetable oils in fattening diets on nonvolatile taste substances of swimming crab (*Portunus trituberculatus*). *J. Food Biochem.* **2020**, *44*, e13345. [CrossRef] [PubMed]
64. Xu, J.; Xu, X.; Hu, J.; Zhou, Z.; Wan, W.; Zhou, Y.; Miao, S. Effects of dietary arachidonic acid on the growth performance, feed utilization and fatty acid metabolism of Chinese mitten crab (*Eriocheir sinensis*). *Aquac. Rep.* **2022**, *24*, 101170. [CrossRef]
65. Bakanev, S.V. Fecundity and some other reproductive parameters of red king crab in the Barents Sea. In *The Red King Crab in the Barents Sea*; Berenboim, B.I., Ed.; PINRO Press: Murmansk, Russia, 2003; pp. 78–88. (In Russian)
66. Al Khawli, F.; Pateiro, M.; Domínguez, R.; Lorenzo, J.M.; Gullón, P.; Kousoulaki, K.; Ferrer, E.; Berrada, H.; Barba, F.J. Innovative green technologies of intensification for valorization of seafood and their by-products. *Mar. Drugs* **2019**, *17*, 689. [CrossRef] [PubMed]
67. Calder, P.C. Omega-3 polyunsaturated fatty acids and inflammatory processes: Nutrition or pharmacology? *Brit. J. Clin. Pharmacol.* **2013**, *83*, 150–162. [CrossRef] [PubMed]
68. Mozaffarian, D.; Wu, J.H.Y. Omega-3 fatty acids and cardiovascular disease: Effects on risk factors, molecular pathways, and clinical events. *J. Am. Coll. Cardiol.* **2011**, *58*, 2047–2067. [CrossRef]
69. Serini, S.; Calviello, G.; Trombino, S.; Piccioni, E.; Maggiano, N. Polyunsaturated fatty acids as promoters of apoptosis: Implications for cancer. *Apoptosis* **2011**, *16*, 115–126.
70. Patten, G.S.; Abeywardena, M.Y. Fish oil and colony-stimulating factor-1 improve insulin sensitivity in the obese insulin-resistant rat: Role of macrophages. *J. Nutr. Biochem.* **2014**, *25*, 1091–1098.
71. Yessoufou, A.; Ple, A.; Moutairou, K.; Hichami, A. N−3 long chain polyunsaturated fatty acids in the prevention and treatment of obesity-induced metabolic disorders. *Lipids Health Dis.* **2011**, *10*, 1–14.
72. Food and Agriculture Organization of the United Nations. *FAO Fats and Fatty Acids in Human Nutrition. Report of an Expert Consultation*; Food and Agriculture Organization of the United Nations: Rome, Italy, 2010.

animals MDPI

Article

Estimate of Growth Parameters of *Penaeus kerathurus* (Forskäl, 1775) (Crustacea, Penaeidae) in the Northern Adriatic Sea

Martina Scanu [1,2,*], Carlo Froglia [2], Fabio Grati [2] and Luca Bolognini [2,*]

[1] Department of Biological, Geological, and Environmental Sciences (BiGeA), Alma Mater Studiorum—University di Bologna, 40126 Bologna, Italy
[2] National Research Council—Institute of Marine Biological Resources and Biotechnologies (CNR IRBIM), 60125 Ancona, Italy; c.froglia@alice.it (C.F.); fabio.grati@cnr.it (F.G.)
* Correspondence: martina.scanu@irbim.cnr.it (M.S.); luca.bolognini@cnr.it (L.B.)

Simple Summary: The study focuses on the caramote prawn in the northern Adriatic Sea, GSA 17, an economically important crustacean species. Despite its increasing landings, there is a lack of comprehensive information in this region on its fishery-dependent data, age, and growth. Using modal progression analysis and the ELEFAN approach with the "TropFishR" R package, the study addresses these gaps, employing new functions with bootstrapping procedures to enhance reliability. One year of monthly length-frequency distributions (LFDs) from commercial bottom trawls reveals sexual dimorphism, with faster growth in females. These findings contribute essential insights for sustainable fisheries management in the northern Adriatic Sea, enriching the understanding of the caramote prawn's growth dynamics.

Abstract: Crustacean fisheries are gaining prominence globally amid a decline in finfish stocks. Some decapod crustacean species have experienced increased landings in response to shifting market demands and environmental dynamics. Notably, the caramote prawn (*Penaeus kerathurus*—Forskål, 1775) in the northern Adriatic Sea, Geographical Sub Area (GSA) 17, has risen in both landings and economic importance in recent years. However, despite its significance, comprehensive information on fishery-dependent data, age, and growth in this region remains lacking. To address this gap, this study employs modal progression analysis and the ELEFAN approach, utilizing the "TropFishR" R package and newly developed functions, including bootstrapping procedures. These advancements aim to overcome issues identified in previous versions and enhance the accuracy and reliability of age and growth estimations. The study leverages one year of monthly length-frequency distributions (LFDs) collected from commercial bottom trawls in the northern Adriatic Sea. The results of the analysis confirm the presence of sexual dimorphism in the caramote prawn species, with females exhibiting faster growth rates compared to males. Additionally, the growth performance index supports this observation, further underscoring the importance of accounting for sexual dimorphism in growth modeling and fisheries management strategies. By contributing to a growing body of knowledge on the growth dynamics of the caramote prawn, this study provides valuable insights for sustainable fisheries management in the northern Adriatic Sea. Understanding the age and growth patterns of key crustacean species is essential for developing effective conservation measures and ensuring the long-term health and productivity of marine ecosystems. The findings of this study serve as a foundation for informed decision-making and proactive management practices aimed at preserving the ecological integrity and economic viability of crustacean fisheries in the region.

Keywords: *Penaeus kerathurus*; growth; Adriatic Sea; caramote prawn

1. Introduction

With finfish stocks declining worldwide, the significance of crustacean fisheries is becoming increasingly pronounced [1,2]. In a countertendency to the general decline observed in marine stocks, certain decapod crustacean species' landings have notably increased over the last decade [3]. Concurrently, the global crustacean market is undergoing substantial growth [2], propelled by rising consumer demand for major crustacean species, particularly shrimps. This trend reflects not only a shift in consumer preferences but also the evolving dynamics of global seafood consumption patterns. As traditional fish stocks face mounting pressure from overfishing and environmental degradation, crustaceans emerge as a viable and increasingly sought-after alternative for seafood consumers worldwide. The inherent versatility, nutritional value, and culinary appeal of crustaceans, including shrimps, make them a staple in diverse cuisines and dining experiences across the globe [4]. Furthermore, the observed increase in decapod crustacean landings underscores the resilience and adaptability of certain marine species to changing environmental conditions and fishing pressures [5]. While challenges persist in ensuring the sustainability and long-term viability of crustacean fisheries, proactive management strategies and conservation measures can help mitigate adverse impacts and safeguard crustacean populations for future generations. As the global crustacean market continues to expand [1], stakeholders across the seafood industry must prioritize responsible harvesting practices, traceability, and transparency in supply chains to ensure the sustainability and integrity of crustacean fisheries. By promoting sustainable fishing practices, supporting ecosystem-based management approaches, and fostering collaboration among industry stakeholders, we can strive towards a more resilient and equitable seafood sector that balances ecological conservation with economic prosperity.

In the northern Adriatic Sea, specifically within Geographical Sub Area (GSA) 17, the primary crustacean species targeted for commercial catches include the Norway lobster (*Nephrops norvegicus*) [6] and the spottail mantis shrimp (*Squilla mantis*) [7]. The Norway lobster, a significant target species, is primarily harvested in the Jabuka/Pomo Pit region. It represents a target species of the fishery activity in this area. Additionally, the spottail mantis shrimp, initially regarded as a secondary catch in the common sole (*Solea solea*) gillnet fishery during the early 2000s, has since evolved into a key target species for small-scale fisheries. These fisheries utilize various types of traps to capture the spottail mantis shrimp, reflecting a shift in targeting practices over recent years.

Nevertheless, there is another crustacean in the area that is worthy of attention: the caramote prawn—*Penaeus kerathurus* (Forskål, 1775). This prawn has shown a sharp increase in landings (from 167 tons in 2004 to 676 tons in 2022 [8]), and thanks to its high commercial value [9,10], it is playing a key role in fishers' revenues [11]. This species was recorded in the Adriatic Sea for the first time only in 1863 [12], and different authors support the hypothesis of the expansion of its available habitat through a gradual meridionalization phenomenon [10] from the southern part of the Mediterranean, where it was particularly abundant [13]. However, these changes in the spatio-temporal dynamics could likely have been generated by the synergic action of multiple factors. Inter alia, the installation of breakwaters to prevent coastal erosion, resulting in the extension of suitable nursery grounds, could have enhanced the recruitment of postlarvae, and the trawling ban during part of the summer season, introduced in 1987, could have delayed the recruitment of juveniles to the fishery.

Implementing science-based management practices is paramount to achieving optimal management of fisheries [14]. To attain this objective, obtaining accurate biological parameters of the target species is crucial. This enables a deeper understanding of the temporal variations in their abundance, distribution, and biology, all of which can undergo significant changes depending on the level of exploitation [15,16].

To ensure the sustainable and profitable exploitation of crustacean species, there is a clear need to initiate a systematic process of collecting, analyzing, and reporting demographic information. These data are essential for determining changes in abundance in response to fishing pressure and for predicting future trends in stock status.

In general, to evaluate and specify the present and potential future condition of a fishery, all this collected information is integrated into a comprehensive stock assessment framework. Mathematical models, which take into account various factors causing changes in harvested fish stocks such as catch, abundance, and biology data, are utilized to calculate fishery reference points. These reference points help in comparing the current status of a stock to a desirable one, thereby aiding in determining the success of a harvest strategy [17].

By adopting such an approach, fisheries can ensure sustainable exploitation practices, maintain healthy stock levels, and promote the long-term viability of crustacean species populations. This concerted effort towards science-based management is crucial for safeguarding marine ecosystems and ensuring the continued availability of crustacean resources for future generations. Unfortunately, excluding total catches, limited information exists on the fishery-dependent data and age and growth of the caramote prawn in the northern Mediterranean region [18–22]. Understanding the biology of a species is a crucial aspect of providing scientific advice [23] to fisheries managers [24] and improving the biological consistency of stock-assessment models.

Modal progression analysis, a technique based on length, has been utilized in fisheries science since its early days to estimate the body growth of fish and aquatic invertebrates [25]. This method involves plotting histograms of fish length collected monthly, referred to as the monthly length-frequency distributions (LFDs), and connecting the peaks (modes) to monitor the progression of each cohort from one month to the next. Typically, the "von Bertalanffy growth function" (VBGF), in its standard and seasonal version, is computed from the modes observed in the monthly LFD data.

The ELEFAN I filter algorithm [26] in the R package TropFishR [27] utilizes a moving average to identify peaks and troughs in original length-frequency distributions (LFDs). Ideally, fitting algorithms should find the optimal growth model regardless of initial values. However, even the most accurate automated algorithms may get trapped in local maxima, posing challenges in locating the overall maximum within multi-dimensional search spaces [23]. To overcome this issue, Schwamborn et al. (2019) presented new algorithms for TropFishR, namely ELEFAN_GA_boot and ELEFAN_SA_boot [23].

Using the cited algorithms [23] on monthly LFDs, this work is aimed at filling in the existing gaps in the growth of the caramote prawn in the northern Adriatic Sea toward a proper evaluation of the status of the stock. The growth patterns of the majority of plant species and ectothermic animals, including fish, and especially crustaceans, which are also subject to molting, exhibit strong seasonality, influenced by factors such as temperature, light, and food availability [28]. This is why in this work both the standard and the seasonal VBGFs will be used.

2. Materials and Methods

2.1. Study Area

The northern Adriatic Sea (Figure 1) is known for its eutrophic shallow waters and expansive continental shelf, which boasts an average depth of 35 m, making it the broadest within the Mediterranean Sea. In contrast, the central part of the basin delves much deeper, plunging to depths of 270 m in the Jabuk/Pomoa Pit. The Adriatic Sea itself exhibits very distinct characteristics between its eastern and western sides. The eastern side is marked by greater depth and rocky terrain, while the western side is predominantly shallow, sandy, and heavily influenced by various river outlets that impact seawater circulation [29,30]. The circulation within the Adriatic follows a cyclonic pattern, governed by two primary currents: the eastern and the western. The eastern current moves in a northward direction along the eastern coastline, featuring three gyres. The northern gyre, affected by strong winds like the Bora and significant freshwater input, primarily from the Po River, generates the western current. This current flows southward along the Italian shores, carrying substantial amounts of nutrients [30]. The considerable nutrient influx from river discharges positions the GSA 17 as one of the most productive zones within the entire Mediterranean Sea [31]. Consequently, it stands as one of the most exploited European basins [32], owing to its high productivity [29,33]. The eutrophic nature of the northern Adriatic Sea supports a diverse array of marine life, ranging from phytoplankton to fish species, and contributes significantly to the regional economy through fisheries and tourism.

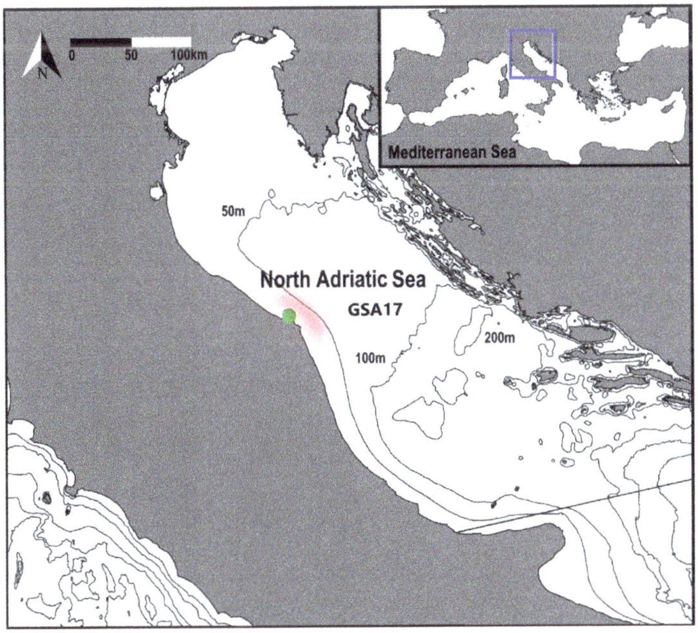

Figure 1. Study area: Northern Adriatic Sea; the green dot represents the sampling harbor, while the light red area represents the fishing ground of the bottom otter trawl and the sampling station of the SoleMon survey taken into account for this study.

Furthermore, the intricate interplay between natural processes and human activities in the Adriatic Sea underscores the importance of sustainable management practices and international cooperation to preserve its ecological integrity and ensure the continued prosperity of coastal communities that rely on its resources. Ongoing research and monitoring efforts are crucial for gaining a comprehensive understanding of the complex dynamics at play and informing effective management strategies for this unique and valuable marine ecosystem.

2.2. Monthly LFDs Collection

The VBGF has emerged as a widely utilized tool for modeling the growth of prawns within the realm of crustacean fisheries [34]. In crustacean fisheries, in particular, life history parameters are often derived from LFDs, which serve as valuable indicators of population structure and dynamics. Indeed, unlike fish, for which age can be derived from otolith reading, these species do not have hard structures from which the same estimates can be derived [35].

A temporal series of size-frequency distributions, collected monthly and stratified by sex, facilitates the tracking of size increments within each age class through modal class progression analysis [26]. This analytical approach enables the estimation of average growth rates, providing invaluable insights into the developmental trajectories of crustacean populations. To facilitate the estimation of life history parameters and growth dynamics, samplings to obtain size-frequency data were conducted in the Ancona harbor region (refer to Figure 1) aboard a commercial bottom trawler working 3–10 nautical miles off the coast (20–45 m depth), spanning from April 2021 to March 2022. The working depth of the vessel was perfectly overlapping with the common habitat of the species, which is more common until 50 m depth [20]. The vessel (F/V "Trionfo"; overall length: 14.4 m; gross tonnage: 25 GT) was equipped with a single cone-shaped net towed on the seabed. The gear was the Italian "Americana" net, having a diamond nominal mesh size of 52 mm at codend and gradually increasing mesh size to the trawl mouth. To improve the fishing efficiency, ahead of the leadline, fishers usually fit at least one metallic tickler chain to facilitate the separation of the demersal species from the sea bottom [36]; in this case, the gear was equipped with 2 chains.

Furthermore, the analysis incorporated LFDs recorded in November 2021 across seven stations within the same area, obtained during the SoleMon survey. This survey is a "rapido" trawl survey performed in GSA17 during fall season, since 2005 [33].

Additionally, LFDs of juvenile specimens sampled in August from the nearshore nursery habitat (7.5 km from the Ancona harbor, depth 0–2 m) using a small-meshed dredge were included in the analysis, enriching the dataset and enhancing the comprehensiveness of the study. The experimental dredge adopted was specifically designed for the purpose of this data collection and was manually towed, employing a rope of known length. The gear consisted of a metal frame of 73 cm width and a net bag to collect the specimens (stretched mesh size 2 mm, total length 300 cm). The frame had two sledge runners to prevent it from digging into the substratum. A small chain (mimicking the tickler one) was placed in front of the net to avoid losing contact with the bottom and to facilitate the entry of the organisms during the tow.

By integrating data collected from diverse sources and sampling methodologies, the analysis endeavors to provide a holistic understanding of the growth patterns, population dynamics, and ecological processes governing prawn populations in the northern Adriatic Sea.

2.3. VBGF Parameters Estimation

While more precise methods based on length-at-age or tagging exist, the length-based approach remains highly pertinent, particularly in circumstances where resources and available data are limited—commonly observed in data-poor situations. Moreover, tagging is impractical with small-sized shrimps, and individual length-at-age determination is unfeasible in shrimps that do not have sclerotized permanent structures keeping growth marks [35].

Electronic length frequency analysis (ELEFAN), initially outlined by Pauly and David [26] and Pauly [37], constitutes a set of fishery assessment procedures that leverage LFD data, commonly obtained from catch records and scientific surveys [38]. It is commonly used to estimate life history parameters related to growth and mortality.

This method sequentially organizes time series data, such as month-to-month LFD samples. A high-pass filter, using a moving average of the LFD, detects peaks, and a VBGF is then applied to these peaks.

Constraints often arise concerning the capacity to import data and conduct automated analyses in fisheries research. However, the "TropFishR" package stands out for its enhanced expansion and adaptability in addressing these challenges [27]. One notable feature is its incorporation of two potent optimization methods that simultaneously explore all parameters. The ELEFAN_SA function employs simulated annealing (SA), while the ELEFAN_GA function utilizes genetic algorithms (GAs) [38].

Simulated annealing (SA) is a probabilistic technique used to approximate the global optimum of a function, making it suitable for scenarios prioritizing an approximate global optimum over a precise local one [34]. On the other hand, genetic algorithms (GAs), inspired by natural selection, generate high-quality solutions to optimization problems through mutation, crossover, and selection processes [34].

This study was conducted using a promising path toward a robust and reliable approach to LFD analysis [23]. The approach was incorporated into a set of ready-to-use R functions and is the basis for the new R package "TropFishR" [39] and its development version (https://github.com/tokami/TropFishR). It contains, as well as the basic version, two optimization routines for the fitting of VBGF: one using the simulated annealing package "GenSA" [40], and the other used the genetic algorithm package "GA" [41], with ELEFAN_SA_boot and ELEFAN_GA_boot functions, respectively.

Compared to the older version of the same methodology, this version overcomes some issues identified [23,38]. Instead of striving for a singular best fit, this new approach involves presenting a range of probable best fits or, more precisely, the range where the likely mean parameters of the population are positioned (i.e., confidence intervals of the parameter estimates) [23]. The given final result is the median above all the estimated fits and can be obtained for both the classical (1) and seasonal (2) Von Bertalanffy models [42]:

$$\mathrm{CL} = \mathrm{CL_{inf}}\left[1 - e^{-K(t-t_0)}\right] \quad (1)$$

$$\mathrm{CL} = \mathrm{CL_{inf}}\left[1 - e^{-K(t-t_0)+S(t)-S(t_0)}\right] \quad (2)$$

where CL is the predicted length at age t (in years), $\mathrm{CL_{inf}}$ is the asymptotic length, K is the rate of growth toward the asymptote, and t_0 is the hypothetical age at zero length. In the seasonal (2) equation,

$$S(t) = \frac{C}{2\pi}\mathrm{Sin}\pi(t - t_0) \quad \text{and} \quad S(t) = \frac{C}{2\pi}\mathrm{Sin}\pi(t_0 - t_s)$$

where C is a parameter that measures the size of the seasonal variation in growth, while the parameter t_s is the time between t = 0 and the start of a growth oscillation.

Different adaptations of Equation (1) have been employed to represent the seasonal fluctuations in growth. Among these adaptations, Equation (2) is the most widely accepted [43].

Particularly in intricate statistical problems where the underlying distributions cannot be predetermined, this is often achieved through comprehensive nonparametric bootstrapping. Since the advent of high-speed computers, bootstrapping has become the standard method for estimating uncertainty or error in numerous methods and models [44]. Bootstrapping proves especially valuable in situations where precise analytical expressions for error terms are challenging to derive [45] or when dealing with intricate nonlinear behaviors involving multiple interacting parameters, as observed in VBGF curve fitting procedures. In addition, the ELEFAN_Boot approach has some direct benefits, among which are the unconstrained search, better reproducibility and accuracy, and assessment of uncertainty (confidence intervals) inherent in all VBGF parameter estimates, as well as the growth performance index (φ'), to also be exported to subsequent analyses [23].

3. Results

In the study area, caramote prawns were consistently landed by the commercial bottom trawler throughout the year, with the exception of August due to the summer fishing ban. A total of 2597 specimens were collected during the study period, comprising 1657 females and 940 males. The CL of the prawns ranged from 19 to 66 mm for females and from 20 to 44 mm for males, reflecting a diverse size distribution within the population.

Analysis of the LFDs revealed distinct patterns within the prawn population. The resulting distributions indicated the presence of two dominant modes, representing animals one and two years old, respectively. Additionally, a smaller group of specimens up to three years old was observed, highlighting the presence of another age class within the population (Figure 2).

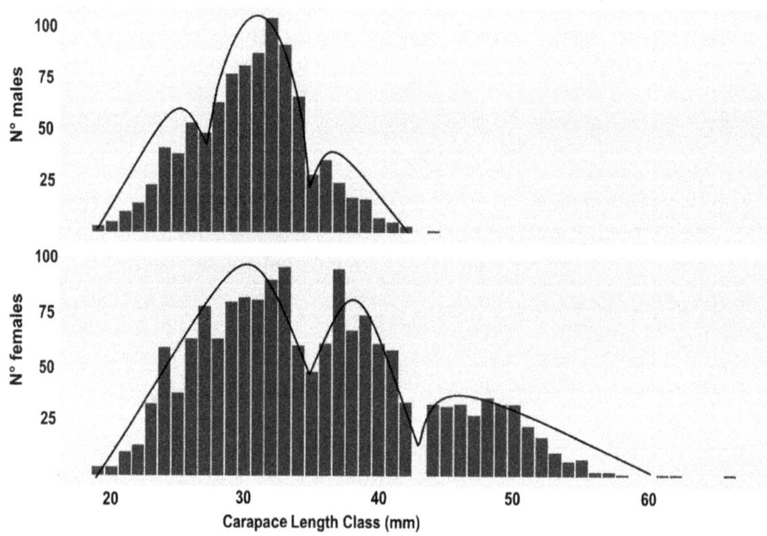

Figure 2. Overall commercial length-frequency distribution of the females and males of *Penaeus kerathurus* in the northern Adriatic Sea.

The LFDs of small specimens, unsorted by sex, obtained from the nearshore nursery ground, exhibited a CL range of 1 to 11 mm. These specimens were integrated into the analysis for both sexes, contributing to a comprehensive understanding of size distribution dynamics within the population. Notably, previous studies have identified that growth differences between sexes may emerge as early as the juvenile stages [19,46], underscoring the importance of considering sex-specific growth patterns in demographic analyses.

Given the significant disparity in size between male and female specimens [47], separate LFDs for each sex were utilized to estimate the VBGF parameters. Employing both simulated annealing and genetic algorithm methodologies, the study sought to identify the optimal fitting parameters for males and females, respectively. The results of these analyses are presented in Tables 1 and 2 for both the classical and seasonal VBGFs, providing insights into the growth dynamics and demographic characteristics of the caramote prawn population in the study area.

Table 1. Parameters result and confidence intervals from ELEFAN_SA_boot available functions, separated for females and males.

Sex	VBGF	CL_{inf}	K	t_anchor [1]	φ′ [2]	C	t_s
Females	Classic	64.18 (58.47–67.18)	0.42 (0.40–0.57)	0.63 (0.58–0.88)	3.23 (3.13–3.41)		
	Seasonal	63.77 (58.22–65.72)	0.43 (0.40–0.71)	0.63 (0.10–0.90)	3.24 (3.11–3.50)	0.41 (0.03–0.96)	0.48 (0.15–0.87)
Males	Classic	39.30 (37.90–47.30)	0.73 (0.43–0.89)	0.47 (0.20–0.75)	3.05 (2.80–3.34)		
	Seasonal	45.74 (38.16–47.27	0.85 (0.42–0.99)	0.62 (0.16–0.72)	3.25 (2.79–3.34)	0.81 (0.23–0.99)	0.60 (0.03–0.9)

[1] t_anchor describes the fraction of the year where yearly repeating growth curves have a cross length equal to zero. [2] φ′ is based on logarithmized mean asymptotic length and relative growth rate obtained from the growth model [48].

Table 2. Parameters result and confidence intervals from ELEFAN_GA_boot available functions, for females and males separated.

Sex	VBGF	CL_{inf}	K	t_anchor [1]	φ′ [2]	C	t_s
Females	Classic	61.76 (57.71–65.63)	0.47 (0.41–0.78)	0.64 (0.18–0.90)	3.25 (3.13–3.57)		
	Seasonal	61.80 (58.22–65.72)	0.49 (0.41–0.74)	0.63 (0.14–0.89)	3.27 (3.15–3.50)	0.46 (0.09–0.88)	0.47 (0.22–0.81)
Males	Classic	41.69 (38.12–46.90)	0.75 (0.50–0.90)	0.44 (0.20–0.85)	3.11 (2.86–3.30)		
	Seasonal	43.57 (38.27–46.63)	0.77 (0.49–0.95)	0.46 (0.15–0.84)	3.16 (2.85–3.31)	0.55 (0.19–0.87)	0.61 (0.10–0.87)

[1] t_anchor describes the fraction of the year where yearly repeating growth curves have a cross length equal to zero. [2] φ′ is based on logarithmized mean asymptotic length and relative growth rate obtained from the growth model [48].

4. Discussion

As reported by other authors [49,50], *P. kerathurus* exhibits a protracted spawning period that results in different microcohorts recruiting to the fishery in slightly different steps and constituting the first cohort. The fluctuating recruitment and the extended spawning period of these short-lived penaeids are reflected in a more difficult modal progression analysis and identification of each 0+ cohort [19]. It was particularly evident for females rather than for males due to relative abundances.

In this comprehensive study, the main recruitment event was identified immediately following the summer fishing ban, between September and October (Figure 3). Across near Mediterranean regions, the migration of young individuals from inshore to deeper waters was observed from September to February [18,51,52], while in the Gulf of Gabes, this migration extended until April [53], possibly influenced by variations in water temperature [54].

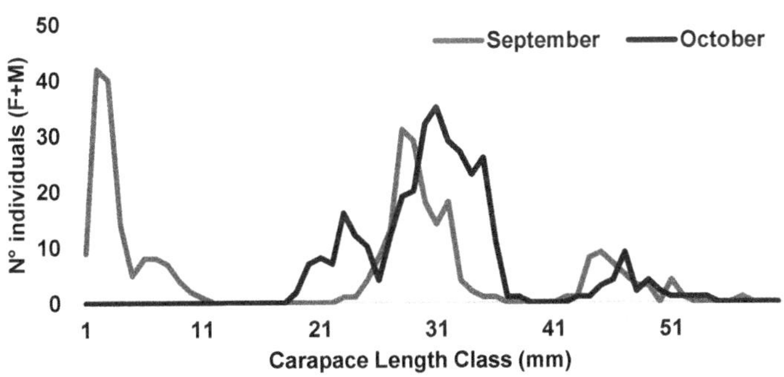

Figure 3. Length-frequency distribution of the female and male individuals of *Penaeus kerathurus* combined underlining the recruitment peak.

In the context of age-classification, both females and males were grouped into one to four age classes per month and one to three age classes per month, respectively. These classifications delineate three distinct generations: 0+, 1+, and 2+. Notably, the final age class typically comprises a minimal number of individuals, a phenomenon attributed to either high natural mortality rates or gear selectivity patterns. This trend is particularly pronounced in areas characterized by intensive fishing activities, such as the northern Adriatic Sea [55,56], where the availability of individuals for capture is typically limited.

The observed growth patterns of the caramote prawn exhibited accelerated growth during the first year of life compared to the second year. Furthermore, the study corroborated the presence of strong sexual dimorphism [47], a finding consistent with data collected and analyzed in other regions [19], irrespective of the algorithm utilized for estimation. Additionally, the growth performance index [48] demonstrated a higher value for females compared to males, a trend observed in prior studies [19,50].

To contextualize these findings, a comparative analysis of growth parameters estimated for the caramote prawn across different geographic areas of the Mediterranean Sea is presented in Table 3. This comparative framework provides valuable insights into the variability of growth patterns exhibited by the species across diverse ecological contexts, shedding light on regional nuances and contributing to our broader understanding of population dynamics and ecological processes in the Mediterranean Sea. The φ′ index, able to provide information about the performance of stocks in different areas and under certain environmental stress conditions [57], highlights more favorable conditions in Greece, followed by Algeria, Italy, and Tunisia.

Table 3. Growth parameters of *P. kerathurus* estimated in the Mediterranean Sea.

Sex	CL_{inf} *	K	φ′	Area	Reference
Females	54.25 *	0.6	3.25^2	Gulf of Gabès (Tunisia)	[53]
	64.14	0.8	3.52	Gulf of Annaba (Algeria)	[53]
	66.15 *	0.57	-	Amvrakikos Gulf (Greece)	[58]
	69.04	1.06	3.71	Amvrakikos Gulf (Greece)	[18]
	62.48	1.15	3.65	Thermaikos Gulf (Greece)	[19]
	64.9	0.7	-	Off N coast Tuscany (Italy)	[59]
	77.5	0.55	3.54	Gulf of Annaba (Algeria)	[50]
	61.45 *	0.7	2.58^2	Gulf of Tunis (Tunisia)	[60]
	60.23 *	0.76	2.64	Gulf of Gabès (Tunisia)	[60]
	62.84^1	0.46^1	3.24	Northern Adriatic Sea (Italy)	Current study
Males	37.46 *	0.78	3.04^2	Gulf of Gabès (Tunisia)	[53]
	45.5	1	3.52	Gulf of Annaba (Algeria)	[53]
	57.53 *	0.47	-	Amvrakikos Gulf (Greece)	[58]
	62.66	1.25	3.69	Amvrakikos Gulf (Greece)	[18]
	47.78	1.28	3.47	Thermaikos Gulf (Greece)	[19]
	46	0.9	-	Off N coast Tuscany (Italy)	[59]
	64.92	0.59	3.40	Gulf of Annaba (Algeria)	[50]
	58.59 *	0.73	2.55^2	Gulf of Tunis (Tunisia)	[60]
	59.61 *	0.76	2.63	Gulf of Gabès (Tunisia)	[60]
	43.27^1	0.77^1	3.16	Northern Adriatic Sea (Italy)	Current study
Combined	72	0.78	3.61	South of Sicily (Italy)	[61]
	60.9	0.24	-	Marmara Sea (Turkey)	[62]

* When the total length at infinitive was estimated, to make it comparable, it was converted to CL using the allometric relationship: females $\log CL = -0.841 + 1.112 \log TL$ and males $\log CL = -0.628 + 1.002 \log$ [63]. [1] Values were averaged between the two algorithms and the two VBGF configurations used. [2] Estimated using total length of the individuals, while monthly LFDs and their fitting to parameters are in Supplementary Materials.

The broad range of variation observed in the estimates of key parameters for the same species across different geographical areas underscores the multifaceted nature of ecological dynamics and scientific inquiry. Several factors contribute to this variability, highlighting the intricate interplay between environmental conditions, sampling methodologies, and analytical techniques.

Firstly, environmental factors such as temperature and prey availability exert a significant influence on the growth and development of marine organisms [19,50]. Variations in habitat characteristics and resource availability across different regions can result in distinct physiological responses and growth trajectories among populations of the same species. Therefore, differences in environmental conditions must be carefully considered when interpreting and comparing parameter estimates across diverse geographic areas.

Secondly, the strategy employed for sample collection and generation of LFDs can significantly impact the accuracy and representativeness of data obtained. Variability in sampling protocols, including sampling frequency, spatial coverage, and gear selectivity, can introduce biases and discrepancies in the resulting datasets, affecting the reliability of parameter estimates derived from them.

Finally, the methodological approach used for estimating and processing key parameters such as growth rates, age-classification, and population dynamics can vary widely among studies. Differences in modeling techniques, statistical methodologies, and data processing procedures can lead to divergent results and interpretations, contributing to the observed variability in parameter estimates across studies and geographic regions.

By acknowledging and addressing these sources of variability, researchers can enhance the robustness and comparability of data generated from different studies, facilitating more accurate assessments of species dynamics and ecosystem health. Collaborative efforts aimed at standardizing sampling protocols, adopting transparent methodologies, and promoting data sharing can foster greater consistency and coherence in scientific research, ultimately advancing our understanding of marine ecosystems and informing evidence-based conservation and management strategies.

5. Conclusions

Understanding the biology of a species is a fundamental and pivotal aspect in providing scientific guidance for effective fisheries management, as underscored by research in this field [24]. Achieving sustainability in the harvesting of fish stocks necessitates the acquisition and utilization of precise information pertaining to population dynamics [64], forming the cornerstone of responsible and well-informed fisheries management practices.

In the quest for sustainable fisheries management, the significance of accurate data cannot be overstated. It serves as the bedrock for making informed decisions that contribute to the preservation of marine ecosystems and the long-term viability of fisheries resources. The reliance on comprehensive information about population dynamics is particularly crucial in assessing and predicting the health of fish stocks, aiding in the development of strategies that balance conservation objectives with the socio-economic needs of communities dependent on these resources.

At the Mediterranean level, the caramote prawn emerges as a crucial fishery resource across the entire basin [49,54,60,65]. Despite facing fishing pressure, it also contends with competition from a very similar and invasive species, namely *Penaeus aztecus* Ives, 1891, native to the NW Atlantic and Gulf of Mexico, which has shown significant dispersion within the region during the last 15 years [65,66]. However, it is concerning that there is still no comprehensive stock assessment evaluating the status of this resource in any of the various management areas (GSAs). This underscores the urgent need for concerted efforts towards assessing and managing caramote prawn populations. It is hoped that this study will contribute valuable insights to initiate such assessments, particularly in the Adriatic region, where the importance of this resource has been steadily increasing, particularly in recent years [11].

In the context of fisheries management, the utilization of algorithms plays a vital role in processing and analyzing data. In instances where differences between two algorithms are found to be negligible, a pragmatic approach emerges, suggesting that average values can be effectively employed for stock assessments [34]. By employing various methodologies to model biological growth, this study aims to hypothetically accommodate all types of stock assessment models, ranging from those that solely consider parameters like growth rate (k) and asymptotic length (L_{inf}) to those capable of incorporating more complex factors such as relative amplitude of the seasonal oscillations (C), with the phase of the seasonal oscillations denoting the time of year corresponding to the start of the convex segment of sinusoidal oscillation (when growth turns positive) (t_s). Through this comprehensive approach, the research endeavors to provide a robust framework capable of capturing the intricacies of biological growth across a spectrum of modeling paradigms, thereby enhancing our understanding and management of natural resources. In fact, this pragmatic strategy not only streamlines the assessment process but also ensures a practical and efficient utilization of resources in fisheries management.

The amalgamation of scientific insights, data-driven algorithms, and pragmatic approaches creates a robust framework for fisheries management, where the dual goals of conservation and sustainable resource utilization are harmonized. It highlights the importance of an interdisciplinary approach that incorporates biological understanding, technological advancements, and adaptive management strategies to address the complex challenges posed by fisheries management in an ever-changing environment.

In summary, the integration of precise biological knowledge, advanced algorithms, and pragmatic methodologies is paramount for achieving effective fisheries management. This holistic approach not only enhances our comprehension of population dynamics but also facilitates the development of strategies that promote sustainable harvesting practices, safeguarding the delicate balance between ecological preservation and human livelihoods.

Supplementary Materials: The following supporting information can be downloaded at https://www.mdpi.com/article/10.3390/ani14071068/s1: Figure S1: Monthly LFDs of female individuals together with the fitting of the averaged seasonal parameters; Figure S2: Monthly LFDs of male individuals together with the fitting of the averaged seasonal parameters.

Author Contributions: Conceptualization, M.S., C.F., F.G. and L.B.; data curation, analysis, M.S.; writing—original draft preparation, M.S.; writing—review and editing, C.F. and F.G.; supervision, L.B. All authors have read and agreed to the published version of the manuscript.

Funding: This research received no external funding.

Institutional Review Board Statement: Ethical review and approval were waived for this study due to the utilization of length data obtained from surveys and commercial catches. No manipulation of the organisms themselves was conducted, as the study solely relied on pre-existing data collected from non-invasive sources.

Informed Consent Statement: Not applicable.

Data Availability Statement: The raw data supporting the conclusions of this article will be made available by the authors on request.

Acknowledgments: The research leading to these results has been conceived under the International Ph.D. Program "Innovative Technologies and Sustainable Use of Mediterranean Sea Fishery and Biological Resources" (www.FishMed-PhD.org). This study represents partial fulfillment of the requirements for the Ph.D. thesis of MS.

Conflicts of Interest: The authors declare no conflicts of interest.

References

1. Boenish, R.; Kritzer, J.P.; Kleisner, K.; Steneck, R.S.; Werner, K.M.; Zhu, W.; Schram, F.; Rader, D.; Cheung, W.; Ingles, J.; et al. The Global Rise of Crustacean Fisheries. *Front. Ecol. Environ.* **2022**, *20*, 102–110. [CrossRef]
2. FAO. *The State of Mediterranean and Black Sea Fisheries 2023*; FAO: Rome, Italy, 2023; ISBN 978-92-5-138411-4.
3. Grati, F.; Fabi, G.; Lucchetti, A.; Consoli, P. Analisi Delle Catture Si Solea Vulgaris Quensel, 1806 Effettuate Con Reti Ad Imbrocco in Adriatico Settentrionale. *Biol. Mar. Mediterr.* **2002**, *90*, 154–160.
4. Dayal, J.S.; Ponniah, A.G.; Khan, H.I.; Babu, E.P.M.; Ambasankar, K.; Kumarguru, K.P. Shrimps—A Nutritional Perspective. *Curr. Sci.* **2013**, *104*, 1487–1491.
5. Champion, C.; Broadhurst, M.K.; Ewere, E.E.; Benkendorff, K.; Butcherine, P.; Wolfe, K.; Coleman, M.A. Resilience to the Interactive Effects of Climate Change and Discard Stress in the Commercially Important Blue Swimmer Crab (*Portunus armatus*). *Mar. Environ. Res.* **2020**, *159*, 105009. [CrossRef] [PubMed]
6. Angelini, S.; Martinelli, M.; Santojanni, A.; Colella, S. Biological Evidence of the Presence of Different Subpopulations of Norway Lobster (*Nephrops norvegicus*) in the Adriatic Sea (Central Mediterranean Sea). *Fish. Res.* **2020**, *221*, 105365. [CrossRef]
7. Grati, F.; Aladzuz, A.; Azzurro, E.; Bolognini, L.; Carbonara, P.; Cobani, M.; Domenichetti, F.; Dragicevic, B.; Dulcic, J.; Đurovic, M.; et al. Seasonal Dynamics of Small-Scale Fisheries in the Adriatic Sea. *Mediterr. Mar. Sci.* **2018**, *19*, 21. [CrossRef]
8. European Commission (EC). *Regulation (EU) 2017/1004 on the Establishment of a Union Framework for the Collection, Management and Use of Data in the Fisheries Sector and Support for Scientific Advice Regarding the Common Fisheries Policy and Repealing Council Regulation (EC) No 199/2008*; European Commission (EC): Brussels, Belgium, 2017.
9. Zitari-Chatti, R.; Chatti, N.; Elouaer, A.; Said, K. Genetic Variation and Population Structure of the Caramote Prawn *Penaeus kerathurus* (Forskäl) from the Eastern and Western Mediterranean Coasts in Tunisia. *Aquac. Res.* **2008**, *39*, 70–76. [CrossRef]

10. Froglia, C.; Scarcella, G.; Lucchetti, A. On the Recent Increase of Penaeus (Melicertus) Kerathurus Stock in Northern and Central Adriatic Sea: Possible Explanations. *Rapp. Comm. Int. Mer Médit.* **2013**, *40*, 780.
11. Lucchetti, A. La Mazancolla-*Melicertus kerathurus* (Forskal, 1755). *Il Pesce* **2006**, 117.
12. Heller, C. *Die Crustaceen Des Südlichen Europa: Crustacea Podophthalmia. Mit Einer Übersicht Über Die Horizontale Verbreitung Sämmtlicher Europäischer Arten, von Camil Heller*; W. Braumüller: Wien, NY, USA, 1863.
13. Jaziri, H.; Khoufi, W.; Meriem, S.B. Assessment Approach of *Melicertus kerathurus* Stock along the North-Eastern Tunisian Coast Using a Surplus Production Model Incorporating Temperature Parameter. *Am. J. Clim. Chang.* **2015**, *4*, 417–430. [CrossRef]
14. Pikitch, E.K.; Santora, C.; Babcock, E.A.; Bakun, A.; Bonfil, R.; Conover, D.O.; Dayton, P.; Doukakis, P.; Fluharty, D.; Heneman, B.; et al. Ecosystem-Based Fishery Management. *Science* **2004**, *305*, 346–347. [CrossRef] [PubMed]
15. Ligas, A.; Sartor, P.; Colloca, F. Trends in Population Dynamics and Fishery of *Parapenaeus longirostris* and *Nephrops norvegicus* in the Tyrrhenian Sea (NW Mediterranean): The Relative Importance of Fishery and Environmental Variables. *Mar. Ecol.* **2011**, *32*, 25–35. [CrossRef]
16. Galimany, E.; Baeta, M.; Durfort, M.; Lleonart, J.; Ramón, M. Reproduction and Size at First Maturity in a Mediterranean Exploited *Callista chione* Bivalve Bed. *Sci. Mar.* **2015**, *79*, 233–242. [CrossRef]
17. Punt, A.E. Those Who Fail to Learn from History Are Condemned to Repeat It: A Perspective on Current Stock Assessment Good Practices and the Consequences of Not Following Them. *Fish. Res.* **2023**, *261*, 106642. [CrossRef]
18. Conides, A.; Glamuzina, B.; Jug-dujakovic, J.; Papaconstantinou, C.; Kapiris, K. Age, Growth, and Mortality of the Karamote Shrimp, *Melicertus kerathurus* (Forskål, 1775), in the East Ionian Sea (Western Greece). *Crustaceana* **2006**, *79*, 33–52. [CrossRef]
19. Kevrekidis, K.; Thessalou-Legaki, M. Population Dynamics of *Melicertus kerathurus* (Decapoda: Penaeidae) in Thermaikos Gulf (N. Aegean Sea). *Fish. Res.* **2011**, *107*, 46–58. [CrossRef]
20. Kevrekidis, K.; Thessalou-Legaki, M. Catch Rates, Size Structure and Sex Ratio of *Melicertus kerathurus* (Decapoda: Penaeidae) from an Aegean Sea Trawl Fishery. *Fish. Res.* **2006**, *80*, 270–279. [CrossRef]
21. Akyol, O.; Ceyhan, T. Catch per Unit Effort of Coastal Prawn Trammel Net Fishery in Izmir Bay, Aegean Sea. *Medit. Mar. Sci.* **2009**, *10*, 19. [CrossRef]
22. Ihsanoglu, M.A.; Daban, I.B.; Işmen, A.; Cabbar, K.; Yığin, C.Ç. Reproductive Biology of *Penaeus kerathurus* (Forskål, 1775) (Decapoda: Penaeidae) in the Sea of Marmara, Turkey. *Oceanol. Hydrobiol. Stud.* **2021**, *50*, 33–37. [CrossRef]
23. Schwamborn, R.; Mildenberger, T.K.; Taylor, M.H. Assessing Sources of Uncertainty in Length-Based Estimates of Body Growth in Populations of Fishes and Macroinvertebrates with Bootstrapped ELEFAN. *Ecol. Model.* **2019**, *393*, 37–51. [CrossRef]
24. Morgan, M.J. Integrating Reproductive Biology into Scientific Advice for Fisheries Management. *J. Northwest Atl. Fish. Sci.* **2008**, *41*, 37–51. [CrossRef]
25. Petersen, C.G.J. Eine Methode zur Bestimmung des Alters und Wuchses der Fische. *Mitteilungen. Dtsch. Seefischerei-Ver.* **1891**, *11*, 226–235.
26. Pauly, D.; David, N. ELEFAN I, a BASIC Program for the Objective Extraction Datal of Growth Parameters from Length-Frequency. In *Berichte der Deutshcen Wissenchaftlichen Kommission für Meeresforschung*; P. Parey: Berlin, Germany, 1981; Volume 28, pp. 205–211.
27. Mildenberger, T.K.; Taylor, M.H.; Wolff, M. TropFishR: An R Package for Fisheries Analysis with Length-Frequency Data. *Methods Ecol. Evol.* **2017**, *8*, 1520–1527. [CrossRef]
28. García-Berthou, E.; Carmona-Catot, G.; Merciai, R.; Ogle, D.H. A Technical Note on Seasonal Growth Models. *Rev. Fish Biol. Fish.* **2012**, *22*, 635–640. [CrossRef]
29. Campanelli, A.; Grilli, F.; Paschini, E.; Marini, M. The Influence of an Exceptional Po River Flood on the Physical and Chemical Oceanographic Properties of the Adriatic Sea. *Dyn. Atmos. Ocean.* **2011**, *52*, 284–297. [CrossRef]
30. Marini, M.; Jones, B.H.; Campanelli, A.; Grilli, F.; Lee, C.M. Seasonal Variability and Po River Plume Influence on Biochemical Properties along Western Adriatic Coast. *J. Geophys. Res. Ocean.* **2008**, *113*, 1–18. [CrossRef]
31. Bastari, A.; Micheli, F.; Ferretti, F.; Pusceddu, A.; Cerrano, C. Large Marine Protected Areas (LMPAs) in the Mediterranean Sea: The Opportunity of the Adriatic Sea. *Mar. Policy* **2016**, *68*, 165–177. [CrossRef]
32. Pellini, G.; Gomiero, A.; Fortibuoni, T.; Ferrà, C.; Grati, F.; Tassetti, A.N.; Polidori, P.; Fabi, G.; Scarcella, G. Characterization of Microplastic Litter in the Gastrointestinal Tract of *Solea solea* from the Adriatic Sea. *Environ. Pollut.* **2018**, *234*, 943–952. [CrossRef] [PubMed]
33. Grati, F.; Scarcella, G.; Polidori, P.; Domenichetti, F.; Bolognini, L.; Gramolini, R.; Vasapollo, C.; Giovanardi, O.; Raicevich, S.; Celić, I.; et al. Multi-Annual Investigation of the Spatial Distributions of Juvenile and Adult Sole (*Solea solea* L.) in the Adriatic Sea (Northern Mediterranean). *J. Sea Res.* **2013**, *84*, 122–132. [CrossRef]
34. Zhou, S.; Hutton, T.; Lei, Y.; Miller, M.; Van Der Velde, T.; Deng, R. Modelling Growth of Red Endeavour Prawns (Metapenaeus ensis) Using New ELEFAN and Bayesian Growth Models. 2021.
35. Kilada, R.; Driscoll, J.G. Age Determination in Crustaceans: A Review. *Hydrobiologia* **2017**, *799*, 21–36. [CrossRef]
36. Lucchetti, A.; Petetta, A.; Bdioui, M.; Gökçe, G.; Saber, M.; Sacchi, J.; Ozbilgin, H.; Carlson, A.; Carpentieri, P. *Catalogue of Fishing Gear in the Mediterranean and Black Sea Region*; FAO Fisher: Rome, Italy, 2023; ISBN 9789251380550.
37. Pauly, D.; Murphy, G.I. *Theory and Management of Tropical Fisheries*; ICLARM: Manila, Philippines, 1982; p. 360.
38. Taylor, M.H.; Mildenberger, T.K. Extending Electronic Length Frequency Analysis in R. *Fish. Manag. Ecol.* **2017**, *24*, 330–338. [CrossRef]

39. Mildenberger, T.; Taylor, M.; Wolff, M.; TropFishR: Tropical Fisheries Analysis with R. R Package Version 1.6.0. Available online: https://cran.r-project.org/package=TropFishR (accessed on 15 December 2023).
40. Xiang, Y.; Gubian, S.; Suomela, B.; Hoeng, J. Generalized Simulated Annealing for Global Optimization: The GenSA Package. *R J.* **2013**, *5*, 13–28. [CrossRef]
41. Scrucca, L. GA: A Package for Genetic Algorithms in R. *J. Stat. Softw.* **2013**, *53*, 1–37. [CrossRef]
42. Von Bertalanffy, L. A Quantitative Theory of Organic Growth (Inquiries on Growth Laws. II). *Hum. Biol.* **1938**, *10*, 181–213.
43. Ogle, D.H. An Algorithm for the von Bertalanffy Seasonal Cessation in Growth Function of Pauly et al. (1992). *Fish. Res.* **2017**, *185*, 1–5. [CrossRef]
44. Cai, B.; Pellegrini, F.; Pang, M.; de Moor, C.; Shen, C.; Charu, V.; Tian, L. Bootstrapping the Cross-Validation Estimate. *arXiv* **2023**, arXiv:2307.00260. [CrossRef]
45. Stine, R. An Introduction to Bootstrap Methods. *Sociol. Methods Res.* **1989**, *18*, 243–291. [CrossRef]
46. Rodríguez, A. Biología Del Langostino *Penaeus kerathurus* (Forskål, 1755) Del Golfo de Cádiz. II: Distribución y Estructura de La Población. *Investig. Pesq.* **1986**, *49*, 581–595.
47. Sardà, F. Bio-Ecological Aspects of the Decapod Crustacean Fisheries in the Western Mediterranean. *Aquat. Living Resour.* **1993**, *6*, 299–305. [CrossRef]
48. Pauly, D.; Munro, J.L. Once More on the Comparison of Growth in Fish and Invertebrates. *Fishbyte* **1984**, *2*, 1–21.
49. Kevrekidis, K. Population Structure, Sex Ratio, CPUE and Reproductive Aspects of *Melicertus kerathurus* and *Penaeus aztecus* in the Thermaikos Gulf, Aegean Sea. *Cah. Biol. Mar.* **2021**, *62*, 268–283.
50. Morghad, H.; Derbal, F.; Rachedi, M. Reproduction, Growth, Mortality and Exploitation of *Penaeus kerathurus* (Forskål, 1775) from the Eastern Coasts of Algeria. *Thalassas* **2023**, 1–12. [CrossRef]
51. Klaoudatos, S.; Tsevis, N.; Conides, A. Studies on Migratory Movements of the Prawn *Penaeus kerathurus* (Forskal, 1775) at Amvrakikos Gulf, Western Greece. *Mar. Ecol.* **1992**, *13*, 133–147. [CrossRef]
52. Kevrekidis, K.; Thessalou-Legaki, M. Reproductive Biology of the Prawn *Melicertus kerathurus* (Decapoda: Penaeidae) in Thermaikos Gulf (N. Aegean Sea). *Helgol. Mar. Res.* **2013**, *67*, 17–31. [CrossRef]
53. Meriem, S.B. Premire Approche de La Croissance de *Penaeus kerathurus* (Decapoda, Penaeidae) Dans le Golfe de Gabès, Tunisie [First Approach to the Growth of *Penaeus kerathurus* (Decapoda, Penaeidae) in the Gulf of Gabes, Tunisia]. *Crustaceana* **2004**, *77*, 277–297. [CrossRef]
54. Bolognini, L.; Froglia, C.; Guicciardi, S.; Scanu, M.; Grati, F. Effects of Breakwater Deployment on the Life-History Traits of the Caramote Prawn *Penaeus kerathurus* (Forskål, 1775) in the Adriatic Sea. In Proceedings of the ECSA 58—EMECS 13. Estuaries and Coastal Seas in the Anthropocene-Structure, Functions, Services and Management, Hull, UK, 7–11 September 2020.
55. Ferrà, C.; Tassetti, A.N.; Grati, F.; Pellini, G.; Polidori, P.; Scarcella, G.; Fabi, G. Mapping Change in Bottom Trawling Activity in the Mediterranean Sea through AIS Data. *Mar. Policy* **2018**, *94*, 275–281. [CrossRef]
56. Amoroso, R.O.; Pitcher, C.R.; Rijnsdorp, A.D.; McConnaughey, R.A.; Parma, A.M.; Suuronen, P.; Eigaard, O.R.; Bastardie, F.; Hintzen, N.T.; Althaus, F.; et al. Bottom Trawl Fishing Footprints on the World's Continental Shelves. *Proc. Natl. Acad. Sci. USA* **2018**, *115*, E10275–E10282. [CrossRef] [PubMed]
57. Pauly, D. Growth Performance in Fishes: Rigorous Description of Patterns as a Basis for Understanding Causal Mechanisms. *Aquabyte* **1991**, *4*, 3–5.
58. Conides, A.; Klaoudatos, S.; Tsevis, N. Study on the Growth Rates of the Prawn *Penaeus kerathurus* in Amvrakikos Gulf. In Proceedings of the 3rd Panhellenic Congress of Oceanography and Fisheries, Hersonissos, Greece, 6–9 May 1990; pp. 610–619.
59. Righini, P.; Baino, R.; Cecchi, A. Note Sulla Biologia e Sui Parametri di Crescita di *Penaeus kerathurus* (Crustacea: Decapoda) Lungo la Costa Toscana. *Biol. Mar. Mediterr.* **1998**, *5*, 836–838.
60. Jaziri, H.; Khoufi, W.; Ben Meriem, S. Characterization of Two Substocks of *Penaeus kerathurus* (Forskål, 1775) Using the Life-History Parameters. *Russ. J. Mar. Biol.* **2020**, *46*, 421–430. [CrossRef]
61. Vitale, S.; Cannizzaro, L.; Lumare, L.; Mazzola, S. Population Parameters of *Melicertus kerathurus* (Decapoda, Penaeidae) in Southwest Sicilian Shallow Waters (Mediterranean Sea) Using Length-Frequency Analysis. *Crustaceana* **2010**, *83*, 997–1007. [CrossRef]
62. İhsanoğlu, M. Less Known Aspects of *Penaeus kerathurus* (Forskål, 1775) (Decapoda, Penaeidae) Obtained from the Fishermen in the Sea of Marmara: Age, Growth, and Mortality Rates. *Crustaceana* **2020**, *93*, 1185–1195. [CrossRef]
63. Rodríguez, A. Biología Del Langostino *Penaeus kerathurus* (Forskål, 1775) Del Golfo de Cádiz. III: Biometría, Edad y Cresimiento. *Investig. Pesq.* **1987**, *51*, 23–37.
64. Kuparinen, A.; Mäntyniemi, S.; Hutchings, J.A.; Kuikka, S. Increasing Biological Realism of Fisheries Stock Assessment: Towards Hierarchical Bayesian Methods. *Environ. Rev.* **2012**, *20*, 135–151. [CrossRef]

65. Spinelli, A.; Sendín Baquero, P.; Tiralongo, F. Westward Expansion of the Brown Shrimp *Penaeus aztecus* Ives 1891 (Decapoda: Penaeidae) in the Mediterranean Sea: A Review on the Mediterranean Distribution and First Record from Spain. *Nat. Hist. Sci.* **2023**, *1891*. [CrossRef]
66. Froglia, C.; Scanu, M. Notes on the Spreading of *Penaeus aztecus* Ives 1891 (Decapoda, Penaeidae) in the Mediterranean Sea and on Its Repeated Misidentifications in the Region. *Biology* **2023**, *12*, 793. [CrossRef]

animals

MDPI

Article

Rose or Red, but Still under Threat: Comparing Microplastics Ingestion between Two Sympatric Marine Crustacean Species (*Aristaeomorpha foliacea* and *Parapenaeus longirostris*)

Laura Ciaralli [1,2], Tommaso Valente [1], Eleonora Monfardini [1,3], Giovanni Libralato [2], Loredana Manfra [1,4], Daniela Berto [5], Federico Rampazzo [5], Giorgia Gioacchini [6], Giulia Chemello [6], Raffaella Piermarini [1], Cecilia Silvestri [1] and Marco Matiddi [1,*]

1 ISPRA, Italian Institute for Environmental Protection and Research, Centro Nazionale Laboratori, Necton Lab, Via del Fosso di Fiorano 64, 00143 Rome, Italy; laura.ciaralli@unina.it (L.C.); tommaso.valente@isprambiente.it (T.V.); eleonoramonfardini94@gmail.com (E.M.); loredana.manfra@isprambiente.it (L.M.); raffaella.piermarini@isprambiente.it (R.P.); cecilia.silvestri@isprambiente.it (C.S.)

2 Department of Biology, University of Naples Federico II, Via Vicinale Cupa Cinthia 26, 80126 Naples, Italy; giovanni.libralato@unina.it

3 PhD Program in Evolutionary Biology and Ecology, Department of Biology, University of Rome 'Tor Vergata', Via della Ricerca Scientifica snc, 00133 Rome, Italy

4 Department of Ecosustainable Marine Biotechnology, Villa Comunale, Stazione Zoologica Anton Dohrn, 80121 Naples, Italy

5 ISPRA, Italian National Institute for Environmental Protection and Research, Via Padre Venturini snc, Loc. Brondolo, 30015 Chioggia, Italy; daniela.berto@isprambiente.it (D.B.); federico.rampazzo@isprambiente.it (F.R.)

6 Department of Life and Environmental Sciences (DiSVA), Polytechnic University of Marche, 60131 Ancona, Italy; giorgia.gioacchini@staff.univpm.it (G.G.); g.chemello@staff.univpm.it (G.C.)

* Correspondence: marco.matiddi@isprambiente.it

Simple Summary: This research investigated microplastic ingestion in two marine crustacean species of high commercial importance, namely the Giant Red Shrimp *Aristaeomorpha foliacea* and the Deep-Water Rose Shrimp *Parapenaeus longirostris*. The primary purpose of this study was to better understand how these species are affected by microplastic pollution, a growing concern in the marine environment worldwide. Based on stable isotope analysis of muscle tissue and on the examination of their gastrointestinal tracts, it was found that the trophic niche of the two species is similar, but the type of particles ingested differed significantly in terms of shape, colour, size class, and polymer. These outcomes shed light on the pervasive impact of microplastic contamination on marine wildlife, emphasising potential threats to consumers at higher levels of the food web. A better knowledge of pathways that microplastics follow through marine food webs is crucial for understanding the ecological implications posed by this emerging contaminant.

Abstract: Increasing plastic contamination poses a serious threat to marine organisms. Microplastics (MPs) ingestion can represent a risk for the organism itself and for the ultimate consumer. Through the analysis of the gastrointestinal tract, coupled with stable isotope analysis on the muscle tissue, this study provides insights into the relationship between MPs pollution and ecology in two commercial marine species caught in the Central Tyrrhenian Sea: *Aristaeomorpha foliacea* and *Parapenaeus longirostris*. Stable isotope analysis was conducted to determine the trophic position and the trophic niche width. The gastrointestinal tracts were processed, and the resultant MPs were analysed under FT-IR spectroscopy to estimate the occurrence, abundance, and typology of the ingested MPs. The trophic level of the species was similar (*P. longirostris* TP = 3 ± 0.10 and *A. foliacea* TP = 3.1 ± 0.08), with an important trophic niche overlap, where 38% and 52% of *P. longirostris* and *A. foliacea* has ingested MPs, respectively. Though species-level differences may not be evident regarding MP's abundance per individual, a high degree of dissimilarity was noted in the typologies of ingested particles. This research provides valuable insights into how MPs enter marine trophic webs, stressing that isotopic niche analysis should be combined with other methods to explain in detail the differences in MPs ingestion.

Keywords: plastic pollution; MPs; trophic ecology; stable isotope analysis; decapoda; shrimps

1. Introduction

Plastics are synthetic polymers mostly derived from petrol-based and non-renewable sources that, once dispersed, become persistent environmental contaminants. Currently, commercial-scale production of plastic generates extensive amounts of polymers [1]. Indeed, synthetic polymers possess a set of technical features that make plastic materials unique, such as lightweight, flexibility, versatility, durability, thermal and electrical insulation, impermeability, and cost-effectiveness, which have greatly contributed to their rapid spreading [2]. Approximately 50% of plastic material is allocated to the production of single-use items, including straws [3], disposable carrier bags [4], food packaging, plastic wrappers, plastic cutlery, and cans [2]. In 2021, it was estimated that plastic waste was being produced at a rate of 300 million tons per year [5] and that plastic degradability in the natural environment can take from 58 up to 1200 years [1]. Therefore, polymers have the tendency to accumulate in ecosystems, leading to multiple negative repercussions for several species [6–8].

Anthropogenic activities in both terrestrial and marine environments contribute to the ongoing discharge of plastic into the marine ecosystem. Several combined factors contribute to the spatial distribution of plastics [9–11], and considering a semi-enclosed basin, such as the Mediterranean Sea, these mechanisms are highly intensified [12]. Notably, the Mediterranean Sea has been recognized as the world's second-largest biodiversity hotspot [13,14]. At the same time, it is a focal point of anthropogenic pressure, also due to the existing plastic pollution hot zone [15–18]. Furthermore, rivers are reported to serve as a channel for poorly managed waste to reach the sea, representing a primary route for plastics to enter the ocean. Subsequently, the presence of the Tiber River in our study area represents the main source of pollutants [19], including plastic waste [20–22]. According to Crosti et al. (2018), 80% of the buoyant litter sampled in the Tiber River estuary was plastic coming from land belonging to the food packaging sector and the cosmetic industry [22].

In the last decade, the scientific community has shown a growing interest in the pollution caused by MPs, generally defined as all sorts of synthetic particles smaller than 5 mm in size [23]. Given the significant impact of plastic pollution, the outcome is the unavoidable interaction between MPs and marine animals. Several studies [24–27] have widely demonstrated that MPs can enter the marine food web, representing a crucial concern for the environment. It has been reported that humans consume a notable quantity of MPs via food, particularly through the consumption of fish, crustaceans, and molluscs [28]. According to Fossi et al. (2018), as MPs frequently cover the size range of prey for various marine species, the ingestion of MPs may occur either accidentally or intentionally, such as mistaking plastic particles for potential prey [24]. Furthermore, trophic transfer, involving the secondary ingestion of MPs already ingested by prey, represents a contributing factor [27,29,30]. In both cases, plastic ingestion has several noxious repercussions. Research activities, both performed under laboratory-controlled conditions [31,32] and in natural environments [33], pointed out that MPs exposure can lead to oxidative stress, tissue damage, increased enzymatic activity, gills obstruction, translocation to other tissue and, in the worst-case scenario, lethal effects [34]. Despite the low toxicity linked to most plastic polymers, the hazard posed by MPs can result from additional substances, degradation byproducts, and adsorbed contaminants [35,36]. Another growing concern relates to the hydrophobic nature of persistent organic pollutants (POPs) and other hazardous substances, which can readily attach to the surfaces of MPs and, thus, due to the transfer provided by the particles, their ingestion is facilitated [37,38]. As a result, MPs ingestion by biota represents a potential transfer of noxious chemicals through marine ecosystems [39]. The integration of these effects makes the study of microplastic ingestion by marine organism of great interest.

According to the FAO [40], shrimps constitute 3.8% of the total global fisheries catch by weight. However, shrimps are considered among the most valuable fishery products worldwide since they contribute significantly, accounting for 11.2% of the global fisheries in monetary value. Particularly, the species selected for this study, the Deep-Water Rose Shrimp *Parapenaeus longirostris* (Lucas, 1846) and the Giant Red Shrimp *Aristaeomorpha foliacea* (Risso, 1827), are ecologically and commercially important targeted fishing species as part of the Penaeidae and Aristeidae families, respectively. The individuals belonging to this species are commonly found on sandy–muddy bottoms in the entire Mediterranean Sea [41,42]. *P longirostris* has an extremely broad bathymetric range, which extends from 20 m up to 750 m [43], whereas *A. foliacea*'s depth distribution is reported between 120 and 1300 m [44], even though it is primarily found within the intermediate range of depths between 450 and 600 m [45]. According to the findings of Cartes et al. (2014) and Kapiris et al. (2004), both *P. longirostris* and *A. foliacea* are active predators that also exhibit secondary scavenging behaviour [46,47]. *P. longirostris*, cycling between a hunting stage and a digging period, shows a strong feeding preference for a diverse range of bathypelagic, benthic, and endobenthic prey, particularly targeting polychaetes, crustaceans, and molluscs, whereas *A. foliacea* displays a highly variegated diet, with crustaceans and fishes representing the two most encountered prey categories [48].

In the field of trophic ecology, the analysis of carbon and nitrogen stable isotope ratios has become a routine analysis [49]. This fast and reliable method is based on the fact that the ratio of stable isotopes in a consumer predictably reflects that of its food sources [50,51]. Stable isotope analysis provides information on trophic niches, feeding interactions, and energy flow. It has been widely used in fish ecology studies [49,50], also in relation to the ingestion of MPs [27], providing interesting results.

MPs ingestion [52–55] has been previously reported in the Mediterranean Sea in both species examined in the present research, as well as in species with a strong trophic relationship with them [56,57]. Indeed, *P. longirostris* and *A. foliacea* are commercially valuable species and serve as the most essential prey for many benthopelagic predators [54,55] and apex demersal predators [56]. Several studies, complied in a review [57], documented the trophic transfer of MPs across multiple trophic levels. The accumulation of MPs at lower levels of the marine food web could potentially trigger cascading effects within marine ecosystems [58,59], further underscoring the relevance of MPs ingestion by shrimp species. In spite of that, the literature on ecology and MPs ingestion by decapods is still scarce [24,60]; this is the first study focusing on the MPs ingestion in *P. longirostris* and *A. foliacea*, collected in sympatry, considering their trophic level, niche metrics, and niche overlap.

The purposes of the present research study are: (i) investigating the ingestion rate of MPs in *P. longirostris* and *A. foliacea* collected in sympatry in the Central Tyrrhenian Sea; (ii) detecting differences between the two species in terms of the quantity and quality of ingested MPs; (iii) deepening our understanding of the ecological factors and the species-specific trophic behaviour, which can affect the interaction with MPs particles. Although previous research has investigated the link between MPs ingestion and the trophic ecology of species through stable isotope analysis [27,61], to the best of our knowledge, this is the first study to specifically involve shrimp species in investigating this issue.

2. Material and Methods

2.1. Study Area

The study area was in the Central Tyrrhenian Sea, precisely facing the mouth of the Tiber River (Figure 1) off the coast of the city of Rome. The Tiber River, with a total length of about 405 km and a hydrographic basin area of 17,375 km^2, is the most important watercourse in Central Italy (third in length) and, additionally, one of the most polluted rivers in the country [19,22]. The study was carried out in a single transect, at around 500 m depth, located north of the Tiber River mouth (Figure 1). Individuals of both species were simultaneously collected on 18 July 2023, during the same haul and at the same distance from the coast (around 26 km) by a professional fishing vessel using bottom trawl techniques.

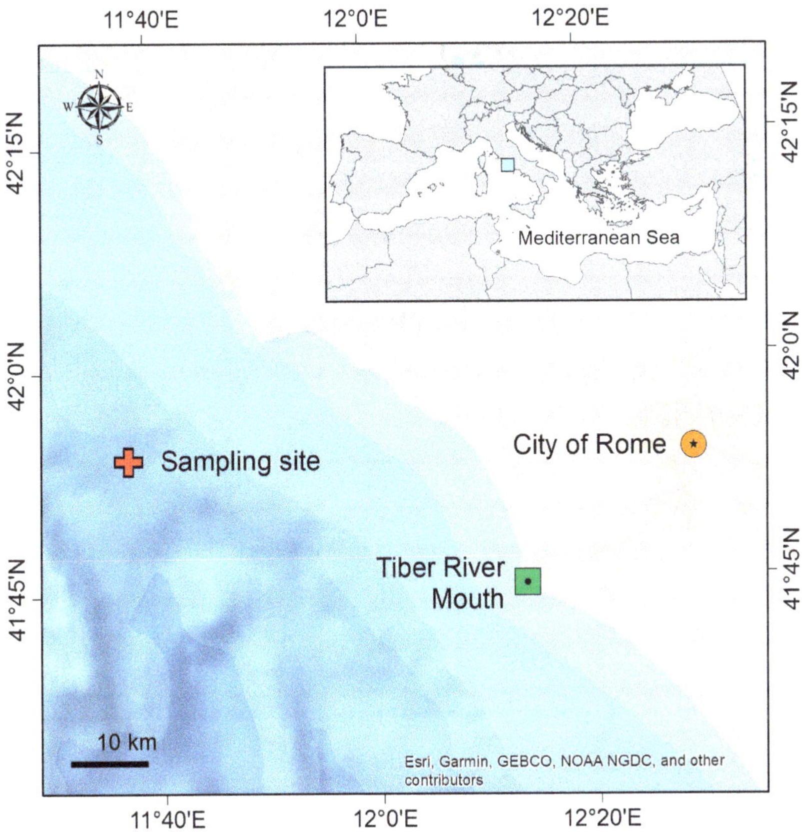

Figure 1. Sampling site. The map represents the sampling site with a spatial point of reference in the top left and the scale in the bottom left. The sampling site is highlighted with a red cross marker, while the Tiber River mouth (one of the most polluted Italian rivers) and the metropolitan city of Rome (the main source of pollution) are indicated with a green and a yellow marker, respectively.

2.2. Sample Collection and Analytical Methods

2.2.1. Samples Processing

A total of 180 individuals (90 per species) were collected, preserved in ice, and wrapped in aluminium to avoid any possibility of secondary contamination, and were immediately transported to the laboratory. Fifty individuals per species, intended for the analysis of ingested MPs, were frozen at −20 °C to be further processed, while 80 individuals (40 per species) were processed immediately after transportation to the laboratory for stable isotope analysis. Prior to dissection, total weight (Tw, g), total length (tL, cm), carapace length (cL, cm), stomach weight (sW, g), intestine weight (iW, g), and hepatopancreas weight (Hw, g) were recorded, using a manual calliper and a precision scale (both with 1 mm of accuracy). The carapace length was determined by measuring the junction line from the right orbital edge to the midpoint of the posterior margin of the carapace. Sex (F = female, M = male) was determined macroscopically through the evolutionary convergence shared by the two species, using sexual dimorphism features, including the presence of the copulatory organ, or petasma, in the male, and differences in rostral length [62]. Ethical review and approval [GG1] were waived for this study due to Italian legislation (D.L. 04/04/14 N.26 art1 a.1), which states that no ethical approval is required for experiments carried out on invertebrates.

2.2.2. Stable Isotope Analysis

Zooplankton samples were collected in the same sampling area during a dedicated sampling campaign in order to establish isotopic baselines. Therefore, stable isotope analysis (SIA) was performed on 40 individuals per species. The muscular segment of the second metamere of the abdomen was sampled from each individual, placed in Falcon tubes, and promptly frozen at −20 °C for preservation. The samples were then freeze-dried, crushed, reduced to powder, and combined in pools of two individuals each. Aliquots of approximately of 0.5 ± 0.1 mg were then placed in a tin capsule (5 mm × 9 mm), oven-dried for at least 24 h at 60 °C and then analysed for the stable isotope ratio of carbon ($\delta^{13}C$) and nitrogen ($\delta^{15}N$) using an elemental analyser Flash 2000 (Thermo Fisher Scientific, Bremen, Germany) coupled with a Delta V Advantage flow mass spectrometer (Thermo Fisher Scientific, Bremen, Germany). The results of $\delta^{13}C$ and $\delta^{15}N$ isotopes were expressed as parts per thousand (‰) on the relative δ-scale referred to the V-PDB (Vienna Pee Dee Belemnite) and N_2 atmospheric air, international standards selected for carbon and nitrogen isotopic ratios, respectively. Ratios were then calculated according to Equation (1):

$$\delta(‰)X = \left(\frac{R\ sample - R\ standard}{R\ standard}\right) \times 1000 \tag{1}$$

where R represents the ratio of abundances of the relative masses of the heavy and light isotopes ($^{13}C/^{12}C$; $^{15}N/^{14}N$). The average analytical reproducibility was ±0.1‰ for $\delta^{13}C$ and ±0.2‰ for $\delta^{15}N$. Samples were in duplicate with a standard deviation on average lower than 0.15‰ for $\delta^{13}C$ and $\delta^{15}N$ measurements. International standards IAEA−600 [63] ($\delta^{13}C$ = −27.77‰ and $\delta^{15}N$ = 1.00‰), USGS40 ($\delta^{13}C$ = −26.39‰ and $\delta^{15}N$ = −4.52‰), IAEA-C-8 ($\delta^{13}C$ = −18.31‰), IAEA-CH6 ($\delta^{13}C$ = −10.45‰), IAEA-N-1 ($\delta^{15}N$ = +0.43‰), and IAEA-N-2 ($\delta^{15}N$ = 20.41) [64–68] were analysed for results calibration and to detect possible drift deviations. The amount of lipid in the muscle tissue can bias the $\delta^{13}C$ values, leading to erroneous diet interpretation. The C:N was examined, and the ratio was below 3.5 for every species, hence, no lipid correction was necessary [69,70].

2.2.3. Microplastics Analysis

The analysis of ingested MPs was performed following the protocol for analysing micro-litter ingestion in fish included in the protocol European Guidelines for Monitoring Marine Litter in European Seas [71,72], adopting specific modifications to customise it to the case of this study. After the dissection, the gastrointestinal tracts were separated into stomachs and intestines, then individually weighed and placed in 250 mL glass beakers. In order to avoid introducing bias into the analyses, all individuals with a completely empty stomach or intestine were excluded for the purposes of the analysis. The stomach wall of each individual was separated from the content, rinsed with distilled water, and placed in labelled glass Petri dish, to be subsequently analysed individually under the stereomicroscope. This step proved to be useful for both the challenging digestion of the stomach wall in H_2O_2, owing to its chitinous nature, and to verify the absence of particles adhered to the walls. With the purpose of removing all organic matter, the stomach contents and the intestines underwent chemical digestion using 15% H_2O_2, and after 5 days at 40 °C, the digestate was filtered onto glass microfibre filter membranes (Whatman GF/B™; pore size: 1.0 µm) employing a vacuum pump system. All microparticles were counted and photographed using a camera-equipped dissecting microscope (ZEISS Stemi 2000-C with Axiocam 208 colour). All the MPs were categorised based on their shapes (fibre, filament, film, fragment, foam, granule, pellet, and bundle), according to their colour (black, blue, green, grey, red, and white) and included into three size class [72] (Size class: 1 from 1 mm to 5 mm; Size class 2: from 1 mm to 330 μm; Size class 3, from 100 μm to 330 μm), fixing the lower limit at 100 μm [72]. To verify the exact chemical composition of each particle, they were individually analysed through μFT-IR spectroscopy (Nicolet iN5 FTIR Microscope, Thermo Fisher Scientific, Madison, WI, USA), using OMNIC™ Series Software (version 9.13.1256) with Aldrich™ Polymers FT-IR Spectral Library. All the particles that had a non-polymeric composition were excluded from the present research.

2.2.4. Quality Control for Microplastics Analysis

To minimise secondary contamination, all analyses were conducted in a clean lab room and under a laminar flow hood. Additionally, 3 Dyson Purifier Cool™ air purifiers were used overnight to purify the laboratory air. During the analyses, staff access to the lab was restricted to a maximum of two people, and only 100% cotton clothes and lab coats were used. All dissecting tools were cleaned with ethanol and distilled water. Every 5 samples, a blank sample was generated to monitor secondary contamination. Following Matiddi et al.'s (2021) protocol [72], if blank contamination occurred, microlitter items sharing characteristics, such as shape, colour, polymer type, and size with the blank contamination were excluded from the specific batch results.

2.3. *Statistical Analysis*

All the analyses included in this study were performed using R studio (Version 4.3.2. 2023)[73] utilising the "stats", "FactoMineR", "factoextra", "ca", packages for statistical analysis, and the "base" and "tidyverse" packages for graphical outputs. A default significance level (p-value < 0.05) was fixed for all the tests performed.

Concerning SIA, the Bayesian Layman's metrics and the computation of standard ellipse areas for the two species were performed through SIBER (Stable Isotope Bayesian Ellipses in R) and represented in δ-space scatterplot [74]. The trophic position (TP) of species was determined using δ^{15}N, incorporating an enrichment factor ΔN of 3.4‰. This value is recommended for constructing food webs when there is no prior knowledge of ΔN. The δ^{15}N of zooplankton (mean ± standard deviation: 4.04 ± 0.65) in the current study ($\delta^{15}N_{zooplankton}$) was designated as the baseline primary consumer for trophic level 2 (Equation (2)) [69]:

$$TP\ consumer = \lambda + \left(\frac{\delta^{15}N\ consumer - \delta^{15}N\ zooplankton}{\Delta N} \right) \quad (2)$$

In Equation (2), λ is assumed to represent the trophic level of the baseline, specifically that of the zooplankton (λ = 2). In order to deepen the ecological dynamics of the two target species, Bayesian Layman metrics [75,76] were calculated, including δ^{13}C range (CR, reflecting the diversity in the sources of the food web) and δ^{15}N range (NR, representing the vertical food web structure tracking trophic dynamics), total area enclosed by the convex hull containing the data (TA, depicting the ecological niche space), and average distance of sample values from the centroids (CD, which describes the diversity characterising the trophic webs). Lastly, with the aim of facilitating direct comparisons of isotopic niches among different communities [74], Bayesian sample-size corrected Standard Ellipse (SEAc) was estimated.

Fullness Index (F.I.%; ${}_{gi}W/{}_{t}W$) and Hepatosomatic Index (H.S.I.%; ${}_{h}W/{}_{t}W$) were respectively calculated as the ratio of gastrointestinal weight to total weight and hepatopancreas weight to total weight. The occurrence of MPs was recorded as a binary presence–absence factor (0–1) and as the number of MPs found in the gastrointestinal tract of each individual involved in this study. Furthermore, the frequency of occurrence (F.O.%), was calculated as the percentage ratio of individuals exhibiting MPs occurrence either in the stomach or in the intestine, divided by the total number of individuals analysed. A preliminary descriptive analysis, comprising bar plots and scatter plots, was conducted to obtain a comprehensive overview of the factors contributing to MPs ingestion. The number of items ingested by each individual was analysed through Kendall's rank correlation test, which measures the strength of monotonic association between two vectors [77], to ascertain the nature of its relationship with a total weight (tW), carapace length (cL), gastrointestinal weight (giW) and hepatopancreas weight (hW). Differences in the abundance of MPs among sex and organs (stomachs and intestines) were analysed using a Wilcox signed-rank test, a non-parametric test useful to compare two groups, especially suitable for small sampling size data containing outliers. In order to analyse MPs typologies, a code was generated for each particle, encompassing size class, shape, colour, and the polymer. With the aim of assessing the diversity of ingested MPs for the two species, a Shannon–Wiener diversity index (H′) was applied to this code. In addition, a Chi-square test was applied to the contingency table to compare the identifying code of each MP between the two species. Eventually, a Correspondence analysis (CA) was used to explore relationships among shape, size, and polymer composition of ingested MPs, visualising data in a plot with reduced dimension, and distinguishing between the stomach and intestine of the two species.

3. Results

3.1. Stable Isotope Analysis

Enrichment of $\delta^{15}N$ highlighted the similarity of the two species, with *P. longirostris* TP = 3 ± 0.10 and *A. foliacea* TP = 3.1 ± 0.08. The results coming from Bayesian Layman metrics are reported in Table 1. Based on the size-corrected standard ellipse areas (SEAc), it turned out that the niche overlap between *P. longirostris* and *A. foliacea*, amounted to 55.20%. As described in Table 1 and shown in Figure 2, *A. foliacea* exhibited a slightly broader trophic niche space compared to that of *P. longirostris*.

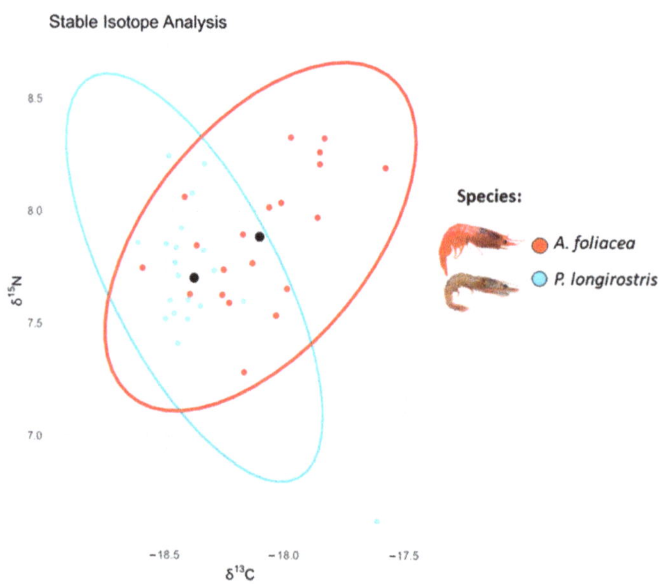

Figure 2. Stable isotope analysis. Scatterplot based on stable isotope analysis conducted on muscular tissue of *Parapenaeus longirostris* (light-blue), and *Aristaeomorpha foliacea* (red) caught off the Rome coast during July 2023. Each point represents a single observation, while the confidence ellipses represent a region within which 95% of the data is expected approximately to fall, providing a visualisation of the variability around the central tendency, highlighted in black, in the $\delta^{13}C$ and $\delta^{15}N$ values for the species.

Table 1. Bayesian Layman metrics. Results of Stable Isotope Analysis conducted on muscular tissue of target species (*Parapenaeus longirostris, Aristaeomorpha foliacea*) caught off the Rome coast during a single sampling trip carried out on July 2023. Mean value with standard error of δ^{15}N and δ^{13}C and Layman Bayesian metrics: the δ^{15}N range (NR), the δ^{13}C range (CR), the total area enclosed by the convex hull containing the data (TA), the average distance of sample values from the centroids (CD), and a sample-size corrected ellipse area (SEAc) metrics.

Species	δ^{15}N	δ^{13}C	NR	CR	TA	CD	SEAc
P. longirostris	7.7 ± 0.07	−18.4 ± 0.04	1.624	1.002	0.502	0.275	0.168
A. foliacea	7.8 ± 0.06	−18.1 ± 0.05	1.046	1.014	0.543	0.339	0.199

3.2. Microplastics Ingestion

MPs ingestion was observed (overall occurrence: 46%) in both examined species, with a F.O.% amounting to 38% for *P. longirostris* and 52% for *A. foliacea*. A total of 72 (78.1% in the stomachs and 21.9% in the intestine) and 128 (39.6% in stomachs and 60.4% in intestines) items have been collected in the gastrointestinal tract of *P. longirostris* and *A. foliacea*, respectively. The data on biometric measurements, biometric indices, and preliminary data on MPs ingestion are reported in Table 2.

Table 2. Morpho-anatomical and biometric indices. Biometric data and indices measured on *Parapenaeus longirostris* and *Aristaeomorpha foliacea* caught off the Rome coast during July 2023. Mean and standard error were calculated for total weight (tW, g) and carapace length (cL, cm). Female individual percentage (Sex (F%)), average of Fullness Index (FI (%): (giW/tW) × 100), average of the Hepatosomatic Index (HSI%: (hW/tW) × 100), and mean ± SE incidence per individual (Items/ind.; microplastics per individual on the whole dataset), were also determined.

Species	tW (g)	cL (cm)	Sex (F%)	FI μ (%)	HSI (%)	Items/Ind.
P. longirostris	11.1 ± 0.44	2.9 ± 0.04	80%	1.9%	2.9%	1.44 ± 0.62
A. foliacea	13.1 ± 0.36	3.3 ± 0.03	56%	4.1%	6.5%	2.56 ± 0.73
All individuals	12.1 ± 0.3	3.1 ± 0.03	68%	2.9%	4.7%	2.04 ± 0.05

The observation under the stereomicroscope of each stomach wall allowed the identification of n = 2 particles within the stomach belonging to individuals of two different species, which were added to the final count. As illustrated in Table 2, the mean number of ingested MPs ± SE per individual was centered on low values (Figure 3A). According to the results, no significant correlation was found between the occurrence and number of ingested items related to the total weight (tW), carapace length (cL), sex (F%), and hepatopancreas weight (GLM, $p > 0.05$). The relationship between the gastrointestinal weight and the number of ingested MPs exhibited a slight positive correlation (Figure 3B) (z = 2.3274; tau = 0.18; p-value < 0.01). Differences in terms of the number of ingested particles between the two parts of the gastrointestinal tract, the stomach (*P. longirostris*: mean ± SE: 0.28 ± 0.06; *A. foliacea*; mean ± SE: 0.39 ± 0.07) and the intestine (*P. longirostris:* mean ± SE: 0.16 ± 0.05; *A. foliacea*: mean ± SE: 0.37 ± 0.07), were detected for both species (p-value < 0.01).

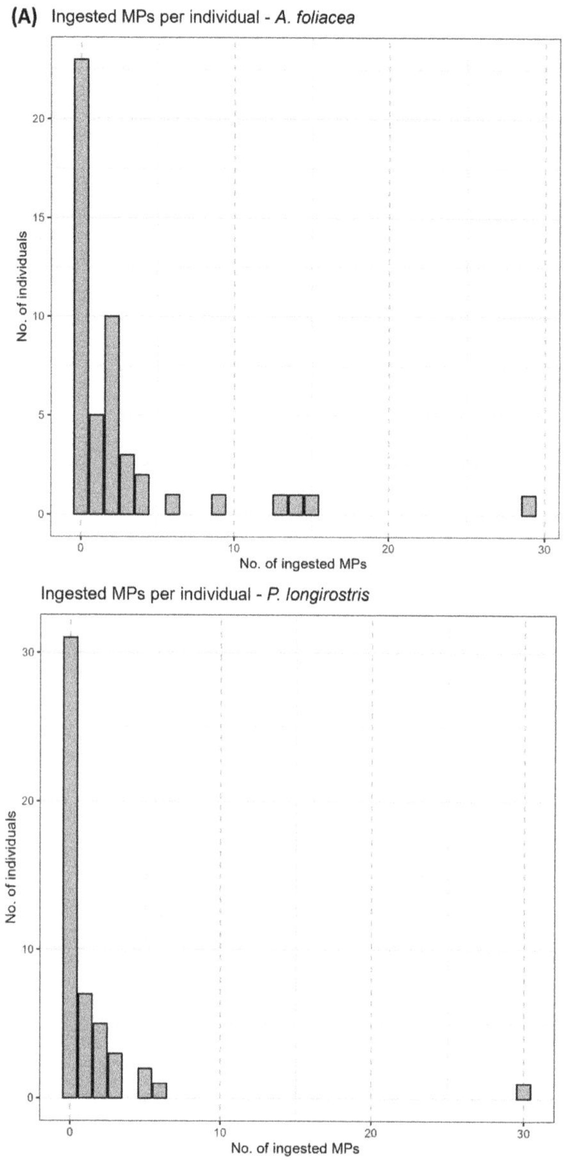

Figure 3. *Cont.*

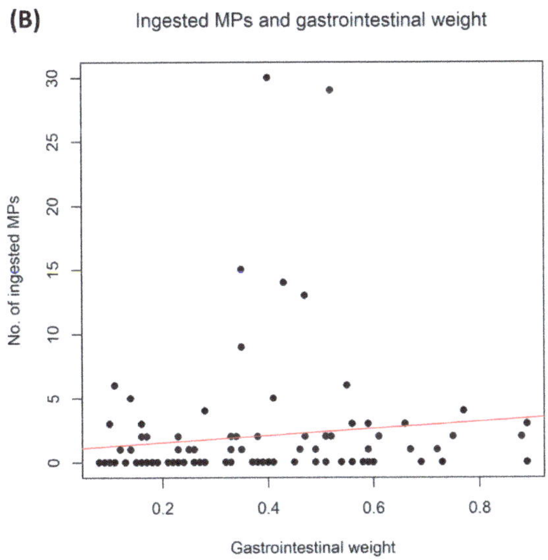

Figure 3. Abundance of ingested microplastics. The number of microplastics per individual and the relationship between the gastrointestinal weight and the number of ingested items are represented with two barplots and a scatterplot, respectively, based on the analysis of the gastrointestinal tract of 100 shrimp individuals (50 *Parapenaeus longirostris* and 50 *Aristaeomorpha foliacea*) caught off theRome coast during July 2023. (**A**) The number of ingested MPs for the two species, with *A. foliacea* on the top and *P. longirostris* on the bottom; (**B**) microplastics ingestion correlated with gastrointestinal weight.

Microplastics Characterisation

All particles were categorised by shape and colour, then grouped into three size classes before being chemically analysed using µFT-IR spectroscopy.

Among the MPs ingested, fibres (82.2%) were prevalent in *P. longirostris*, followed by filaments (11%) and fragments (5.5%). One single fibre bundle was found in *P. longirostris* stomach together with several single fibres. In contrast, *A. foliacea* exhibited a lower occurrence of fibres (20.2%), while predominantly ingesting fragments (76.7%). Only 3.1% of the overall MPs ingestion in *A. foliacea* was constituted by filaments. Particularly, fragment ingestion abundance comparison highlighted a significant difference between the two species (p-value < 0.01).

Overall, the colour analysis of the ingested MPs (Figure 4) revealed that the most abundant colours detected were green (47%), black (18.5%), and white (18%). *P. longirostris* predominantly ingested white (44.5%), red (33.5%), and black (16.7%) microparticles. In contrast, *A. foliacea* showed an evident preference for green (72.6%) MPs, followed by black (19.5%) and red particles (5.5%). Additionally, grey MPs were exclusively ingested by *P. longirostris*.

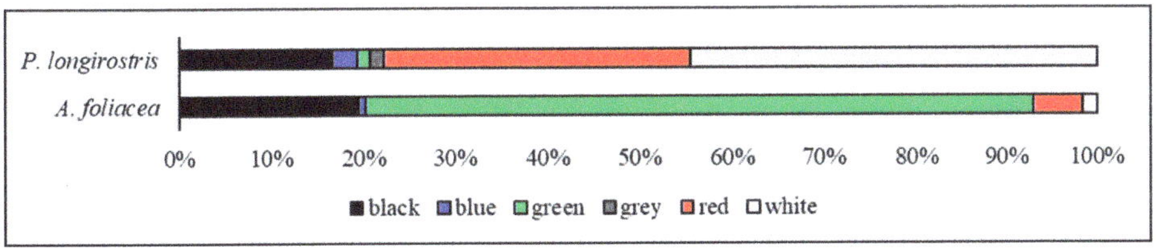

Figure 4. Microplastics colour composition. Colours of microplastic particles (N = 200) extracted from the gut of *Parapenaeus longirostris* (above) and *Aristaeomorpha foliacea* (below) caught off the Rome coast during July 2023.

Concerning the size class, the 74% of the items ingested by *P. longirostris* belonged to size class 1, the 23% to size class 2, and only the 3% to size class 3. By contrast, *A. foliacea* showed percentages of 8.5%, 12.5%, and 79%, respectively (p-value < 0.01).

With regard to the chemical characterisation of the ingested MPs, *P. longirostris* showed a high percentage of polyethylene terephthalate (PET; 59.7%) and polyacrylic (PC; 20.8%) ingestion, with a significant presence of anthropogenic resinous compounds (RES; 11.1%) and a slightly greater variety of ingested polymers, including polyethylene (PE; 4.2%), polyamide (PA; 1.39%), polystyrene (PS; 1.4%) and polyvinyl-chloride (PVC; 1.4%). Conversely, in *A. foliacea* was quite evident a substantial predominance of epoxy resin compounds (RES; 89.8%) and a moderate presence of polyamide (PA; 3.1%), polyethylene terephthalate (PET; 3.9%), polyacrylic (PC; 2.3%)and polypropylene (PP; 0.8%). The Shannon–Wiener Diversity Index H differed between the two species, amounting to 2.588 for *P. longirostris* and 1.477 for *A. foliacea.* Moreover, a strong difference of MP's code between the two species has been detected (X-squared = 166.74, df = 37, p-value < 0.001), enhancing a significant difference in the types of MPs swallowed from the two species. Representative images of the MPs detected in the two species, along with their respective chemical spectra, are presented in Figure 5.

Parapenaeus longirostris

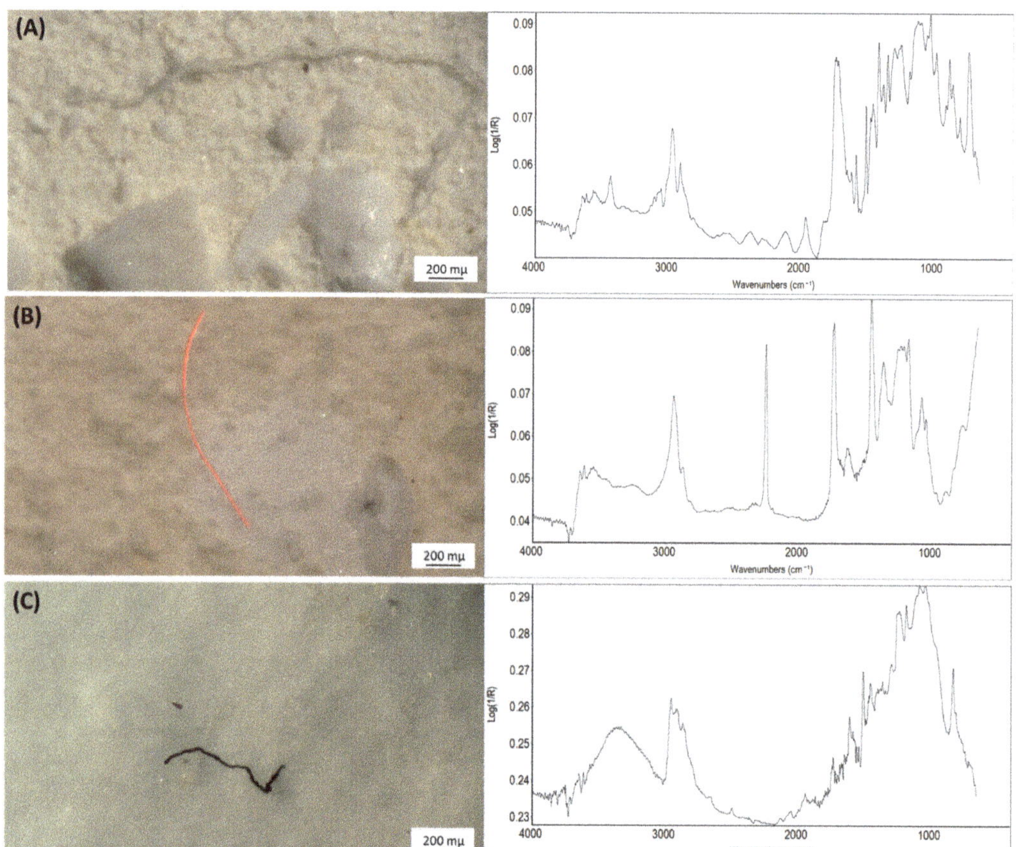

Figure 5. *Cont.*

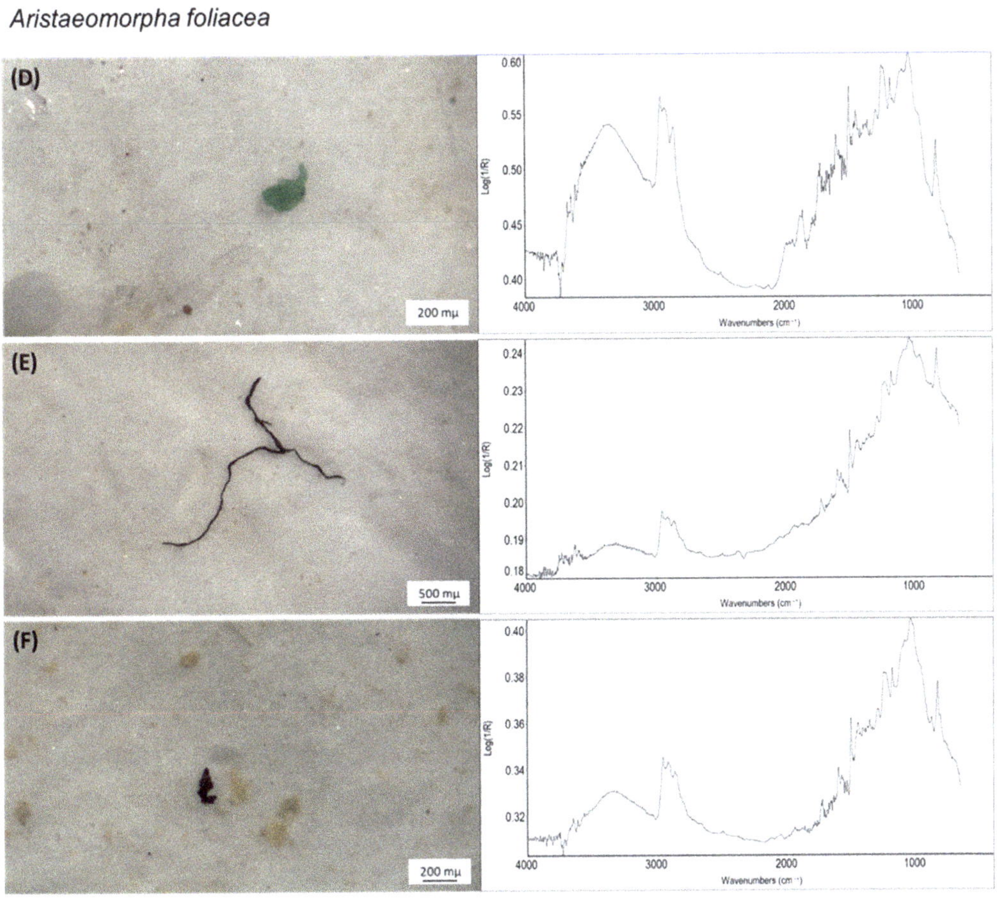

Figure 5. Most ingested microplastic types in the two target species. Types of the microplastics most observed in the gastrointestinal tract of *P. longirostris* (above) and *A. foliacea* (below) caught off the Rome coast during July 2023, along with their respective chemical spectra (y-axis: absorbance; x-axis: wavenumbers (cm^{-1})) to the right. (**A**) white fibre, polyethylene terephthalate (match: 91%); (**B**) red fibre, polyacrylic (match: 92%); (**C**) black fibre, resinous compounds (match: 75%); (**D**) green fragment, resinous compounds (match: 70%); (**E**) black fibre, resinous compounds (match: 81%); (**F**) black fragment, resinous compounds (match: 77%).

Finally, correspondence analysis biplots are reported in Figure 6. Dimension 1 explains most of the variance in the data across all three plots, highlighting a separation between the two species among the different categories. Conversely, the stomach and intestines of the two species are close in all the plots.

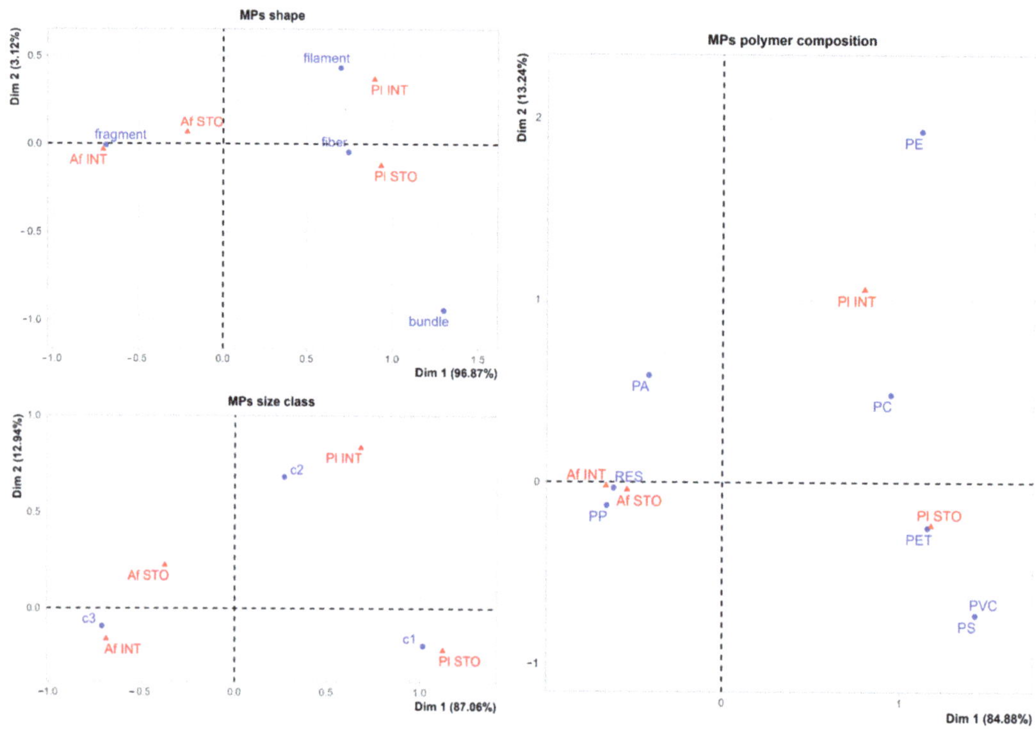

Figure 6. Biplot on microplastics characterisation. A biplot performed on the contingency table of the correspondence analysis of shape, size class, and polymer of the particles ingested by target species (*Parapenaeus longirostris* and *Aristaeomorpha foliacea*) caught off the Rome coast during July 2023, is provided here. Pl = *Parapenaeus longirostris*, Af = *Aristaeomorpha foliacea*, STO = stomach, INT = intestine. Size class: c1: 1 from 1 mm to 5 mm; c2: from 1 mm to 330 µm; c3, from 100 µm to 330 µm. Polymer: PA: polyamide; PC: acrylic; PE: polyethylene; PET: polyethylene terephthalate; PP: polypropylene; PS: polystyrene; PVC: polyvinyl chloride; RES: resinous compounds).

4. Discussion

Stable isotope analysis has disclosed that the two species exhibit a substantial trophic niche overlap. Although the literature on feeding habits reports a slightly more basal diet for *P. longirostris* (polychaetes, molluscs, and crustaceans) [47,78] and a more predatory one for *A. foliacea*, (crustaceans, fishes, and macroplankton euphausiids) [45,48], according to our findings, the trophic positions of the two species are remarkably similar. As a result, differences in MPs ingestion observed in this study are likely due to species–specific behaviour occurring during the feeding phase rather than the trophic position of the two species.

Both species ingest MPs at comparable rates, and no significant differences in the average number of particles per individual have been detected. Although the literature on this topic is not particularly abundant, research conducted in other regions of the Mediterranean Sea [79–82] has previously observed the ingestion of microplastics in *P. longirostris* and *A. foliacea*. Despite the present study area being quite polluted [19,21,22], the percentages of particle ingestion are lower compared to those reported in D'Iglio et al. (2022) (*P. longirostris*: 76%; *A. foliacea*: 83%) and Yücel (2023) (*P. longirostris*: 100%), conducted in the Southwestern Ionian Sea and Northeastern Mediterranean Sea, respectively [80,82]. On the other hand, the opposite trend was evidenced for both species by Leila et al. (2023) in the Eastern Ionian Sea (*A. foliacea*: 14.65%) and Bono et al. (2022) in the Strait of Sicily (*P. longirostris*:

21%) [79,81]. The overall ingestion rate is low since most individuals ingest a few particles. Ingested MPs can be expelled or retained in the gastrointestinal tract [24,83,84]. Despite the FO% being relatively high in both species, the average number of items per individual is low in both species, so it is likely that shrimps expel the ingested MPs after a short period of time [85]. As excretion occurs, MPs do not accumulate over time within the gastrointestinal tract of shrimps. This is further supported by the positive relationship between gastrointestinal weight and the number of ingested items, which could indicate that the retention time of MPs in the target species has a duration comparable to that of their dietary intake. This result aligns with what has been previously observed in various species; even in regions characterised by significant plastic contamination, the quantity of ingested MPs remains relatively low [25,27,85,86]. Moreover, the short retention time of MPs in organisms and their high frequency of occurrence make shrimps potential small-scale bioindicators of microplastic pollution [87].

The differences in MPs distribution observed in the two organs of the two species' gastrointestinal tract (i.e., stomach and intestine) may be explained by a difference in the periods of the foraging activity. For *P. longirostris*, MPs retained in the intestinal tract are lower both in terms of abundance and numerical variability compared to those contained in the stomach, while in the case of *A. foliacea*, MPs abundance exhibits an opposite distribution, potentially indicating a temporal delay in feeding activity, compared to *P. longirostris*.

Statistical analysis revealed significant differences between the two species in terms of chemical-physical characteristics of the ingested microparticles, underlying interesting insights regarding selective ingestion. The shape of the ingested MPs is remarkably different between the two species. As previously observed [80–82], *P. longirostris* mostly ingested fibres. The presence of fibre bundles is frequent in this species [82] but also in *A. foliacea* [79] and in *Aristeus antennatus* (Risso, 1816) [88]. In relation to *A. foliacea*, a distinct prevalence of fragments is observed. Fragment ingestion was also documented by D'Iglio et al. (2022), contrasting with Leila et al. (2023), who, using a different method of MPs identification, reported 100% of plastic particles as fibres [79,80]. The different types of MPs ingested by the two species could be explained by diverse weights combined with the shape-linked behaviour of MPs in the marine environment [89,90], as well as by MP's patchy distributions and different availability in marine sediments [91,92]. Furthermore, it can also be related to differences in prey-searching modalities [87,88]. Both species exhibit active predatory behaviour [46,47]. Predatory behaviour could lead to greater interest in particles that tend to move under the effect of the marine current. Therefore, these particles could be actively selected by the target species, according to their feeding habits [93]. In this view, as fragments tend to float [89], the ingestion of large fragments could be the result of intentional ingestion triggered by the resemblance of these particles to natural prey [93,94]. Moreover, since the size of the fragments often resembles that of sand, they are frequently available to the invertebrates at the base of the food web [95,96], which are also consumed by the target species of this study [46–48].

The size class of ingested particles shows an opposite trend between the two species. *P. longirostris* showed a remarkable preference for larger items. This result perfectly aligns with findings highlighted by Yücel (2023), where most particles found fell within the 1 to 2.5 mm size range, and D'Iglio et al. (2022), where most of the MPs are in the range of 1 to 5 mm, but contrasts with what was found by Bono et al. (2020), where particles fell within the range of 100 to 300 µm [81,82]. By contrast, *A. foliacea* displayed a marked preference for small-sized (size class 3) items, as previously observed [80].

Differences or affinities found with related studies may be significantly influenced by variations in sampling setup and sample processing, type of analysis, and size choice.

The colour distribution provides insights into the selective ingestion of MPs by species, enhancing a noteworthy colour preference. Overall, green-coloured MPs are the most ingested particles, consistent with what was reported by Sbrana et al. (2020) in *Boops boops*, collected in shallow water in the same area [25]. According to Denton, 1990, at depths exceeding 200 metres in sea waters, daylight appears blue and maintains a nearly constant spectral composition and angular distribution [97]. Moreover, the intensity of daylight decreases ten times, approximately every 75 metres of depth. However, shrimps possess a well-developed visual system composed of compound eyes that consist of a multifaceted cornea situated on mobile peduncles, which allows them to have a complete 360° view of their surroundings [98]. Especially in species that live in low-light habitats, the presence of a layer of reflective pigment, known as *tapetum*, helps to detect light and movements [98]. In the present case study, it can be assumed that, at a depth ranging from 368 to 582 metres, sunlight is still partially visible by shrimps [97] and, therefore, it is possible that the two species recognise colours. In this view, the high frequency of green MPs might result from inadvertently ingesting MPs that resemble natural food sources [88,94]. By contrast, the ingestion of black and white particles could occur accidentally due to the lower visibility of these colours in depth.

FT-IR analysis revealed that polymer ingestion differs between the two species. PET and PA are the two polymers most widely used as raw materials in the textile and apparel industry [99]. Their high presence in *P. longirostris* is in line with the higher ingestion of fibres. Resin-based plastics are commonly used in packaging materials, including bottles and containers, as well as painting and coatings [100]. The high percentage of resinous fragments in *A. foliacea* may be linked to this type of source. According to Lithner et al. (2011), each polymer is associated with a specific risk level, which can be calculated as Polymer Hazard Index (PHI) [101]. In this case study, the range of the hazard score associated with resinous compounds (hazard score range: 7139–4226), mostly ingested by *A. foliacea*, is definitely higher than that associated with polyethylene terephthalate and other polyesters (hazard score range: 4–1414), preferred by *P. longirostris*. Therefore, the risk of exposure could be higher for *A. foliacea*. However, it was impossible to perform an exact calculation of the PHI because polymers, especially resin-based, might have variable concentrations of different substances, each associated with different hazards. The integration of FT-IR analysis with other techniques, such as using elemental analyser isotope ratio mass spectrometry (EA/IRMS) to characterise plastic polymers, could be interesting both for obtaining a more accurate response and for interpreting ecological influence on MPs ingestion more precisely [102]. Indeed, our study points out that, at the same trophic level, there are no significant differences in the ingestion of MPs. By delving deeper into both trophic ecology (e.g., diet or feeding rate studies based on other techniques, such as stomach content analysis [46], DNA metabarcoding [103], and fluorescence [104]) and the nature of plastics (exact concentration of substances), we could gain a further interpretation of our results.

Interesting insights regarding the trophic transfer of MPs can be obtained by comparing the current results with those of studies conducted on species that prey on *P. longirostris* and *A. foliacea*. *Galeus melastomus* (Rafinesque, 1810), *Etmopterus spinax* (Linnaeus, 1758), and *Scyliorhinus canicula* (Linnaeus, 1758), as well as *Merluccius merluccius* (Linnaeus, 1758) and *Mullus barbatus* (Linnaeus, 1758) are widespread species in the Mediterranean Sea that actively feed on crustaceans [105–108], and in some case specifically on *P. longirostris* [109,110] and on *A. foliacea* [111]. More than one study reported MPs ingestion by these elasmobranchs and fish in the Mediterranean Sea [109,112–115]. As highlighted by Alomar and Deudero (2017) [114], a bioaccumulation between prey and predators is likely to occur in the case of *G. melastomus*. Specifically, the positive correlation between the stomach weight and the abundance of plastic particles, MPs ingestion is likely to occur via secondary ingestion through prey ingestion [114]. Also, Valente et al. (2019) [59] and Zicarelli et al. (2023) [107], hypothesises that these elasmobranch species may facilitate the transfer of plastics up the trophic chain, indicating potential complex dynamics within the food web. Similarly, the same can be hypothesised for the prey of *P. longirostris* and *A. foliacea*. Literature concerning MPs ingestion by invertebrates is even more scarce than the one on crustaceans [113], and also studies on the specific diet of the target species is quite obsolete [44,47,48]. A more recent study has reported that *A. foliacea* in the Tyrrhenian

Sea primarily feeds on other crustaceans (*Plesionika martia*, *Meganyctiphanes norvegica*) and various species of Myctophidae. To the best of our knowledge, no specific studies have been conducted on the prey of the present target species, except from *M. norvegica*, whose ingestion is related to MPs uptake in *Balaenoptera physalus* in the North Atlantic Sea [114] and Myctophidae [116]. Anyway, studies conducted on benthic organisms [117] and molluscs [59,96] report ingestion of microplastics. Consequently, it can be hypothesised that there is an uptake beginning from the prey of the target species. Moreover, as previously suggested, prey-searching modalities, foraging depths, nictemeral migrations and feeding habits of organisms can potentially alter the bioavailability of different MP types [89].

5. Conclusions

In conclusion, findings from this investigation contribute not only to a better comprehension of MPs exposure of two commercially important crustacean species but also underline how variable MPs ingestion could be among different species and ecological niches. Indeed, this study clearly shows that even when comparing species with overlapping trophic niches, the distinction in the types of ingested MPs might remain evident. Considering this, the contrasting trend in MPs typologies ingestion may be explained by differences in feeding modalities and foraging timing of the two species. Although MPs ingestion by marine species is nowadays well-documented, the research focused on deep-sea organisms, particularly crustaceans, remains limited. Compared to fishes and molluscs, crustaceans have received less attention in this context. Nonetheless, given the significant interest of crustaceans for human consumption, understanding their role is crucial for unravelling the dynamics of trophic webs and interactions with MPs and, therefore, for humans' well-being. In addition, the currently available literature strictly regarding decapods reports highly variable research settings and a remarkable difference in the obtained results, emphasising the necessity of the identification of a unique and standardised method that allows the comparison of results. A deeper analysis of the diet of these two species could provide interesting insights regarding the influence of the trophic habits on MPs ingestion.

Author Contributions: Conceptualisation, L.C., T.V., F.R., C.S. and M.M.; data curation, L.C., T.V., D.B. and F.R.; formal analysis, L.C., T.V., D.B., F.R. and M.M.; funding acquisition, L.M. and M.M.; investigation, T.V. and M.M.; methodology, L.C., T.V., D.B., F.R. and M.M.; project administration, L.M. and M.M.; resources, L.C., T.V. and M.M.; software, L.C. and T.V.; supervision, T.V., G.L., L.M., D.B. and M.M.; validation, T.V., D.B., F.R., G.G. and M.M.; visualisation, T.V., E.M., G.L., L.M., D.B., F.R., G.G., G.C., R.P., C.S. and M.M.; writing—original draft, L.C.; writing—review and editing, L.C., T.V., E.M., G.L., L.M., D.B., F.R., G.G., R.P., C.S. and M.M. All authors have read and agreed to the published version of the manuscript.

Funding: This work was realised in the framework of BIOPLAST4SAFE project (code: PREV-B-2022-12377008): Biomonitoring of biodegradable micro and nanoplastics: from the environment to humans in a one health perspective (Biomonitoraggio di micro e nanoplastiche biodegradabili: dall'ambiente all'uomo in una prospettiva one health) with the technical and economic support of the Italian Ministry of Health–PNC funds. Laura Ciaralli was supported by a PhD fellowship (PhD in Biology, University of Naples Federico II) funded by the ISPRA (Rome, Italy). Laura Ciaralli was supported by PhD fellowship (PhD in Biology, University of Naples Federico II) co-funded by ISPRA (Rome, Italy).

Institutional Review Board Statement: Ethical review and approval were not required for this study in accordance with Italian legislation (D.L. 04/04/14 N.26 art1 a.1), which specifies that ethical approval is not necessary for experiments conducted on invertebrates.

Informed Consent Statement: Not applicable.

Data Availability Statement: The data presented in this study are available at the request of the corresponding author for reasons related to the project's privacy policy.

Acknowledgments: The authors also wish to thank the reviewers for their valuable comments and suggestions, which have helped improve the paper.

Conflicts of Interest: The authors report the absence of conflicts of interest.

References

1. Chamas, A.; Moon, H.; Zheng, J.; Qiu, Y.; Tabassum, T.; Jang, J.H.; Abu-Omar, M.; Scott, S.L.; Suh, S. Degradation Rates of Plastics in the Environment. *ACS Sustain. Chem. Eng.* **2020**, *8*, 3494–3511. [CrossRef]
2. Bergmann, M.; Gutow, L.; Klages, M. *Marine Anthropogenic Litter*, 1st ed.; Springer Nature: Berlin, Germany, 2015; pp. 29–75.
3. Gao, A.L.; Wan, Y. Life Cycle Assessment of Environmental Impact of Disposable Drinking Straws: A Trade-off Analysis with Marine Litter in the United States. *Sci. Total Environ.* **2022**, *817*, 153016. [CrossRef] [PubMed]
4. Chen, Y.; Awasthi, A.K.; Wei, F.; Tan, Q.; Li, J. Single-Use Plastics: Production, Usage, Disposal, and Adverse Impacts. *Sci. Total Environ.* **2021**, *752*, 141772. [CrossRef] [PubMed]
5. IUCN: International Union for Conservation of Nature. *Marine Plastic Pollution*; IUCN: Gland, Switzerland, 2021.
6. Ciaralli, L.; Rotini, A.; Scalici, M.; Battisti, C.; Chiesa, S.; Christoforou, E.; Libralato, G.; Manfra, L. The Under-Investigated Plastic Threat on Seagrasses Worldwide: A Comprehensive Review. *Environ. Sci. Pollut. Res.* **2024**, *31*, 8341–8353. [CrossRef] [PubMed]
7. Gouin, T. Toward an Improved Understanding of the Ingestion and Trophic Transfer of Microplastic Particles: Critical Review and Implications for Future Research. *Environ. Toxicol. Chem.* **2020**, *39*, 1119–1137. [CrossRef] [PubMed]
8. Thompson, R.C.; Olsen, Y.; Mitchell, R.P.; Davis, A.; Rowland, S.J.; John, A.W.G.; McGonigle, D.; Russell, A.E. Lost at Sea: Where Is All the Plastic? *Science* **2004**, *304*, 838. [CrossRef]
9. Aigars, J.; Barone, M.; Suhareva, N.; Putna-Nimane, I.; Dimante-Deimantovica, I. Occurrence and Spatial Distribution of Microplastics in the Surface Waters of the Baltic Sea and the Gulf of Riga. *Mar. Pollut. Bull.* **2021**, *172*, 112860. [CrossRef] [PubMed]
10. Lorenz, C.; Roscher, L.; Meyer, M.S.; Hildebrandt, L.; Prume, J.; Löder, M.G.J.; Primpke, S.; Gerdts, G. Spatial Distribution of Microplastics in Sediments and Surface Waters of the Southern North Sea. *Environ. Pollut.* **2019**, *252*, 1719–1729. [CrossRef]
11. Wagner, S.; Klöckner, P.; Stier, B.; Römer, M.; Seiwert, B.; Reemtsma, T.; Schmidt, C. Relationship between Discharge and River Plastic Concentrations in a Rural and an Urban Catchment. *Environ. Sci. Technol.* **2019**, *53*, 10082–10091. [CrossRef] [PubMed]
12. Cózar, A.; Sanz-Martín, M.; Martí, E.; González-Gordillo, J.I.; Ubeda, B.; Gálvez, J.Á.; Irigoien, X.; Duarte, C.M. Plastic Accumulation in the Mediterranean Sea. *PLoS ONE* **2015**, *10*, e0121762. [CrossRef]
13. Coll, M.; Piroddi, C.; Steenbeek, J.; Kaschner, K.; Ben Rais Lasram, F.; Aguzzi, J.; Ballesteros, E.; Bianchi, C.N.; Corbera, J.; Dailianis, T.; et al. The Biodiversity of the Mediterranean Sea: Estimates, Patterns, and Threats. *PLoS ONE* **2010**, *5*, e11842. [CrossRef] [PubMed]
14. Cuttelod, A.; García, N.; Malak Abdul, D.; Temple, H.; Katariya, V. The Mediterranean: A Biodiversity Hotspot under Threat. In *Wildlife in a Changing World–An Analysis of the 2008 IUCN Red List of Threatened Species*; IUCN: Gland, Switzerland, 2008.
15. Sharma, S.; Sharma, V.; Chatterjee, S. Microplastics in the Mediterranean Sea: Sources, Pollution Intensity, Sea Health, and Regulatory Policies. *Front. Mar. Sci.* **2021**, *8*, 634934. [CrossRef]
16. Suaria, G.; Avio, C.G.; Mineo, A.; Lattin, G.L.; Magaldi, M.G.; Belmonte, G.; Moore, C.J.; Regoli, F.; Aliani, S. The Mediterranean Plastic Soup: Synthetic Polymers in Mediterranean Surface Waters. *Sci. Rep.* **2016**, *6*, 37551. [CrossRef] [PubMed]
17. Boudouresque, C.F.; Bernard, G.; Pergent, G.; Shili, A.; Verlaque, M. Regression of Mediterranean Seagrasses Caused by Natural Processes and Anthropogenic Disturbances and Stress: A Critical Review. *Bot. Mar.* **2009**, *52*, 395–418. [CrossRef]
18. Danovaro, R. Pollution Threats in the Mediterranean Sea: An Overview. *Chem. Ecol.* **2003**, *19*, 15–32. [CrossRef]
19. Inghilesi, R.; Ottolenghi, L.; Orasi, A.; Pizzi, C.; Bignami, F.; Santoleri, R. Fate of River Tiber Discharge Investigated through Numerical Simulation and Satellite Monitoring. *Ocean Sci.* **2012**, *8*, 773–786. [CrossRef]
20. Cesarini, G.; Crosti, R.; Secco, S.; Gallitelli, L.; Scalici, M. From City to Sea: Spatiotemporal Dynamics of Floating Macrolitter in the Tiber River. *Sci. Total Environ.* **2023**, *857*, 159713. [CrossRef] [PubMed]
21. De Lucia, G.; Vianello, A.; Camedda, A.; Vani, D.; Tomassetti, P.; Coppa, S.; Palazzo, L.; Amici, M.; Romanelli, G.; Zampetti, G.; et al. Sea Water Contamination in the Vicinity of the Italian Minor Islands Caused by Microplastic Pollution. *Water* **2018**, *10*, 1108. [CrossRef]
22. Crosti, R.; Arcangeli, A.; Campana, I.; Paraboschi, M.; González-Fernández, D. 'Down to the River': Amount, Composition, and Economic Sector of Litter Entering the Marine Compartment, through the Tiber River in the Western Mediterranean Sea. *Rend. Lincei Sci. Fis. Nat.* **2018**, *29*, 859–866. [CrossRef]
23. Arthur, C.; Baker, J.; Bamford, H. Occurrence, Effects, and Fate of Microplastic Marine Debris. In Proceedings of the International Research Workshop, Tacoma, WA, USA, 9–11 September 2009.
24. Fossi, M.C.; Pedà, C.; Compa, M.; Tsangaris, C.; Alomar, C.; Claro, F.; Ioakeimidis, C.; Galgani, F.; Hema, T.; Deudero, S.; et al. Bioindicators for Monitoring Marine Litter Ingestion and Its Impacts on Mediterranean Biodiversity. *Environ. Pollut.* **2018**, *237*, 1023–1040. [CrossRef]
25. Sbrana, A.; Valente, T.; Scacco, U.; Bianchi, J.; Silvestri, C.; Palazzo, L.; De Lucia, G.A.; Valerani, C.; Ardizzone, G.; Matiddi, M. Spatial Variability and Influence of Biological Parameters on Microplastic Ingestion by *Boops boops* (L.) along the Italian Coasts (Western Mediterranean Sea). *Environ. Pollut.* **2020**, *263*, 114429. [CrossRef] [PubMed]
26. Abidli, S.; Pinheiro, M.; Lahbib, Y.; Neuparth, T.; Santos, M.M.; Trigui El Menif, N. Effects of Environmentally Relevant Levels of Polyethylene Microplastic on *Mytilus galloprovincialis* (Mollusca: Bivalvia): Filtration Rate and Oxidative Stress. *Environ. Sci Pollut. Res.* **2021**, *28*, 26643–26652. [CrossRef]

27. Valente, T.; Costantini, M.L.; Careddu, G.; Berto, D.; Piermarini, R.; Rampazzo, F.; Sbrana, A.; Silvestri, C.; Ventura, D.; Matiddi, M. Tracing the Route: Using Stable Isotope Analysis to Understand Microplastic Pathways through the Pelagic-Neritic Food Web of the Tyrrhenian Sea (Western Mediterranean). *Sci. Total Environ.* **2023**, *885*, 163875. [CrossRef] [PubMed]
28. Yuan, Z.; Nag, R.; Cummins, E. Human Health Concerns Regarding Microplastics in the Aquatic Environment—From Marine to Food Systems. *Sci. Total Environ.* **2022**, *823*, 153730. [CrossRef] [PubMed]
29. Hasegawa, T.; Nakaoka, M. Trophic Transfer of Microplastics from Mysids to Fish Greatly Exceeds Direct Ingestion from the Water Column. *Environ. Pollut.* **2021**, *273*, 116468. [CrossRef]
30. Nelms, S.E.; Galloway, T.S.; Godley, B.J.; Jarvis, D.S.; Lindeque, P.K. Investigating Microplastic Trophic Transfer in Marine Top Predators. *Environ. Pollut.* **2018**, *238*, 999–1007. [CrossRef] [PubMed]
31. Paul-Pont, I.; Tallec, K.; Gonzalez-Fernandez, C.; Lambert, C.; Vincent, D.; Mazurais, D.; Zambonino-Infante, J.-L.; Brotons, G.; Lagarde, F.; Fabioux, C.; et al. Constraints and Priorities for Conducting Experimental Exposures of Marine Organisms to Microplastics. *Front. Mar. Sci.* **2018**, *5*, 252. [CrossRef]
32. Oliveira, M.; Ribeiro, A.; Hylland, K.; Guilhermino, L. Single and Combined Effects of Microplastics and Pyrene on Juveniles (0+ Group) of the Common Goby *Pomatoschistus microps* (Teleostei, Gobiidae). *Ecol. Indic.* **2013**, *34*, 641–647. [CrossRef]
33. Li, B.; Su, L.; Zhang, H.; Deng, H.; Chen, Q.; Shi, H. Microplastics in Fishes and Their Living Environments Surrounding a Plastic Production Area. *Sci. Total Environ.* **2020**, *727*, 138662. [CrossRef]
34. Timilsina, A.; Adhikari, K.; Yadav, A.K.; Joshi, P.; Ramena, G.; Bohara, K. Effects of Microplastics and Nanoplastics in Shrimp: Mechanisms of Plastic Particle and Contaminant Distribution and Subsequent Effects after Uptake. *Sci. Total Environ.* **2023**, *894*, 164999. [CrossRef]
35. Gallo, F.; Fossi, C.; Weber, R.; Santillo, D.; Sousa, J.; Ingram, I.; Nadal, A.; Romano, D. Marine Litter Plastics and Microplastics and Their Toxic Chemicals Components: The Need for Urgent Preventive Measures. *Environ. Sci. Eur.* **2018**, *30*, 13. [CrossRef] [PubMed]
36. Rodrigues, M.O.; Abrantes, N.; Gonçalves, F.J.M.; Nogueira, H.; Marques, J.C.; Gonçalves, A.M.M. Impacts of Plastic Products Used in Daily Life on the Environment and Human Health: What Is Known? *Environ. Toxicol. Pharmacol.* **2019**, *72*, 103239. [CrossRef] [PubMed]
37. Alberghini, L.; Truant, A.; Santonicola, S.; Colavita, G.; Giaccone, V. Microplastics in Fish and Fishery Products and Risks for Human Health: A Review. *Int. J. Environ. Res. Public Health* **2022**, *20*, 789. [CrossRef] [PubMed]
38. Rochman, C.M.; Hoh, E.; Kurobe, T.; Teh, S.J. Ingested Plastic Transfers Hazardous Chemicals to Fish and Induces Hepatic Stress. *Sci. Rep.* **2013**, *3*, 3263. [CrossRef] [PubMed]
39. Rocha-Santos, T.; Costa, M.; Mouneyrac, C. (Eds.) *Handbook of Microplastics in the Environment*; Springer International Publishing: Cham, Switzerland, 2020; ISBN 978-3-030-10618-8.
40. FAO. *The State of World Fisheries and Aquaculture 2020. Sustainability in Action*; FAO: Rome, Italy, 2021. [CrossRef]
41. Cau, A.; Carbonell, A.; Follesa, M.C.; Mannini, A.; Orsi Relini, L.; Politou, C.Y.; Ragonese, S.; Rinelli, P. MEDITS-Based Information on the Deep Water Red Shrimps *Aristaeomorpha foliacea* and *Aristeus antennatus* (Crustacea: Decapoda: Aristeidae). *Sci. Mar.* **2002**, *66*, 103. [CrossRef]
42. Fischer, W.; Bauchot, M.L.; Schneider, M. *Fiches FAO d'identification Des Espèces Pour Les Besoins de La Pêche. (Révision 1)*; Méditerranée et Mer Noire. Zone de Pêche 37. Végétaux et Invertébrés. Publication Préparée Par La FAO, Résultat d'un Accord Entre La FAO et La Commission Des Communautés Européennes; Département Des pêches de la FAO (Food and Agriculture Organization of the United Nations): Rome, Italy, 1987.
43. Tom, M.; Goren, M.; Ovadia, M. The Benthic Phase of the Life Cycle of *Parapenaeus longirostris* (Crustacea, Decapoda, Penaeidae) along the Mediterranean Coast of Israel. *Hydrobiologia* **1988**, *169*, 339–352. [CrossRef]
44. Bianchini, M.; Ragonese, S. Life Cycles and Fisheries of the Deep Water Shrimps *Aristaeomorpha foliacea* and *Aristeus antennatus*. In Proceedings of the International Workshop Held in the Istituto Tecnologia Pesca e Pescato, Mazara, Italy, 28–30 April 1994; Volume 3, pp. 1–88.
45. Cartes, J. Diets of, and Trophic Resources Exploited by, Bathyal Penaeoidean Shrimps from the Western Mediterranean. *Mar. Freshw. Res.* **1995**, *46*, 889. [CrossRef]
46. Cartes, J.E.; Fanelli, E.; Kapiris, K.; Bayhan, Y.K.; Ligas, A.; López-Pérez, C.; Murenu, M.; Papiol, V.; Rumolo, P.; Scarcella, G. Spatial Variability in the Trophic Ecology and Biology of the Deep-Sea Shrimp *Aristaeomorpha foliacea* in the Mediterranean Sea. *Deep Sea Res. Part I Oceanogr. Res. Pap.* **2014**, *87*, 1–13. [CrossRef]
47. Kapiris, K. Feeding Ecology of *Parapenaeus longirostris* (Lucas, 1846) (Decapoda: Penaeidae) from the Ionian Sea (Central and Eastern Mediterranean Sea). *Sci. Mar.* **2004**, *68*, 247–256. [CrossRef]
48. Chartosia, N.; Tzomos, T.H.; Kitsos, M.S.; Karani, I.; Tselepides, A.; Koukouras, A. Karani Diet Comparison of the Bathyal Shrimps, *Aristeus antennatus* (Risso, 1816) and *Aristaeomorpha foliacea* (Risso, 1827) (Decapoda, Aristeidae) in the Eastern Mediterranean. *Crustaceana* **2005**, *78*, 273–284. [CrossRef]
49. O'Reilly, C.M.; Hecky, R.E.; Cohen, A.S.; Plisnier, P.-D. Interpreting Stable Isotopes in Food Webs: Recognizing the Role of Time Averaging at Different Trophic Levels. *Limnol. Oceanogr.* **2002**, *47*, 306–309. [CrossRef]
50. Peterson, B.J.; Fry, B. STABLE ISOTOPES IN ECOSYSTEM STUDIES. *Annu. Rev. Ecol. Syst.* **1987**, *18*, 293–320. [CrossRef]
51. Schoeninger, M.J.; DeNiro, M.J. Nitrogen and Carbon Isotopic Composition of Bone Collagen from Marine and Terrestrial Animals. *Geochim. Cosmochim. Acta* **1984**, *48*, 625–639. [CrossRef]

52. Vizzini, S.; Mazzola, A. Stable Isotopes and Trophic Positions of Littoral Fishes from a Mediterranean Marine Protected Area. *Environ. Biol. Fish.* **2009**, *84*, 13–25. [CrossRef]
53. Fanelli, E.; Azzurro, E.; Bariche, M.; Cartes, J.E.; Maynou, F. Depicting the Novel Eastern Mediterranean Food Web: A Stable Isotopes Study Following Lessepsian Fish Invasion. *Biol. Invasions* **2015**, *17*, 2163–2178. [CrossRef]
54. D'Iglio, C.; Famulari, S.; Albano, M.; Giordano, D.; Rinelli, P.; Capillo, G.; Spanò, N.; Savoca, S. Time-Scale Analysis of Prey Preferences and Ontogenetic Shift in the Diet of European Hake *Merluccius merluccius* (Linnaeus, 1758) in Southern and Central Tyrrhenian Sea. *Fishes* **2022**, *7*, 167. [CrossRef]
55. Ricci, P.; Carlucci, R.; Capezzuto, F.; Carluccio, A.; Cipriano, G.; D'Onghia, G.; Maiorano, P.; Sion, L.; Tursi, A.; Libralato, S. Contribution of Intermediate and High Trophic Level Species to Benthic-Pelagic Coupling: Insights From Modelling Analysis. *Front. Mar. Sci.* **2022**, *9*, 887464. [CrossRef]
56. Ricci, P.; Libralato, S.; Capezzuto, F.; D'Onghia, G.; Maiorano, P.; Sion, L.; Tursi, A.; Solidoro, C.; Carlucci, R. Ecosystem Functioning of Two Marine Food Webs in the North-Western Ionian Sea (Central Mediterranean Sea). *Ecol. Evol.* **2019**, *9*, 10198–10212. [CrossRef]
57. Carbery, M.; O'Connor, W.; Palanisami, T. Trophic Transfer of Microplastics and Mixed Contaminants in the Marine Food Web and Implications for Human Health. *Environ. Int.* **2018**, *115*, 400–409. [CrossRef]
58. Frias, J.P.G.L.; Otero, V.; Sobral, P. Evidence of Microplastics in Samples of Zooplankton from Portuguese Coastal Waters. *Mar. Environ. Res.* **2014**, *95*, 89–95. [CrossRef]
59. Farrell, P.; Nelson, K. Trophic Level Transfer of Microplastic: *Mytilus edulis* (L.) to *Carcinus maenas* (L.). *Environ. Pollut.* **2013**, *177*, 1–3. [CrossRef] [PubMed]
60. Cau, A.; Avio, C.G.; Dessì, C.; Follesa, M.C.; Moccia, D.; Regoli, F.; Pusceddu, A. Microplastics in the Crustaceans *Nephrops norvegicus* and *Aristeus antennatus*: Flagship Species for Deep-Sea Environments? *Environ. Pollut.* **2019**, *255*, 113107. [CrossRef] [PubMed]
61. Garcia, F.; De Carvalho, A.R.; Riem-Galliano, L.; Tudesque, L.; Albignac, M.; Ter Halle, A.; Cucherousset, J. Stable Isotope Insights into Microplastic Contamination within Freshwater Food Webs. *Environ. Sci. Technol.* **2021**, *55*, 1024–1035. [CrossRef] [PubMed]
62. Falciai, L.; Minervini, R. *Guida Dei Crostacei Decapodi d'Europa*; Franco Muzzio: Padova, Italy, 1992; ISBN 978-88-7021-557-1.
63. IAEA−600. Available online: https://analytical-reference-materials.iaea.org/iaea-600 (accessed on 23 November 2023).
64. USGS40. Available online: https://analytical-reference-materials.iaea.org/usgs40 (accessed on 23 November 2023).
65. IAEA-C-8. Available online: https://analytical-reference-materials.iaea.org/iaea-c-8 (accessed on 23 November 2023).
66. IAEA-CH6. Available online: https://analytical-reference-materials.iaea.org/iaea-ch-6 (accessed on 23 November 2023).
67. IAEA-N-1. Available online: https://analytical-reference-materials.iaea.org/iaea-n-1 (accessed on 23 November 2023).
68. IAEA-N-2. Available online: https://analytical-reference-materials.iaea.org/iaea-n-2 (accessed on 23 November 2023).
69. Post, D.M. Using Stable Isotopes To Estimate Trophic Position: Models, Methods, and Assumptions. *Ecology* **2002**, *83*, 703–718. [CrossRef]
70. Skinner, M.M.; Martin, A.A.; Moore, B.C. Is Lipid Correction Necessary in the Stable Isotope Analysis of Fish Tissues? *Rapid Commun. Mass Spectrom.* **2016**, *30*, 881–889. [CrossRef] [PubMed]
71. European Commission; Joint Research Centre; MSFD Technical Group on Marine Litter. *Guidance on the Monitoring of Marine Litter in European Seas: An Update to Improve the Harmonised Monitoring of Marine Litter under the Marine Strategy Framework Directive*; Publications Office: Luxembourg, 2023.
72. Matiddi, M.; Pham, C.K.; Anastasopoulou, A.; Andresmaa, E.; Avio, C.G.; Bianchi, J.; Chiaeb, O.; Palazzo, L.; Darmon, G.; De Lucia, G.A.; et al. Monitoring Micro-Litter Ingestion in Marine Fish: A Harmonized Protocol for MSFD and RSCs Areas. INDICIT II Project 2021. Available online: https://accedacris.ulpgc.es/bitstream/10553/114417/1/Report_Monitoring-microlitter-ingestion-in-marine-fish.pdf (accessed on 23 November 2023).
73. RStudio Team. RStudio: Integrated Development Environment for R. Version 4.3.2. 2023. Available online: https://www.rstudio.com/ (accessed on 23 November 2023).
74. Jackson, A.L.; Inger, R.; Parnell, A.C.; Bearhop, S. Comparing Isotopic Niche Widths among and within Communities: SIBER—Stable Isotope Bayesian Ellipses in R: Bayesian Isotopic Niche Metrics. *J. Anim. Ecol.* **2011**, *80*, 595–602. [CrossRef] [PubMed]
75. Layman, C.A.; Araujo, M.S.; Boucek, R.; Hammerschlag-Peyer, C.M.; Harrison, E.; Jud, Z.R.; Matich, P.; Rosenblatt, A.E.; Vaudo, J.J.; Yeager, L.A.; et al. Applying Stable Isotopes to Examine Food-web Structure: An Overview of Analytical Tools. *Biol. Rev.* **2012**, *87*, 545–562. [CrossRef] [PubMed]
76. Layman, C.A.; Arrington, D.A.; Montaña, C.G.; Post, D.M. CAN STABLE ISOTOPE RATIOS PROVIDE FOR COMMUNITY-WIDE MEASURES OF TROPHIC STRUCTURE? *Ecology* **2007**, *88*, 42–48. [CrossRef] [PubMed]
77. McLeod, A.I. Kendall Rank Correlation and Mann-Kendall Trend Test. *R Package Kendall* **2005**, *602*, 1–10.
78. Iitembu, J.A.; Dalu, T. Patterns of Trophic Resource Use among Deep-Sea Shrimps in the Northern Benguela Current Ecosystem, Namibia. *Food Webs* **2018**, *16*, e00089. [CrossRef]
79. Leila, B.; Sedláček, P.; Anastasopoulou, A. Plastic Pollution in the Deep-Sea Giant Red Shrimp, *Aristaeomorpha foliacea*, in the Eastern Ionian Sea; an Alarm Point on Stock and Human Health Safety. *Sci. Total Environ.* **2023**, *877*, 162783. [CrossRef]
80. D'Iglio, C.; Di Fresco, D.; Spanò, N.; Albano, M.; Panarello, G.; Laface, F.; Faggio, C.; Capillo, G.; Savoca, S. Occurrence of Anthropogenic Debris in Three Commercial Shrimp Species from South-Western Ionian Sea. *Biology* **2022**, *11*, 1616. [CrossRef] [PubMed]

81. Bono, G.; Scannella, D.; Falsone, F.; Falco, F.; Maio, F.D.; Gabriele, M.; Gancitano, V.; Geraci, M.L.; Mancuso, M.; Okpala, C.; et al. Microplastics and Alien Black Particles as Contaminants of Deep-Water Rose Shrimp (Parapenaeus Longistroris Lucas, 1846) in the Central Mediterranean Sea. *J. Adv. Biotechnol. Bioeng.* **2020**, *8*, 23–28. [CrossRef]
82. Yücel, N. Detection of Microplastic Fibers Tangle in Deep-Water Rose Shrimp (*Parapenaeus longirostris*, Lucas, 1846) in the Northeastern Mediterranean Sea. *Environ. Sci. Pollut. Res.* **2022**, *30*, 10914–10924. [CrossRef] [PubMed]
83. Yin, L.; Chen, B.; Xia, B.; Shi, X.; Qu, K. Polystyrene Microplastics Alter the Behavior, Energy Reserve and Nutritional Composition of Marine Jacopever (*Sebastes schlegelii*). *J. Hazard. Mater.* **2018**, *360*, 97–105. [CrossRef] [PubMed]
84. Grigorakis, S.; Mason, S.A.; Drouillard, K.G. Determination of the Gut Retention of Plastic Microbeads and Microfibers in Goldfish (*Carassius auratus*). *Chemosphere* **2017**, *169*, 233–238. [CrossRef] [PubMed]
85. Foekema, E.M.; De Gruijter, C.; Mergia, M.T.; Van Franeker, J.A.; Murk, A.J.; Koelmans, A.A. Plastic in North Sea Fish. *Environ. Sci. Technol.* **2013**, *47*, 8818–8824. [CrossRef] [PubMed]
86. Neves, D.; Sobral, P.; Ferreira, J.L.; Pereira, T. Ingestion of Microplastics by Commercial Fish off the Portuguese Coast. *Mar. Pollut. Bull.* **2015**, *101*, 119–126. [CrossRef] [PubMed]
87. Valente, T.; Pelamatti, T.; Avio, C.G.; Camedda, A.; Costantini, M.L.; De Lucia, G.A.; Jacomini, C.; Piermarini, R.; Regoli, F.; Sbrana, A.; et al. One Is Not Enough: Monitoring Microplastic Ingestion by Fish Needs a Multispecies Approach. *Mar. Pollut. Bull.* **2022**, *184*, 114133. [CrossRef]
88. Carreras-Colom, E.; Constenla, M.; Soler-Membrives, A.; Cartes, J.E.; Baeza, M.; Padrós, F.; Carrassón, M. Spatial Occurrence and Effects of Microplastic Ingestion on the Deep-Water Shrimp *Aristeus antennatus*. *Mar. Pollut. Bull.* **2018**, *133*, 44–52. [CrossRef]
89. Chubarenko, I.; Bagaev, A.; Zobkov, M.; Esiukova, E. On Some Physical and Dynamical Properties of Microplastic Particles in Marine Environment. *Mar. Pollut. Bull.* **2016**, *108*, 105–112. [CrossRef] [PubMed]
90. Hidalgo-Ruz, V.; Gutow, L.; Thompson, R.C.; Thiel, M. Microplastics in the Marine Environment: A Review of the Methods Used for Identification and Quantification. *Environ. Sci. Technol.* **2012**, *46*, 3060–3075. [CrossRef] [PubMed]
91. Van Cauwenberghe, L.; Devriese, L.; Galgani, F.; Robbens, J.; Janssen, C.R. Microplastics in Sediments: A Review of Techniques, Occurrence and Effects. *Mar. Environ. Res.* **2015**, *111*, 5–17. [CrossRef] [PubMed]
92. Erni-Cassola, G.; Zadjelovic, V.; Gibson, M.I.; Christie-Oleza, J.A. Distribution of Plastic Polymer Types in the Marine Environment; A Meta-Analysis. *J. Hazard. Mater.* **2019**, *369*, 691–698. [CrossRef] [PubMed]
93. Lopes, C.; Raimundo, J.; Caetano, M.; Garrido, S. Microplastic Ingestion and Diet Composition of Planktivorous Fish. *Limnol. Ocean. Lett.* **2020**, *5*, 103–112. [CrossRef]
94. Ory, N.C.; Sobral, P.; Ferreira, J.L.; Thiel, M. Amberstripe Scad *Decapterus muroadsi* (Carangidae) Fish Ingest Blue Microplastics Resembling Their Copepod Prey along the Coast of Rapa Nui (Easter Island) in the South Pacific Subtropical Gyre. *Sci. Total Environ.* **2017**, *586*, 430–437. [CrossRef]
95. Murray, F.; Cowie, P.R. Plastic Contamination in the Decapod Crustacean *Nephrops norvegicus* (Linnaeus, 1758). *Mar. Pollut. Bull.* **2011**, *62*, 1207–1217. [CrossRef] [PubMed]
96. Browne, M.A.; Dissanayake, A.; Galloway, T.S.; Lowe, D.M.; Thompson, R.C. Ingested Microscopic Plastic Translocates to the Circulatory System of the Mussel, *Mytilus edulis* (L.). *Environ. Sci. Technol.* **2008**, *42*, 5026–5031. [CrossRef] [PubMed]
97. Denton, E.J. *Light and Life in the Sea*; Cambridge University Press: Cambridge, UK, 1990; ISBN 0-521-39207-1.
98. Bauer, R.T. *Shrimps: Their Diversity, Intriguing Adaptations and Varied Lifestyles*; Fish & Fisheries Series; Springer International Publishing: Cham, Switzerland, 2023; Volume 42, ISBN 978-3-031-20965-9.
99. Deopura, B.L.; Alagirusamy, R.; Textile Institute (Eds.) *Polyesters and Polyamides*; Woodhead Publishing in Textiles; CRC Press: Boca Raton, FL, USA, 2008; ISBN 978-1-84569-298-8.
100. Turner, A. Paint Particles in the Marine Environment: An Overlooked Component of Microplastics. *Water Res. X* **2021**, *12*, 100110. [CrossRef]
101. Lithner, D.; Larsson, Å.; Dave, G. Environmental and Health Hazard Ranking and Assessment of Plastic Polymers Based on Chemical Composition. *Sci. Total Environ.* **2011**, *409*, 3309–3324. [CrossRef]
102. Berto, D.; Rampazzo, F.; Gion, C.; Noventa, S.; Ronchi, F.; Traldi, U.; Giorgi, G.; Cicero, A.M.; Giovanardi, O. Preliminary Study to Characterize Plastic Polymers Using Elemental Analyser/Isotope Ratio Mass Spectrometry (EA/IRMS). *Chemosphere* **2017**, *176*, 47–56. [CrossRef] [PubMed]
103. Shum, P.; Wage-Recchioni, J.; Sellers, G.S.; Johnson, M.L.; Joyce, D.A. DNA Metabarcoding Reveals the Dietary Profiles of a Benthic Marine Crustacean, Nephrops Norvegicus. *PLoS ONE* **2023**, *18*, e0289221. [CrossRef] [PubMed]
104. Mackas, D.; Bohrer, R. Fluorescence Analysis of Zooplankton Gut Contents and an Investigation of Diel Feeding Patterns. *J. Exp. Mar. Biol. Ecol.* **1976**, *25*, 77–85. [CrossRef]
105. Kousteni, V.; Karachle, P.K.; Megalofonou, P. Diet of the Small-Spotted Catshark *Scyliorhinus canicula* in the Aegean Sea (Eastern Mediterranean). *Mar. Biol. Res.* **2017**, *13*, 161–173. [CrossRef]
106. Fanelli, E.; Rey, J.; Torres, P.; Gil De Sola, L. Feeding Habits of Blackmouth Catshark *Galeus melastomus* Rafinesque, 1810 and Velvet Belly Lantern Shark *Etmopterus spinax* (Linnaeus, 1758) in the Western Mediterranean. *J. Appl. Ichthyol.* **2009**, *25*, 83–93. [CrossRef]
107. Anastasopoulou, A.; Mytilineou, C.; Lefkaditou, E.; Dokos, J.; Smith, C.J.; Siapatis, A.; Bekas, P.; Papadopoulou, K.-N. Diet and Feeding Strategy of Blackmouth Catshark *Galeus melastomus*. *J. Fish Biol.* **2013**, *83*, 1637–1655. [CrossRef]

108. Esposito, V.; Andaloro, F.; Bianca, D.; Natalotto, A.; Romeo, T.; Scotti, G.; Castriota, L. Diet and Prey Selectivity of the Red Mullet, *Mullus barbatus* (Pisces: Mullidae), from the Southern Tyrrhenian Sea: The Role of the Surf Zone as a Feeding Ground. *Mar. Biol. Res.* **2014**, *10*, 167–178. [CrossRef]
109. Zicarelli, G.; Romano, C.; Gallo, S.; Valentino, C.; Pepe Bellomo, V.; Leonetti, F.L.; Giglio, G.; Neri, A.; Marsili, L.; Milazzo, C.; et al. Diet and Plastic Ingestion in the Blackmouth Catshark Galeus Melastomus, Rafinesque 1810, in Italian Waters. *Animals* **2023**, *13*, 1039. [CrossRef] [PubMed]
110. Carpentieri, P.; Colloca, F.; Cardinale, M.; Belluscio, A.; Ardizzone, G. Feeding Habits of European Hake (*Merluccius merluccius*) in the Central Mediterranean Sea. *Fish. Bull.* **2005**, *103*, 411–416.
111. Bello, G. *The Feeding Ecology of the Velvet Belly, Etmopterus spinax (Chondrichthyes: Squalidae), of the Adriatic*; Società Italiana Scienze Naturali: Milano, Italy, 1998.
112. Valente, T.; Sbrana, A.; Scacco, U.; Jacomini, C.; Bianchi, J.; Palazzo, L.; De Lucia, G.A.; Silvestri, C.; Matiddi, M. Exploring Microplastic Ingestion by Three Deep-Water Elasmobranch Species: A Case Study from the Tyrrhenian Sea. *Environ. Pollut.* **2019**, *253*, 342–350. [CrossRef]
113. Wang, W.; Ge, J.; Yu, X. Bioavailability and Toxicity of Microplastics to Fish Species: A Review. *Ecotoxicol. Environ. Saf.* **2020**, *189*, 109913. [CrossRef] [PubMed]
114. Alomar, C.; Deudero, S. Evidence of Microplastic Ingestion in the Shark Galeus Melastomus Rafinesque, 1810 in the Continental Shelf off the Western Mediterranean Sea. *Environ. Pollut.* **2017**, *223*, 223–229. [CrossRef] [PubMed]
115. Giani, D.; Baini, M.; Galli, M.; Casini, S.; Fossi, M.C. Microplastics Occurrence in Edible Fish Species (*Mullus barbatus* and *Merluccius merluccius*) Collected in Three Different Geographical Sub-Areas of the Mediterranean Sea. *Mar. Pollut. Bull.* **2019**, *140*, 129–137. [CrossRef] [PubMed]
116. Ferreira, G.V.B.; Justino, A.K.S.; Eduardo, L.N.; Schmidt, N.; Martins, J.R.; Ménard, F.; Fauvelle, V.; Mincarone, M.M.; Lucena-Frédou, F. Influencing Factors for Microplastic Intake in Abundant Deep-Sea Lanternfishes (Myctophidae). *Sci. Total Environ.* **2023**, *867*, 161478. [CrossRef]
117. Vecchi, S.; Bianchi, J.; Scalici, M.; Fabroni, F.; Tomassetti, P. Field Evidence for Microplastic Interactions in Marine Benthic Invertebrates. *Sci. Rep.* **2021**, *11*, 20900. [CrossRef]

animals

Article

Effects of Intermittent and Chronic Hypoxia on Fish Size and Nutrient Metabolism in Tiger Puffer (*Takifugu rubripes*)

Qiang Ma [1], Renxiao Zhang [1,3], Yuliang Wei [1,2], Mengqing Liang [1,2] and Houguo Xu [1,2,*]

[1] State Key Laboratory of Mariculture Biobreeding and Sustainable Goods, Yellow Sea Fisheries Research Institute, Chinese Academy of Fishery Sciences, Qingdao 266071, China; maqiang@ysfri.ac.cn (Q.M.); renxiaoz2022@163.com (R.Z.); weiyl@ysfri.ac.cn (Y.W.); liangmq@ysfri.ac.cn (M.L.)

[2] Laboratory for Marine Fisheries Science and Food Production Processes, Qingdao Marine Science and Technology Center, Qingdao 266237, China

[3] Key Laboratory of Aquaculture Nutrition and Feed, Ministry of Agriculture, Ocean University of China, 5 Yushan Road, Qingdao 266003, China

* Correspondence: xuhg@ysfri.ac.cn; Tel.: +86-532-85822914

Simple Summary: Intermittent and chronic hypoxia are common and harmful to marine animals. The tiger puffer is a representative species of the family Tetraodontidae, which has only slit-like gill openings and is susceptible to hypoxia stress. Both intermittent and chronic hypoxia decrease the growth and visceral weight of tiger puffer but increase the feed conversion ratio and blood hemoglobin content. Chronic hypoxia but not intermittent hypoxia promoted protein synthesis and the glycolysis pathway. Intermittent hypoxia but not chronic hypoxia decreased lipid synthesis and monounsaturated fatty acids content but increased n-3 polyunsaturated fatty acids levels. These changes promoted the adaption of tiger puffer to intermittent and chronic hypoxia.

Abstract: Intermittent and chronic hypoxia are common stresses to marine fish, but the different responses of fish to intermittent and chronic hypoxia have not been well-known. In this study, tiger puffers were farmed in normoxia conditions (NO, 6.5 ± 0.5 mg/L), intermittent hypoxia (IH, 6.5 ± 0.5 mg/L in the day and 3.5 ± 0.5 mg/L in the night), or choric hypoxia (CH, 3.5 ± 0.5 mg/L) conditions for 4 weeks, after which the growth, nutrient metabolism and three *hifα* isoforms expression were measured. Both intermittent and chronic hypoxia decreased the fish growth and visceral weight but increased the feed conversion ratio and blood hemoglobin content. Chronic hypoxia but not intermittent hypoxia promoted protein synthesis and whole-fish protein content by activating *mtor* gene expression and promoted the glycolysis pathway by activating gene expression of *hif1α* and *hif2α*. Intermittent hypoxia but not chronic hypoxia decreased the hepatic lipid synthesis by inhibiting *fasn* and *srebf1* gene expression. Meanwhile, intermittent hypoxia reduced the monounsaturated fatty acid content but increased the n-3 polyunsaturated fatty acids percentage. The results of this study clarified the adaptive mechanism of tiger puffer to intermittent and chronic hypoxia, which provides important information about mechanisms of hypoxia adaption in fish.

Keywords: intermittent hypoxia; chronic hypoxia; nutrients metabolism; hypoxia-inducible factor; marine deoxygenation

1. Introduction

Dissolved oxygen (DO), expressed as mg/L or saturation (%), is the molecular oxygen (O_2) dissolved in the water environment, which plays an important role in the survival of aquatic organisms and the homeostasis of water ecosystems [1,2]. Hypoxia stress of aquatic animals usually refers to the DO level < 2 mg/L [3], but some fish species, such as turbot (*Scophthalmus maximus*) and rainbow trout (*Oncorhynchus mykiss*), are intolerant to hypoxia and have severe stress responses when the DO level falls below 4 mg/L. The O_2 content in the air is stable at about 21%, but the DO level of the natural water environment is affected by water quality and always fluctuates. Therefore, unlike terrestrial animals, aquatic animals are more susceptible to hypoxia stress [4]. According to the duration of hypoxia, hypoxia can be divided into three major types: acute hypoxia, intermittent hypoxia, and chronic hypoxia. In aquaculture, factors such as high temperature, extreme weather, power failure, and long-distance transportation can all reduce the DO level of the water environment, resulting in acute hypoxia stress for aquatic animals. The high-density aquaculture mode of Atlantic salmon needs high DO levels in Norway and Chile, and uncontrollable ocean currents caused acute hypoxia, leading to massive deaths of fish and severe economic losses [5]. In most natural waters, DO level varies with photoperiod, with high DO during the day and low DO at night, resulting in intermittent hypoxia in aquatic animals [6,7]. During the daytime, aquatic plants and algae can perform photosynthesis and produce O_2 continuously, and the water DO level reaches its highest level before sunset; while at night, aquatic plants, animals, and microorganisms perform respiration, and the water DO content is continuously decreasing, and generally reaches its lowest level before sunrise. Therefore, intermittent hypoxia is a common phenomenon for aquatic animals [8]. Ocean warming, eutrophication, and high-density aquaculture all keep the water DO at low levels for a long time and cause chronic hypoxia, resulting in death or growth inhibition of aquatic animals [9]. With the excessive discharge of greenhouse gases (CO_2) and wastewater containing nitrogen and phosphorus, the DO content of the Gulf of Mexico and Amazon rivers has been declining, leading to a dramatic decrease in aquatic animal populations and serious damage to the ecosystem [3,10,11]. Therefore, acute, intermittent, and chronic hypoxia are three common types of hypoxia stress in both natural water environments and aquaculture activities.

During oxidative phosphorylation, O_2 receives electrons from $FADH_2$ and NADH of the respiratory chain to produce adenosine triphosphate (ATP) [12]. In hypoxia conditions, the efficiency of electron transport in the respiratory chain would decrease, which would suppress the synthesis of ATP and cause incomplete reduction of O_2 to generate the reactive oxygen species (ROS) that are harmful to the organism [13,14]. In mammals and fish, the hypoxia-inducible factor (HIF) signaling pathway plays key roles in cellular response to hypoxia stress. The lack of oxygen inhibits the breakdown of HIFα (including HIF1α, HIF2α/EPAS1, and HIF3α/IPAS), and accumulated HIFα could enter into the cell nucleus and bind with the HIFβ. Then, the dimer would combine with the hypoxia-responsive element (HRE) and promote the transcription of target genes [15,16]. Studies in mammals have shown that HIF1α and HIF2α played a major role in hypoxia, activating numerous downstream target genes, such as key genes of glycolysis (GLUT1, HK, and LDH, etc.), vascular endothelial growth factor (VEGF), and erythropoietin (EPO) [17,18]. However, some HIF3α transcripts in mammals are negative regulators of HIF1α and HIF2α by competitively binding to HIFβ to repress the target genes expression under hypoxia [19]. Unlike mammalian HIF3α, zebrafish Hif3α is also an oxygen-dependent transcriptional activator and plays a similar role as Hif1α under hypoxia [20,21]. Oscar (*Astronotus ocellatus*) is a hypoxia-tolerant fish in the Amazon River, and acute hypoxia (DO = 0.5 mg/L for 3 h) increased the red blood cells, hematocrit, hemoglobin, and glucose levels, as well as genes expression of *hif1α* and *vegf* in the liver [22]. Similar results were found in Nile tilapia (*Oreochromis niloticus*) [23,24]. In largemouth bass (*Micropterus salmoides*), acute hypoxia (DO = 1.2–3 mg/L for 4–8 h) significantly increased the serum glucose content, liver contents of lactic acid, triglycerides, and non-esterified fatty acids, as well as the expression of *hif1α*, *glut1*, and glycolysis-related genes [25,26]. These studies all proved that acute hypoxia activated *hif1α* and promoted glucose catabolism by the anaerobic glycolysis pathway in fish. However, the differences in nutrient metabolism patterns and expression of three *hifα* isoforms between intermittent hypoxia and chronic hypoxia are unknown in marine fish species to date.

Tiger puffer (*Takifugu rubripes*) is a representative species of the family Tetraodontidae and is distributed in the northwestern Pacific [27]. Tetraodontids are characterized by a loose abdomen that can be inflated with air or water, a beak-like dental plate divided by a median suture, and tetrodotoxin and saxitoxin in the tissues [28]. The family Tetraodontidae also have the smallest vertebrate genomes yet measured. Tiger puffers have become a model organism for genomics studies, and Japan has approved the sale of CRISPR-edited tiger puffers [29,30]. Tiger puffers are one of the most valuable commercial fish species because of their high nutritional value and delicious flavor. However, the capture production of wild tiger puffers has greatly decreased since the 1980s because of the higher fishing pressure and deteriorating marine environment [31]. Therefore, tiger puffers are widely farmed in China, Japan, and Korea in both floating cages and industrial recirculating aquaculture systems with high density. China is the largest tiger puffer and obscure puffer (*Takifugu obscurus*) producer in the world, and the production was 16,612 and 14,434 tons in 2022, respectively [32]. The tiger puffer is a demersal species with only a slit-like gill opening anterior to the base of the pectoral fin, making them susceptible to hypoxia stress. Therefore, in this study, tiger puffers were reared in normoxia, intermittent, and chronic hypoxia environments for 4 weeks, and then blood, liver, and muscle were collected for indicators measurement. The results of the study illustrated the effects of intermittent and chronic hypoxia on growth and nutrient metabolism in tiger puffers for the first time, as well as improved our understanding of the adaptive mechanisms of different hypoxia patterns in fish.

2. Materials and Methods

2.1. Fish Culture and Hypoxia Conditions Management

Juvenile tiger puffers (three-months-old) were purchased from Huanghai Aquaculture Co. Ltd. (Yantai, China) and were cultured in a flow-through seawater system. A total of 180 experimental tiger puffers with an average initial body weight of appr. 37 g were distributed into three groups: normoxia group (NO), intermittent hypoxia group (IH), and choric hypoxia group (CH). Each group had triplicate polyethylene tanks (300 L), and each tank had 20 fish. The water dissolved oxygen level of the NO group was kept at 6.5 ± 0.5 mg/L, and the DO level in the CH group was kept at 3.5 ± 0.5 mg/L. In the IH group, the DO level was kept at 6.5 ± 0.5 mg/L during the daytime (from 8:00 to 20:00), but the DO level was kept at 3.5 ± 0.5 mg/L in the night (from 20:00 to 8:00). An air pump was used to control the DO level. All fish were fed the experimental diets following the standard procedures. The formula and proximate compositions of the diets are listed in Table 1. All fish were fed twice per day at 7:00 and 19:00, and the feeding rate was 3–4% of body weight per tank. The water salinity, pH, temperature, and total ammonia nitrogen were maintained at 26.5 ± 2.5‰, 7.5 ± 0.5, 22.5 ± 1 °C, and <0.02 mg/L, respectively.

Table 1. Formula and proximate compositions of the experimental diets used for tiger puffer (*Takifugu rubripes*).

Dietary Ingredient (% of Dry Matter)	
Fishmeal	41
Soybean protein concentrate	25
Wheat meal	21
Fish oil	6
Soybean lecithin	1.5
Vitamin premix [a]	0.4
Mineral premix [b]	0.8
Choline chloride	0.5
Butylated hydroxytoluene	0.02
Dimethyl-beta-propiothetin	0.1
$Ca(H_2PO_4)_2$	1.5
Vitamin C	0.5
Carboxymethyl cellulose	1.68
Total	100

Table 1. *Cont.*

Dietary Ingredient (% of Dry Matter)	
Proximate compositions:	
Moisture (%)	9.31
Crude fat (%)	10.62
Crude protein (%)	49.46
Ash (%)	8.65

[a] Vitamin premix and [b] mineral premix, designed for marine fish, were purchased from Qingdao Master Biotech Co., Ltd., Qingdao, China.

2.2. Fish Sampling and Indicators Calculating

After four weeks of farming in normoxia, intermittent, or chronic hypoxia conditions, all fish were fasted for 12 h, then weighed and counted. After anesthesia with eugenol, nine fish in each group (three fish in each tank) were selected for the collection of blood, liver, and muscle samples. The blood was kept at 4 °C for 2 h, then centrifuged at 3000 rpm for 10 min for serum collection. The serum, liver, and muscle samples were frozen with liquid nitrogen immediately and stored at −80 °C for further analysis. Six fish in each group (two fish in each tank) were selected for the measurement of body weight and length, the weight of the liver and viscera, based on which the hepatosomatic index (HSI), viscerosomatic index (VSI), and condition factor (K) were calculated. Six fish per group (two fish per tank) were used for the analysis of the proximate composition of the whole fish body.

2.3. Proximate Analysis of Diet and Whole Fish Compositions

Feed and whole-fish proximate compositions were assayed using the standard methods of AOAC. Weighed feed and whole-fish samples were dried at 105 °C for 24 h to calculate dry matter and moisture levels. The crude protein content was assayed using the Kjeldahl method (N × 6.25). The crude lipid content was detected by the Soxtec 2050 Soxhlet extractor with petroleum ether (Foss, Hilleroed, Denmark). Samples were weighed and put into a muffle furnace at 550 °C for 4 h to calculate the ash content.

2.4. Biochemical Indexes Assays

The contents of glucose (F006-1-1), pyruvate (A081-1-1), lactate (A019-2-1), glycogen (A043-1-1), total soluble protein (A045-2-2), and triglyceride (A110-1-1) in the serum, liver, and muscle were determined using commercial kits. All measurement steps refer to relevant kit protocols at http://www.njjcbio.com/ (accessed on 1 August 2024), and the absorbance was read using a microplate reader (Tecan infinite M200, Männedorf, Switzerland). The blood hemoglobin level was assayed using a portable hemoglobin monitor (electrochemistry method, Taiwan BeneCheck, Taipei City, China).

2.5. Extraction of RNA and qPCR

The total RNA of the muscle and liver was isolated using an RNAiso Plus kit (Takara, Kyoto, Japan), and the quality and quantity were measured by a Titertek-Berthold Colibri spectrometer (Colibri, Berlin, Germany). The absorbance ratios in 260/280 nm of RNA solution were from 1.9 to 2.0, which suggested high purity of the RNA samples. cDNA was synthesized using a reverse-transcribed kit with gDNase (Tiangen, Beijing, China). The primers of reference genes (*β-actin* and *rpl19*) and target genes in Table 2 were designed in NCBI. The reaction system of qPCR was 10 μL, including 1 μL cDNA (20 ng), 5 μL 2 × SYBR Mixture, 0.5 μL qPCR primers (10 μM), and 3.5 μL nuclease-free water. The qPCR program had 95 °C for 30 s, 40 cycles of 94 °C for 5 s, and 60 °C for 30 s, and was carried out in 96-well plates on a Roche LightCycler 96 system (Roche, Basel, Switzerland). The specificity of amplified products was detected by the melting curve at the end of the PCR. The amplification efficiency of primers was between 90% and 110%. The amplification efficiency was calculated using the following equation: $E = 10^{(-1/Slope)} - 1$ [33], and the target genes expression was calculated using the $2^{-\Delta\Delta Ct}$ method [34].

Table 2. Primers used for the analysis of gene expression for tiger puffer (*Takifugu rubripes*).

Gene Name	Sequences (5′ to 3′) Forward and Reverse	Product Length	GenBank NO.
hif3α/hif1αl	AAGCATCAGCATCAAACGGAG	111	XM_011608719.2
(hypoxia-inducible factor 1-alpha-like)	GTGTGGGCGAGCTCATAAAAC		
hif2α/epas1b	CACATGTGCAGAATCCCCCT	191	XM_011603052.2
(endothelial PAS domain-containing protein 1)	ATGGGGTATGCTCTGTTGGC		
hif1α	CCCCCTTCAGCTTACCAGAC	157	XM_003962474.3
(hypoxia-inducible factor 1 subunit alpha)	CTTTTGCGTGGGGTGTTCTG		
vegfa	CATATCACGATGCCGTTTGTG	162	XM_029849217.1
(vascular endothelial growth factor A)	CACATTTGCAGGTCCGTTCG		
gck	GAGGACTGTGGAACTGGTGG	81	XM_029829322.1
(glucokinase)	TCTCCATTGTCCCCGAATGC		
hk1	GGTTGAGGACCACCATAGGC	100	XM_003969460.3
(hexokinase 1)	ATATCCGGGACCAAACGACG		
pfk	GAATGGGCATCTACGTGGGG	140	XM_029844357.1
(ATP-dependent 6-phosphofructokinase, liver type)	ACCTATCATGGTCCCACCCT		
ldha	GCGTCACCGCTAATTCCAAG	193	XM_003967364.3
(L-lactate dehydrogenase A chain)	CAGGCCACGTAGGTCAGAAT		
srebf1	CGAGTGTGGAGCAGCCTAAA	170	XM_011603881.2
(sterol regulatory element binding transcription factor 1)	AGGGCTCTGGGTCTGAATCT		
fasn	GGAGCTGACTACAAGCTGGG	81	XM_011619859.2
(fatty acid synthase)	CAGGAAGGTTCGGTGGTCTC		
cpt1ab	CCTGATGGATGAAGAGCGGT	111	XM_011607269.2
(carnitine O-palmitoyltransferase 1, liver isoform-like)	GAGGCCCACCAGGATTTGAG		
cpt1b	TCTATCCCGCCAGTCCATCT	115	XM_029841851.1
(carnitine O-palmitoyltransferase 1, muscle isoform)	GGCAGGTTCTCCTTCATTGC		
atgl/pnpla2	CGCCGTGGAACATTTCGTTT	84	XM_003967696.3
(patatin-like phospholipase domain containing 2)	GCGCTTGTTCCAACAGACAG		
uba1	TTTCATTGGCGGTTTGGCTG	156	XM_029834111.1
(ubiquitin-like modifier activating enzyme 1)	CGGCAGTTTCTAGGAGCACA		
mtor	CGCCTTCCTCTCTTGTTGGT	162	XM_011621515.2
(mechanistic target of rapamycin kinase)	AGGGGTAGAGGACCCTTGTC		
ir	AGCAAGGACATCCGGAACAG	113	XM_003975383.3
(insulin receptor)	CGGAAATCCTCTGGCTTGGT		
akt1	GGAGACGGACACGCGATATT	139	XM_029832808.1
(AKT serine/threonine kinase 1)	ACTGGCGGAGTAGGAGAACT		
gsk3β	ATCAAGGTTCTGGGCACACC	126	XM_029839479.1
(glycogen synthase kinase-3 beta)	TGGTCGGAATACCTGCTGAC		
rpl19	GATCCCAACGAGACCAACGA	191	XM_003964816.3
(60S ribosomal protein L19)	CGAGCATTGGCTGTACCCTT		
β-actin	GGAAGATGAAATCGCCGCAC	196	XM_003964421.3
(actin beta)	GGTCAGGATACCCCTCTTGC		

2.6. Fatty Acid Compositions Analysis

The liver was weighed and broken, and the total lipid was extracted using the chloroform–methanol (2:1, v/v) method. A total of 5 mg lipid was esterified with KOH–methanol and Boron trifluoride–methanol in 65 °C and 75 °C water baths for 30 min, respectively. Then, fatty acid methyl esters were extracted with hexane and assayed by gas chromatography (GC-2010 pro, Shimadzu, Kyoto, Japan) with a fused silica capillary column (SH-RT-2560, 100 m × 0.25 mm × 0.20 μm). The temperature programming of the column was listed as follows: 150 °C to 200 °C for 15 min and 200 °C to 250 °C for 2 min. Injector and detector temperatures were 250 °C. The contents of fatty acids were expressed as a percentage of each fatty acid with respect to the total fatty acids (%).

2.7. Statistical Analysis

All data are presented as means ± standard error of means (SEM) and were analyzed using the SPSS Statistics 21.0 software (IBM corporation, Armonk, NY, USA). All data were tested for normality and homogeneity of variances using Shapiro–Wilk and Levene's tests. One-way analysis of variance (ANOVA) was performed to determine significant differences among the three treatments, followed by a Duncan's multiple comparison test. Significant differences were set at $p < 0.05$.

3. Results

3.1. Effect of Intermittent and Chronic Hypoxia on Fish Size and Organ Weight

Compared with the normoxia group (NO), the final body weight, weight gain, and viscerasomatic index of the fish in the intermittent hypoxia group (IH) and chronic hypoxia group (CH) were decreased significantly (Figure 1A,B,D). Meanwhile, the feed conversion ratio in the IH and CH groups was significantly higher than in the NO group (Figure 1C). Compared with the NO group, the hepatosomatic index and condition factor were reduced in the IH and CH groups, and significance was found in the hepatosomatic index of the CH group (Figure 1E,F). IH and CH (DO = 3.5 ± 0.5 mg/L) did not affect the survival of the tiger puffer during the 4-week farming (Figure 1G). These data proved that intermittent and chronic hypoxia conditions decreased fish growth and visceral weight.

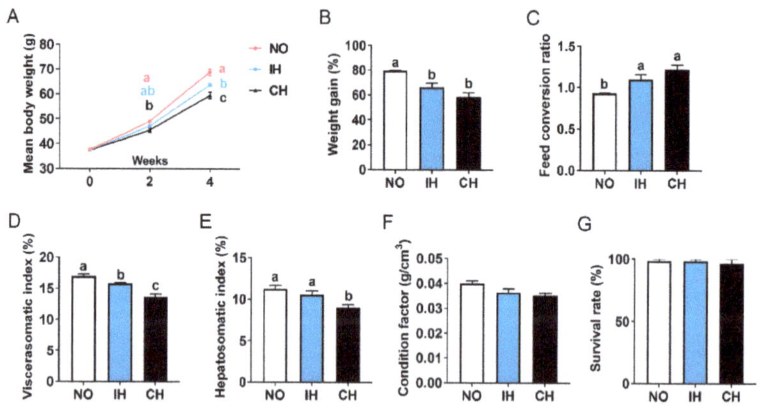

Figure 1. Effect of intermittent and chronic hypoxia on growth and visceral weight in tiger puffer (*Takifugu rubripes*). (**A**) Final body weight; (**B**) Weight gain; (**C**) Feed conversion ratio; (**D**) Viscerasomatic index; (**E**) Hepatosomatic index; (**F**) Condition factor; (**G**) Survival. Note: NO: normoxia group; IH: intermittent hypoxia group; CH: chronic hypoxia group. Different letters above the bars suggest significant differences ($p < 0.05$) among the three groups (mean ± SEM, n = 3–6).

3.2. Effects of Intermittent and Chronic Hypoxia on Whole Fish Compositions

Compared with the NO group, the whole-fish moisture and crude lipid contents in the IH and CH groups showed an increased trend and a decreased trend, respectively (Figure 2A,B). The crude protein content of whole fish in the CH group was significantly higher than the NO and IH groups (Figure 2C). These results suggest that chronic hypoxia decreased the crude lipid content and increased the crude protein of whole fish.

3.3. Effects of Intermittent and Chronic Hypoxia on Glucose Metabolism

The CH group had significantly higher contents of glucose and lactate contents in the serum, as well as higher lactate content in the liver than the NO group, but these indices were not significantly different between the IH and NO groups (Figure 3A,C,F). In addition, IH and CH had no significant effects on the pyruvate, glycogen, and lactate contents in the serum liver or muscle (Figure 3B,D,E,G,H). These data indicate that chronic hypoxia, but not intermittent hypoxia, promoted the conversion of glucose to lactate.

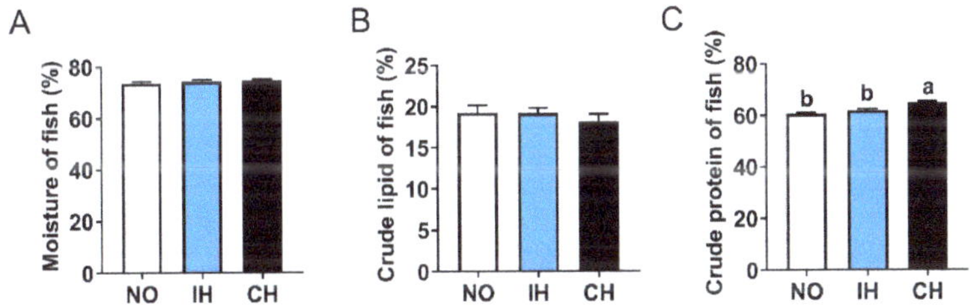

Figure 2. Effects of intermittent and chronic hypoxia on whole-fish proximate compositions in tiger puffer (*Takifugu rubripes*). (**A**) Moisture; (**B**) Crude lipid; (**C**) Crude protein. Note: NO: normoxia group; IH: intermittent hypoxia group; CH: chronic hypoxia group. Different letters above the bars suggest significant differences ($p < 0.05$) among the three groups (mean ± SEM, n = 6).

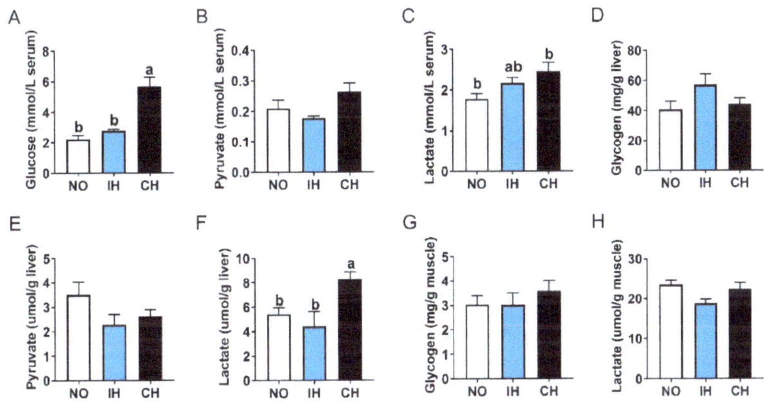

Figure 3. Effects of intermittent and chronic hypoxia on glucose metabolism in tiger puffer (*Takifugu rubripes*). (**A**) Serum glucose content; (**B**) Serum pyruvate content; (**C**) Serum lactate content; (**D**) Liver glycogen content; (**E**) Liver pyruvate content; (**F**) Liver lactate content; (**G**) Muscle glycogen content; (**H**) Muscle lactate content. Note: NO: normoxia group; IH: intermittent hypoxia group; CH: chronic hypoxia group. Different letters above the bars suggest significant differences ($p < 0.05$) among the three groups (mean ± SEM, n = 6).

3.4. *Effects of Intermittent and Chronic Hypoxia on Hifα and Glycolysis Pathway*

Compared with the NO and IH groups, the gene expressions of hypoxia-inducible factor 1 subunit alpha a (*hif1α*), *hif2α*, glucokinase (gck), phosphofructokinase (pfk), lactate dehydrogenase A4 (ldha), and gsk3β in the liver, as well as hif3α in the muscle, were upregulated significantly in the CH group (Figure 4A,B,E–H). However, there was no significant difference in these indicators between the NO and IH groups. The IH and CH groups all had significantly lower *hif1α*, *hk1*, and *gsk3β* gene expressions in the muscle than the NO group (Figure 4I,M,O). The expression of the *vegfa* gene in the liver was significantly lower in the IH group than that in the NO group (Figure 4D). In addition, IH and CH did not affect the expression of *hif3α* in the liver and *hif2α*, *vegfa*, and *ldha* in the muscle (Figure 4C,J,L,N). All these results indicate that chronic hypoxia promoted the expression of *hif1α*, *hif2α*, and glycolysis-related genes in the liver.

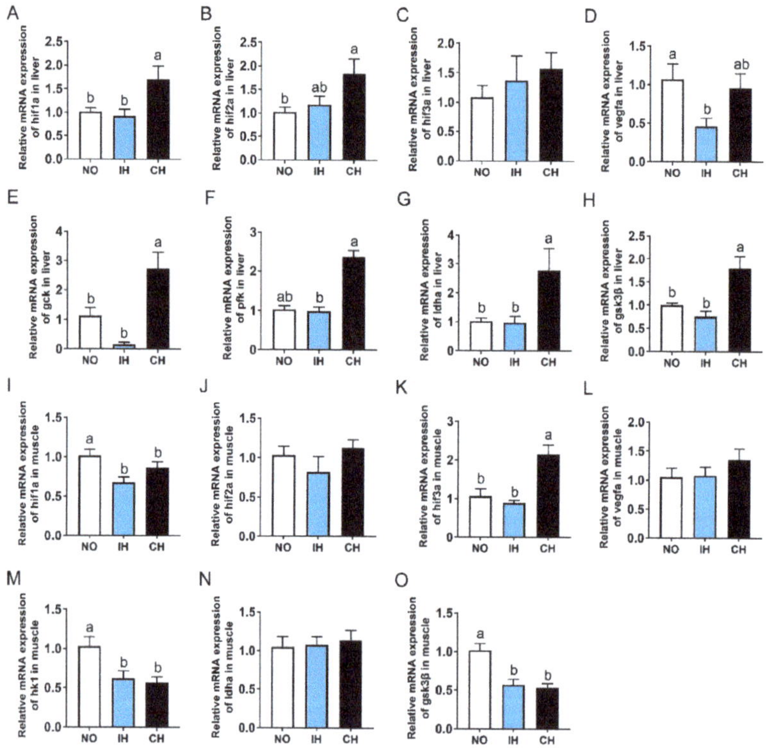

Figure 4. Effects of intermittent and chronic hypoxia on *hifα* and glycolysis pathway in tiger puffer (*Takifugu rubripes*). (**A–D**) Liver *hif1α*, *hif2α*, *hif3α* and *vegfa* expression; (**E–H**) Liver *gck*, *pfk*, *ldha* and *gsk3β* expression; (**I–L**) Muscle *hif1α*, *hif2α*, *hif3α* and *vegfa* expression; (**M–O**) Muscle *hk1*, *ldha* and *gsk3β* expression. Note: NO: normoxia group; IH: intermittent hypoxia group; CH: chronic hypoxia group. Different letters above the bars suggest significant differences ($p < 0.05$) among the three groups (mean ± SEM, n = 6).

3.5. Effects of Intermittent and Chronic Hypoxia on Lipid Metabolism

Compared with the NO group, both IH and CH decreased the crude lipid and triglyceride levels in the liver, and a significant difference was found in the crude lipid content of the IH group (Figure 5B,C). However, there was no significant difference in the triglyceride level of serum and muscle between NO and hypoxia groups (Figure 5A,D). Compared with the NO group, the expression of lipid synthesis-related genes (*fasn* and *srebf1*) in the liver showed a reducing trend in the IH group but was upregulated in the CH group (Figure 5E,F). Both IH and CH did not affect the expression of lipid catabolism-related genes (*atgl* and *cpt1ab*) in the liver and lipid synthesis-related genes (*fasn* and *srebf1*) in the muscle (Figure 5G–J). Compared with the NO group, the expression of lipid catabolism-related genes (*atgl* and *cpt1b*) in the muscle in the IH and CH groups were downregulated significantly (Figure 5K,L). These data suggest that intermittent hypoxia and chronic hypoxia can reduce the lipid content of tiger puffer.

3.6. Effects of Intermittent and Chronic Hypoxia on Fatty Acids Compositions

Compared with the NO group, the C16:0 level was reduced in the IH and CH groups (Table 3). IH decreased the monounsaturated fatty acid (MUFA) content but increased the contents of n-3 polyunsaturated fatty acid (n-3PUFA), including C20:5n-3 (EPA), C22:5n-3

(DPA), and C22:6n-3 (DHA). IH and CH did not affect the levels of saturated fatty acid (SFA) and n-6 polyunsaturated fatty acid (n-6PUFA).

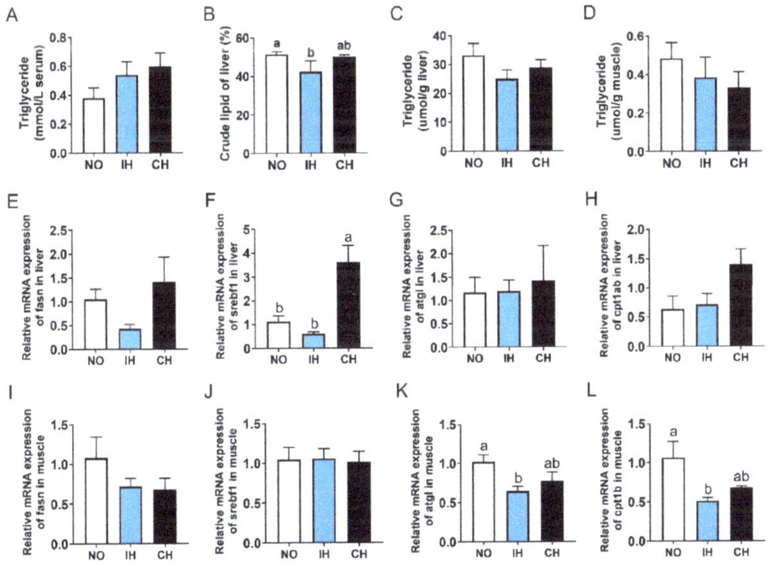

Figure 5. Effects of intermittent and chronic hypoxia on lipid metabolism in tiger puffer (*Takifugu rubripes*). (**A**) Serum triglyceride content; (**B**) Liver crude lipid content; (**C**) Liver triglyceride content; (**D**) Muscle triglyceride content; (**E**) Liver *fasn* expression; (**F**) Liver *srebf1* expression; (**G**) Liver *atgl* expression; (**H**) Liver *cpt1ab* expression; (**I**) Muscle *fasn* expression; (**J**) Muscle *srebf1* expression; (**K**) Muscle *atgl* expression; (**L**) Muscle *cpt1b* expression. Note: NO: normoxia group; IH: intermittent hypoxia group; CH: chronic hypoxia group. Different letters above the bars suggest significant differences ($p < 0.05$) among the three groups (mean ± SEM, n = 6).

Table 3. Liver fatty acid compositions of tiger puffer (*Takifugu rubripes*) (% total fatty acids).

Fatty Acid	NO	IH	CH
C14:0	3.68 ± 0.04	3.62 ± 0.09	3.63 ± 0.08
C16:0	20.48 ± 0.32	19.34 ± 0.24	19.6 ± 0.46
C17:0	0.40 ± 0.01	0.4 ± 0.02	0.41 ± 0.02
C18:0	7.47 ± 0.20	7.36 ± 0.06	7.39 ± 0.20
C20:0	0.26 ± 0.02	0.24 ± 0.01	0.26 ± 0.01
∑SFA	32.29 ± 0.32	30.95 ± 0.29	31.77 ± 0.30
C14:1n-5	0.43 ± 0.01	0.42 ± 0.02	0.42 ± 0.01
C16:1n-7	7.89 ± 0.22	7.76 ± 0.07	7.75 ± 0.12
C17:1n-7	0.38 ± 0.01	0.42 ± 0.01	0.4 ± 0.01
C18:1n-9	24.71 ± 0.23	23.89 ± 0.32	24.66 ± 0.37
C20:1n-9	1.61 ± 0.06	1.70 ± 0.08	1.56 ± 0.05
C22:1n-9	0.16 ± 0.02	0.18 ± 0.01	0.20 ± 0.01
C24:1n-9	0.29 ± 0.04	0.26 ± 0.02	0.30 ± 0.02
∑MUFA	35.41 ± 0.13 ^{a}	34.58 ± 0.27 ^{b}	35.21 ± 0.24 ^{ab}
C18:2n-6	11.04 ± 0.18	11.27 ± 0.29	11.11 ± 0.15
C18:3n-6	0.40 ± 0.02	0.39 ± 0.02	0.38 ± 0.03
C20:2n-6	0.62 ± 0.02	0.72 ± 0.03	0.71 ± 0.04
C20:3n-6	0.71 ± 0.04	0.65 ± 0.03	0.71 ± 0.04

Table 3. *Cont.*

Fatty Acid	NO	IH	CH
C22:2n-6	0.43 ± 0.02	0.47 ± 0.02	0.39 ± 0.07
∑n-6PUFA	13.2 ± 0.16	13.5 ± 0.35	13.3 ± 0.26
C20:3n-3	0.58 ± 0.02 ^{b}	0.55 ± 0.02 ^{b}	0.68 ± 0.03 ^{a}
C20:5n-3	5.50 ± 0.13	6.10 ± 0.13	5.95 ± 0.21
C22:5n-3	4.05 ± 0.16 ^{b}	4.63 ± 0.09 ^{a}	4.00 ± 0.17 ^{b}
C22:6n-3	8.91 ± 0.12 ^{b}	9.69 ± 0.09 ^{a}	9.55 ± 0.24 ^{ab}
∑n-3PUFA	19.03 ± 0.33 ^{b}	20.97 ± 0.23 ^{a}	20.17 ± 0.58 ^{ab}

SFA: saturated fatty acid; MUFA, monounsaturated fatty acid; PUFA: polyunsaturated fatty acid. Different letters above the bars suggest significant differences ($p < 0.05$) among the three groups (mean ± SEM, n = 6).

3.7. Effects of Intermittent and Chronic Hypoxia on Protein Metabolism

The total soluble protein level in the serum and liver, as well as blood hemoglobin content in the IH and CH groups, was higher than that in the NO group (Figure 6A–C). However, the total soluble protein content in the muscle was not affected by IH and CH (Figure 6D). Compared with the NO and IH groups, the expression of *ir*, *akt1*, mechanistic target of rapamycin kinase (*mtor*), and ubiquitin-like modifier activating enzyme 1 (*uba1*) in the liver of the CH group were upregulated, and significant differences were found in *mtor* and *uba1* (Figure 6E–H). However, the expression of *akt1* and *uba1* in the muscle of both IH and CH groups was lower than that in the NO group, and a significant difference was found in *uba1* (Figure 6J,L). In addition, there were no significant differences in *ir* and *mtor* expression in the muscle between the NO and hypoxia groups (Figure 6I,K). All these data indicate that intermittent and chronic hypoxia could promote the synthesis of protein in the liver.

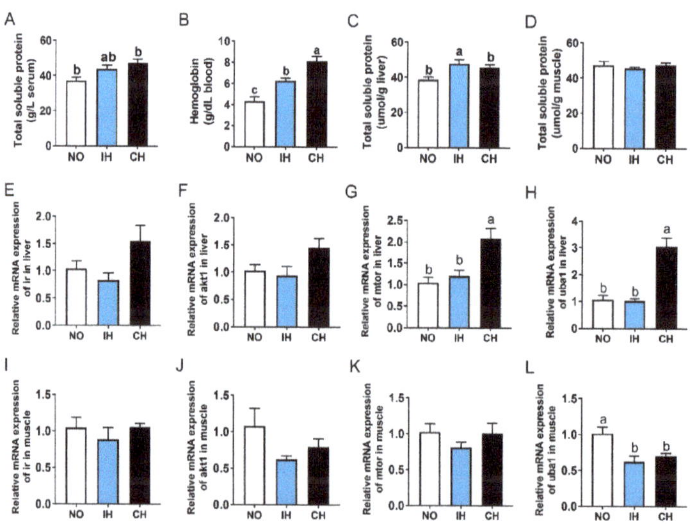

Figure 6. Effects of intermittent and chronic hypoxia on protein metabolism in tiger puffer (*Takifugu rubripes*). (**A**) Serum total soluble protein content; (**B**) Blood hemoglobin content; (**C**) Liver total soluble protein content; (**D**) Muscle total soluble protein content; (**E–H**) Liver *ir*, *akt1*, *mtor*, and *uba1* genes expression; (**I–L**) Muscle *ir*, *akt1*, *mtor*, and *uba1* genes expression. Note: NO: normoxia group; IH: intermittent hypoxia group; CH: chronic hypoxia group. Different letters above the bars suggest significant differences ($p < 0.05$) among the three groups (mean ± SEM, n = 6).

4. Discussion

In natural environments and aquaculture activities, hypoxia (oxygen deficiency) widely exists and is harmful to aquatic animals. Since the middle of the 20th century, eutrophication and global warming caused by excess nutrient loading and fossil fuel burning generated more chronic hypoxia and anoxia environments in shallow coastal and estuarine areas [35,36]. The northern Gulf of Mexico, as well as the Black Sea and Baltic Sea, are the typical ecosystems that were burdened with severe seasonal hypoxia, leading to alteration of food webs and loss of fisheries and biodiversity [37]. In Snug Harbor (MA, USA) and Sanya Bay (Hainan, China), DO and pH changes both showed a distinct diurnal cycle caused by photosynthesis of benthic microalgae and macroalgae, with DO and pH rising at daytime and falling at night [7,8]. A similar phenomenon was also found in freshwater ponds and lakes [6]. So many aquatic animals live in intermittent hypoxia environments, and the physiological strategies for adapting to intermittent hypoxia of tiger puffer are poorly understood. In addition, the different responses of fish to intermittent hypoxia and chronic hypoxia are not clear.

Previous studies have shown that in intermittent hypoxia, the oxygen consumption rate of estuarine killifish (*Fundulus heteroclitus*) significantly decreased in the nighttime-hypoxia phase with 0.8 mg/L O_2, then increased in the daytime-reoxygenation phase with 8 mg/L O_2, but the oxygen consumption rate was consistently at a low level in the chronic hypoxia condition [38]. For largemouth bass, chronic hypoxia (17–26% DO saturation) for 6 weeks significantly decreased the weight gain, specific growth rate, and feed intake [39]. In European sea bass (*Dicentrarchus labrax*), chronic moderate hypoxia (40% DO saturation) for 38 days reduced the fish size and protein digestibility by decreasing the trypsin and amylase activities in the liver, as well as alkaline phosphatase and aminopeptidase-N activities in the intestine [40]. Similar results were also found in zebrafish (*Danio rerio*) [33] and oriental river prawn (*Macrobrachium nipponense*) [41]. In Nile tilapia, compared with normoxia group (pO_2 = 17.4 ± 0.4 kPa), chronic hypoxia (pO_2 = 8.1 ± 0.6 kPa) for 9 weeks reduced the feed intake, apparent digestibility of nutrients, weight gain, hepatosomatic index and viscerosomatic index, but intermittent hypoxia (6 h of hypoxia every night, pO_2 = 0.4 ± 1.0 kPa) significantly promoted the growth and weight gain by increasing feed intake compensatorily [42]. In juvenile qingbo (*Spinibarbus sinensis*), intermittent hypoxia (DO = 7.0 mg/L from 7:00 to 21:00; DO = 3.0 mg/L from 21:00 to 7:00) for 30 days also significantly increased the weight gain and specific growth rate [43]. However, in the present study, both the intermittent hypoxia group and chronic hypoxia group reduced the final mean body weight, weight gain, viscerasomatic index, hepatosomatic index, and condition factor of tiger puffer compared to the normoxia group. Meanwhile, compared with the normoxia condition, chronic hypoxia for 4 weeks had more significant negative effects on fish growth and visceral weight than intermittent hypoxia. The possible reason could be that chronic hypoxia leads to lower oxygen consumption and metabolic rate. Intermittent hypoxia (44–65% DO saturation, average 4 h every day) for 38 days also decreased the feed intake and weight gain in Atlantic salmon (*Salmo salar*) [44]. The possible reasons for the different growth performances of different fish species in intermittent hypoxia could be that Nile tilapia and qingbo fish are very tolerant to hypoxia, but tiger puffer and Atlantic salmon are intolerant and sensitive to hypoxia. Nile tilapia and qingbo fish can recover from intermittent hypoxia quickly, using the saved basal energy expenditure and active metabolic expenditure for growth, and increase feed intake compensatorily from intermittent hypoxia to normoxia condition.

Non-cyprinidae fish species and mammals have three *hifα* isoforms, including *hif1α*, *hif2α*, and *hif3α*. The partial functions of these isoforms have been identified in several fish species, such as zebrafish [45], Nile tilapia [46], blunt snout bream (*Megalobrama amblycephala*) [47], turbot [48], estuarine fish (*Fundulus heteroclitus*) [49], Amur minnow (*Phoxinus lagowskii*) [50], and thick-lipped grey mullet (*Chelon labrosus*) [51]. However, the functions of the three *hifα* isoforms of tiger puffer in intermittent hypoxia and chronic hypoxia are still unclear. Studies have found that the *hif1α* and *hif2α* expression in the liver was increased several-fold in dragonet (*Callionymus valenciennei*) that lived in hypoxic areas of Tokyo Bay, and the *hif1α* and *hif2α* expressions were decreased after reoxygenation for 24 h [52]. In Korean rockfish (*Sebastes schlegeli*), acute hypoxia for 30–60 min significantly increased the *hif1α* mRNA expression in the ovary and the *hif2α* mRNA expression in the gill and spleen but decreased the *hif2α* mRNA expression in the gonads and liver [53]. In C57BL/6 male mice and human HepG2 cells, inhibition of *hif2α* reversed the liver fat accumulation induced by hypoxia, which indicates that *hif2α* plays an important role in lipid synthesis under hypoxia conditions [54]. Our previous studies in Nile tilapia and turbot have shown that the hepatic expression of *hif3α* but not *hif1α* was significantly upregulated under acute hypoxia, which suggests the *hif3α* is a marker gene for fish response to acute hypoxia [46,48]. In the present study, the CH group, but not the IH group, had significantly higher *hif1α*, *hif2α*, *gck*, *pfk*, *ldha*, and *gsk3β* expression in the liver than the NO group. However, the IH and CH groups had significantly lower *hif1α*, *hk1*, and *gsk3β* expression in the muscle than the NO group, and IH and CH did not affect the *hif3α* expression in the liver. These results showed that chronic hypoxia had more significant effects than intermittent hypoxia on fish, and the liver is more sensitive to hypoxia than the muscle. Moreover, *hif1α* and *hif2α* play key roles in the adaptation of tiger puffer to chronic hypoxia.

Many studies have described the changes in nutrient metabolism under chronic hypoxia or intermittent hypoxia, but the differences in nutrient metabolism between intermittent hypoxia and chronic hypoxia have not been known in marine fish species. In largemouth bass, chronic hypoxia (17–26% DO saturation) for 6 weeks significantly decreased the myofiber diameter and protein synthesis by inhibiting the HIF1α/REDD1/mTOR pathway and increased the protein degradation and SOD, GPX, and CAT enzymes activities in the serum and muscle [39]. In juvenile turbot and European sea bass, chronic hypoxia (DO = 3.2–4.5 mg/L) for 42 days reduced the liver glycogen content and increased the serum lactate and blood hematocrit levels but did not change the serum glucose content [55]. In killifish (*Fundulus heteroclitus*), chronic hypoxia (pO_2 = 5 kPa for 28 days) but not intermittent hypoxia (pO_2 = 5 kPa at night, pO_2 = 20 kPa at day for 28 days) reduced the total length of gill filaments and the number of oxidative fibers of the swimming muscle and increased the blood hemoglobin level [56]. However, intermittent hypoxia, not constant hypoxia, increased the capillary density in the glycolytic muscle, the protein content, and the cytochrome c oxidase activity in the liver [56]. In high-latitude fish (*Phoxinus lagowskii*), the metabolomics showed that compared with normoxia, both chronic hypoxia (DO = 3.0–4.0 mg/L for 28 days) and intermittent hypoxia (DO = 6.0–7.0 mg/L from 07:00 to 21:00, DO = 3.0–4.0 mg/L from 21:00 to 07:00 for 28 days) changed the starch and sucrose metabolism, carbohydrate digestion and absorption, and ATP binding cassette transporters pathways, but the significantly changed pathways between the chronic hypoxia and intermittent hypoxia were biosynthesis of secondary metabolites, ATP binding cassette transporters, and bile secretion [57]. In *Phoxinus lagowskii*, intermittent hypoxia increased the TG level, PK and LDH activities, *hif1α*, *hif2α*, *hif3α*, *vhl*, *fih* expression in the heart, as well as the MDA level and the SOD, CAT, LDH activities in the brain [58]. However, chronic hypoxia (DO = 3–4 mg/L) increased the *hif1α*, *hif2α*, *hif3α*, *vhl*, and *fih* expression in the brain and did not change the oxidative stress biomarkers in the brain [58]. Some different results were found in the present study. Chronic hypoxia increased the blood hemoglobin content, the total soluble protein level in the serum and liver, *mtor* (protein synthesis-related) expression, and crude protein of whole fish. Meanwhile, chronic hypoxia increased the lactate content in the serum and liver and promoted the conversion of glucose to lactate. However, intermittent hypoxia significantly reduced the crude lipid and triglyceride contents, expression of lipid synthesis-related genes (*fasn* and *srebf1*) in the liver, and crude lipid content of tiger puffer fish but elevated the n-3 polyunsaturated fatty acid (PUFA) proportion and reduced the monounsaturated fatty acid (MUFA) proportion. Therefore, fish species, stress time and degree, and tissue difference all affect the physiological and metabolic responses of fish to hypoxia.

5. Conclusions

Intermittent and chronic hypoxia are common and harmful to marine animals. The present study suggested that both intermittent and chronic hypoxia decreased fish growth and visceral weight but increased the feed conversion ratio and blood hemoglobin content. Chronic hypoxia but not intermittent hypoxia promoted the protein synthesis by activating *mtor* expression and promoted the glycolysis pathway by activating *hif1α* and *hif2α* expression. Intermittent hypoxia but not chronic hypoxia decreased the lipid synthesis by inhibiting *fasn* and *srebf1* expression. Meanwhile, intermittent hypoxia reduced the monounsaturated fatty acid content but increased the contents of n-3 polyunsaturated fatty acid (EPA, DPA, and DHA). These changes promoted the adaption of tiger puffer to intermittent and chronic hypoxia.

Author Contributions: Conceptualization, methodology, writing—original draft preparation, project administration, funding acquisition, Q.M.; investigation, writing—review and editing, H.X.; software, formal analysis, data curation, visualization, R.Z., M.L. and Y.W. All authors have read and agreed to the published version of the manuscript.

Funding: This research was funded by the National Natural Science Foundation of China (32202950), China Postdoctoral Science Foundation (2022M713471), the Central Public-interest Scientific Institution Basal Research Fund, CAFS (2023TD52), and the earmarked fund for China Agriculture Research System (CARS47-G15).

Institutional Review Board Statement: All experimental procedures and animal care were conducted under a protocol approved by experimental animal care, ethics, and safety inspection from the Yellow Sea Fisheries Research Institute, Chinese Academy of Fishery Sciences (IACUC-20210721001).

Informed Consent Statement: Not applicable.

Data Availability Statement: The datasets generated during the current study are available from the corresponding author upon reasonable request.

Acknowledgments: We thank Zhangbin Liao for the technical assistance with the study.

Conflicts of Interest: The authors declare no conflicts of interest.

References

1. Ekau, W.; Auel, H.; Portner, H.O.; Gilbert, D. Impacts of hypoxia on the structure and processes in pelagic communities (zooplankton, macro-invertebrates and fish). *Biogeosciences* **2010**, *7*, 1669–1699. [CrossRef]
2. Karim, M.R.; Sekine, M.; Ukita, M. Simulation of eutrophication and associated occurrence of hypoxic and anoxic condition in a coastal bay in Japan. *Mar. Pollut. Bull.* **2002**, *45*, 280–285. [CrossRef] [PubMed]
3. Rabalais, N.N.; Turner, R.E.; Wiseman, W.J. Gulf of Mexico hypoxia, A.K.A. "The dead zone". *Annu. Rev. Ecol. Evol. Syst.* **2002**, *33*, 235–263. [CrossRef]
4. Ma, Q.; Hu, C.T.; Yue, J.; Luo, Y.; Qiao, F.; Chen, L.Q.; Zhang, M.L.; Du, Z.Y. High-carbohydrate diet promotes the adaptation to acute hypoxia in zebrafish. *Fish Physiol. Biochem.* **2020**, *46*, 665–679. [CrossRef] [PubMed]
5. Martínez, D.; De Lázaro, O.; Cortés, P.; Oyarzún-Salazar, R.; Paschke, K.; Vargas-Chacoff, L. Hypoxia modulates the transcriptional immunological response in Oncorhynchus kisutch. *Fish Shellfish Immun.* **2020**, *106*, 1042–1051. [CrossRef]
6. Phan-Van, M.; Rousseau, D.; De Pauw, N. Effects of fish bioturbation on the vertical distribution of water temperature and dissolved oxygen in a fish culture-integrated waste stabilization pond system in Vietnam. *Aquaculture* **2008**, *281*, 28–33. [CrossRef]
7. Chen, X.; Wei, G.; Xie, L.; Deng, W.; Sun, Y.; Wang, Z.; Ke, T. Biological controls on diurnal variations in seawater trace element concentrations and carbonate chemistry on a coral reef. *Mar. Chem.* **2015**, *176*, 1–8. [CrossRef]
8. Howarth, R.W.; Hayn, M.; Marino, R.M.; Ganju, N.; Foreman, K.; McGlathery, K.; Giblin, A.E.; Berg, P.; Walker, J.D. Metabolism of a nitrogen-enriched coastal marine lagoon during the summertime. *Biogeochemistry* **2013**, *118*, 1–20. [CrossRef]
9. Gattuso, A.; Garofalo, F.; Cerra, M.C.; Imbrogno, S. Hypoxia Tolerance in Teleosts: Implications of Cardiac Nitrosative Signals. *Front. Physiol.* **2018**, *9*, 366. [CrossRef]
10. Breitburg, D. Effects of hypoxia, and the balance between hypoxia and enrichment, on coastal fishes and fisheries. *Estuar. Coast.* **2002**, *25*, 767–781. [CrossRef]
11. Keeling, R.F.; Kortzinger, A.; Gruber, N. Ocean deoxygenation in a warming world. *Annu. Rev. Mar. Sci.* **2010**, *2*, 199–229. [CrossRef] [PubMed]
12. Saraste, M. Oxidative phosphorylation at the fin de siecle. *Science* **1999**, *283*, 1488–1493. [CrossRef] [PubMed]

13. Goda, N.; Kanai, M. Hypoxia-inducible factors and their roles in energy metabolism. *Int. J. Hematol.* **2012**, *95*, 457–463. [CrossRef] [PubMed]
14. Majmundar, A.J.; Wong, W.J.; Simon, M.C. Hypoxia-inducible factors and the response to hypoxic stress. *Mol. cell* **2010**, *40*, 294–309. [CrossRef]
15. Gaspar, J.M.; Velloso, L.A. Hypoxia inducible factor as a central regulator of metabolism—Implications for the development of obesity. *Front. Neurosci.* **2018**, *12*, 813. [CrossRef]
16. Schwab, L.P.; Peacock, D.L.; Majumdar, D.; Ingels, J.F.; Jensen, L.C.; Smith, K.D.; Cushing, R.C.; Seagroves, T.N. Hypoxia-inducible factor 1 alpha promotes primary tumor growth and tumor-initiating cell activity in breast cancer. *Breast Cancer Res.* **2012**, *14*, R6. [CrossRef]
17. Koukourakis, M.I.; Giatromanolaki, A.; Sivridis, E.; Simopoulos, C.; Turley, H.; Talks, K.; Gatter, K.C.; Harris, A.L.; Tumour Angiogenesis Res, G. Hypoxia-inducible factor (HIF1A and HIF2A), angiogenesis, and chemoradiotherapy outcome of squamous cell head-and-neck cancer. *Int. J. Radiat. Oncol.* **2002**, *53*, 1192–1202. [CrossRef]
18. Koukourakis, M.I.; Giatromanolaki, A.; Skarlatos, J.; Corti, L.; Blandamura, S.; Piazza, M.; Gatter, K.C.; Harris, A.L. Hypoxia inducible factor (HIF-1a and HIF-2a) expression in early esophageal cancer and response to photodynamic therapy and radiotherapy. *Cancer Res.* **2001**, *61*, 1830–1832.
19. Makino, Y.; Cao, R.; Svensson, K.; Bertilsson, G.; Asman, M.; Tanaka, H.; Cao, Y.; Berkenstam, A.; Poellinger, L. Inhibitory PAS domain protein is a negative regulator of hypoxia-inducible gene expression. *Nature* **2001**, *414*, 550–554. [CrossRef]
20. Zhang, P.; Lu, L.; Yao, Q.; Li, Y.; Zhou, J.; Liu, Y.; Duan, C. Molecular, functional, and gene expression analysis of zebrafish hypoxia-inducible factor-3α. *Am. J. Physiol.—Regul. Integr. Comp. Physiol.* **2012**, *303*, R1165–R1174. [CrossRef]
21. Zhang, P.; Yao, Q.; Lu, L.; Li, Y.; Chen, P.J.; Duan, C. Hypoxia-inducible factor 3 is an oxygen-dependent transcription activator and regulates a distinct transcriptional response to hypoxia. *Cell Rep.* **2014**, *6*, 1110–1121. [CrossRef] [PubMed]
22. Baptista, R.B.; Souza-Castro, N.; Almeida-Val, V.M.F. Acute hypoxia up-regulates HIF-1α and VEGF mRNA levels in Amazon hypoxia-tolerant Oscar (*Astronotus ocellatus*). *Fish Physiol. Biochem.* **2016**, *42*, 1307–1318. [CrossRef] [PubMed]
23. Li, M.; Wang, X.; Qi, C.; Li, E.; Du, Z.; Qin, J.G.; Chen, L. Metabolic response of Nile tilapia (*Oreochromis niloticus*) to acute and chronic hypoxia stress. *Aquaculture* **2018**, *495*, 187–195. [CrossRef]
24. Bergstedt, J.H.; Pfalzgraff, T.; Skov, P.V. Hypoxia tolerance and metabolic coping strategies in *Oreochromis niloticus*. *Comp. Biochem. Physiol. Part A Mol. Integr. Physiol.* **2021**, *257*, 110956. [CrossRef]
25. Yang, S.; Yan, T.; Wu, H.; Xiao, Q.; Fu, H.M.; Luo, J.; Zhou, J.; Zhao, L.L.; Wang, Y.; Yang, S.Y.; et al. Acute hypoxic stress: Effect on blood parameters, antioxidant enzymes, and expression of HIF-1alpha and GLUT-1 genes in largemouth bass (*Micropterus salmoides*). *Fish Shellfish Immun.* **2017**, *67*, 449–458. [CrossRef]
26. Sun, J.L.; Zhao, L.L.; Wu, H.; Liu, Q.; Liao, L.; Luo, J.; Lian, W.Q.; Cui, C.; Jin, L.; Ma, J.D.; et al. Acute hypoxia changes the mode of glucose and lipid utilization in the liver of the largemouth bass (*Micropterus salmoides*). *Sci. Total Environ.* **2020**, *713*, 135157. [CrossRef]
27. Stump, E.; Ralph, G.M.; Comeros-Raynal, M.T.; Matsuura, K.; Carpenter, K.E. Global conservation status of marine pufferfishes (Tetraodontiformes: Tetraodontidae). *Glob. Ecol. Conserv.* **2018**, *14*, 10. [CrossRef]
28. Noguchi, T.; Arakawa, O.; Takatani, T. TTX accumulation in pufferfish. *Comp. Biochem. Phys. Part D Genom. Proteom.* **2006**, *1*, 145–152. [CrossRef]
29. Japan embraces CRISPR-edited fish. *Nat. Biotechnol.* **2022**, *40*, 10. [CrossRef]
30. Guo, B.; Zou, M.; Gan, X.; He, S. Genome size evolution in pufferfish: An insight from BAC clone-based Diodon holocanthus genome sequencing. *BMC Genom.* **2010**, *11*, 396. [CrossRef]
31. Ogawa, K.; Matsushita, Y.; Sawada, S.; Amimoto, T. Spatiotemporal distribution of the harmful red tide of *Karenia mikimotoi* and damage to cultured tiger puffer fisheries in southern Harima-nada, eastern Seto Inland Sea, Japan in the summer of 2022. *Nippon Suisan Gakk.* **2023**, *89*, 424–437. [CrossRef]
32. Hou, H.; Zhang, Y.; Ma, Z.; Wang, X.; Su, P.; Wang, H.; Liu, Y. Life cycle assessment of tiger puffer (*Takifugu rubripes*) farming: A case study in Dalian, China. *Sci. Total Environ.* **2022**, *823*, 153522. [CrossRef]
33. Ma, Q.; Wang, X.; Li, L.Y.; Qiao, F.; Zhang, M.L.; Du, Z.Y. High protein intake promotes the adaptation to chronic hypoxia in zebrafish (*Danio rerio*). *Aquaculture* **2021**, *535*, 736356. [CrossRef]
34. Livak, K.J.; Schmittgen, T.D. Analysis of relative gene expression data using real-time quantitative PCR and the $2^{-\Delta\Delta CT}$ method. *Methods* **2001**, *25*, 402–408. [CrossRef]
35. Diaz, R.J.; Rosenberg, R. Spreading Dead Zones and Consequences for Marine Ecosystems. *Science* **2008**, *321*, 926–929. [CrossRef] [PubMed]
36. Breitburg, D.; Levin, L.A.; Oschlies, A.; Gregoire, M.; Chavez, F.P.; Conley, D.J.; Garcon, V.; Gilbert, D.; Gutierrez, D.; Isensee, K.; et al. Declining oxygen in the global ocean and coastal waters. *Science* **2018**, *359*, 46. [CrossRef] [PubMed]
37. Diaz, R.J. Overview of hypoxia around the world. *J. Environ. Qual.* **2001**, *30*, 275–281. [CrossRef] [PubMed]
38. Borowiec, B.G.; McClelland, G.B.; Rees, B.B.; Scott, G.R. Distinct metabolic adjustments arise from acclimation to constant hypoxia and intermittent hypoxia in estuarine killifish (*Fundulus heteroclitus*). *J. Exp. Biol.* **2018**, *221*, jeb190900. [CrossRef]
39. He, Y.; Yu, H.; Zhang, Z.; Zhang, J.; Kang, S.; Zhang, X. Effects of chronic hypoxia on growth performance, antioxidant capacity and protein turnover of largemouth bass (*Micropterus salmoides*). *Aquaculture* **2022**, *561*, 738673. [CrossRef]

40. Vanderplancke, G.; Claireaux, G.; Quazuguel, P.; Huelvan, C.; Corporeau, C.; Mazurais, D.; Zambonino-Infante, J.L. Exposure to chronic moderate hypoxia impacts physiological and developmental traits of European sea bass (*Dicentrarchus labrax*) larvae. *Fish Physiol. Biochem.* **2015**, *41*, 233–242. [CrossRef]
41. Sun, S.; Yang, M.; Fu, H.; Ge, X.; Zou, J. Altered intestinal microbiota induced by chronic hypoxia drives the effects on lipid metabolism and the immune response of oriental river prawn Macrobrachium nipponense. *Aquaculture* **2020**, *526*, 735431. [CrossRef]
42. Obirikorang, K.A.; Acheampong, J.N.; Duodu, C.P.; Skov, P.V. Growth, metabolism and respiration in Nile tilapia (*Oreochromis niloticus*) exposed to chronic or periodic hypoxia. *Comp. Biochem. Phys. Part A Mol. Integr. Physiol.* **2020**, *248*, 110768. [CrossRef] [PubMed]
43. Dan, X.M.; Yan, G.J.; Zhang, A.J.; Cao, Z.D.; Fu, S.J. Effects of stable and diel-cycling hypoxia on hypoxia tolerance, postprandial metabolic response, and growth performance in juvenile qingbo (*Spinibarbus sinensis*). *Aquaculture* **2014**, *428*, 21–28. [CrossRef]
44. Burt, K.; Hamoutene, D.; Perez-Casanova, J.; Gamperl, A.K.; Volkoff, H. The effect of intermittent hypoxia on growth, appetite and some aspects of the immune response of Atlantic salmon (*Salmo salar*). *Aquac. Res.* **2013**, *45*, 124–137. [CrossRef]
45. Cai, X.; Zhou, Z.; Zhu, J.; Liao, Q.; Zhang, D.; Liu, X.; Wang, J.; Ouyang, G.; Xiao, W. Zebrafish Hif3α modulates erythropoiesis via regulation of *gata1* to facilitate hypoxia tolerance. *Development* **2020**, *147*, dev185116. [CrossRef]
46. Ma, Q.; Luo, Y.; Zhong, J.; Limbu, S.M.; Li, L.Y.; Chen, L.Q.; Qiao, F.; Zhang, M.L.; Lin, Q.; Du, Z.Y. Hypoxia tolerance in fish depends on catabolic preference between lipids and carbohydrates. *Zool. Res.* **2023**, *44*, 954. [CrossRef]
47. Liu, Z.; Zhao, X.; Jiang, X.; Zou, S. Transcription of blunt snout bream (*Megalobrama amblycephala*) HIF3α and its localization in the nucleus under both normoxic and hypoxic conditions. *Biochem. Biophys. Res. Commun.* **2018**, *500*, 443–449. [CrossRef]
48. Ma, Q.; Xu, H.; Wei, Y.; Liang, M. Effects of acute hypoxia on nutrient metabolism and physiological function in turbot, *Scophthalmus maximus*. *Fish Physiol. Biochem.* **2023**, *50*, 367–383. [CrossRef]
49. Townley, I.K.; Karchner, S.I.; Skripnikova, E.; Wiese, T.E.; Hahn, M.E.; Rees, B.B. Sequence and functional characterization of hypoxia-inducible factors, HIF1α, HIF2α a, and HIF3α, from the estuarine fish, *Fundulus heteroclitus*. *Am. J. Physiol.—Regul. Integr. Comp. Physiol.* **2017**, *312*, R412–R425. [CrossRef]
50. Yang, Y.; Dong, Z.; Chen, X.; Wang, Z.; Zhang, D.; Liang, L.; Mu, W. Molecular characterization and expression analysis of hypoxia-inducible factor-1 alpha, factor-2 alpha, and factor-3 alpha and physiological response to hypoxia exposure in Amur minnow (*Phoxinus lagowskii*). *Aquacult. Int.* **2022**, *30*, 607–632. [CrossRef]
51. Garcia-Marquez, J.; Alvarez-Torres, D.; Cerezo, I.M.; Dominguez-Maqueda, M.; Figueroa, F.L.; Alarcon, F.J.; Acien, G.; Martinez-Manzanares, E.; Abdala-Diaz, R.T.; Bejar, J.; et al. Combined dietary administration of *Chlorella fusca* and ethanol-inactivated *Vibrio proteolyticus* modulates intestinal microbiota and gene expression in *Chelon labrosus*. *Animals* **2023**, *13*, 3325. [CrossRef] [PubMed]
52. Kodama, K.; Rahman, M.S.; Horiguchi, T.; Thomas, P. Upregulation of hypoxia-inducible factor (HIF)-1α and HIF-2α mRNA levels in dragonet *Callionymus valenciennei* exposed to environmental hypoxia in Tokyo Bay. *Mar. Pollut. Bull.* **2012**, *64*, 1339–1347. [CrossRef]
53. Mu, W.; Wen, H.; Li, J.; He, F. HIFs genes expression and hematology indices responses to different oxygen treatments in an ovoviviparous teleost species *Sebastes schlegelii*. *Mar. Environ. Res.* **2015**, *110*, 142–151. [CrossRef]
54. Cao, R.; Zhao, X.; Li, S.; Zhou, H.; Chen, W.; Ren, L.; Zhou, X.; Zheng, H.; Shi, R. Hypoxia Induces Dysregulation of Lipid Metabolism in HepG2 Cells via Activation of HIF-2α. *Cell. Physiol. Biochem.* **2014**, *34*, 1427–1441. [CrossRef] [PubMed]
55. Pichavant, K.; Person-Le-Ruyet, J.; Le Bayon, N.; Severe, A.; Le Roux, A.; Boeuf, G. Comparative effects of long-term hypoxia on growth, feeding and oxygen consumption in juvenile turbot and European sea bass. *J. Fish Biol.* **2001**, *59*, 875–883. [CrossRef]
56. Borowiec, B.G.; Darcy, K.L.; Gillette, D.M.; Scott, G.R. Distinct physiological strategies are used to cope with constant hypoxia and intermittent hypoxia in killifish (*Fundulus heteroclitus*). *J. Exp. Biol.* **2015**, *218*, 1198–1211. [CrossRef] [PubMed]
57. Wang, S.; Sun, M.; Ning, Z.; Chen, Y.; Zhou, H.; Mu, W. The effects of sustained and diel-cycling hypoxia on high-latitude fish Phoxinus lagowskii. *Comp. Biochem. Phys. Part D Genom. Proteom.* **2023**, *45*, 101059. [CrossRef]
58. Yao, T.; Wang, S.; Xu, K.; Zhang, X.; Zhang, T.; Wang, J.; Wang, Z.; Mu, W. Biochemical, histological, and gene expression analysis in brain and heart in *Phoxinus lagowskii* under sustained and diel-cycling hypoxia. *J. World Aquacult. Soc.* **2022**, *53*, 860–878. [CrossRef]

animals

MDPI

Article

Trophic Ecology of the Pyjama Shark *Poroderma africanum* (Gmelin, 1789) Elucidated by Stable Isotopes

Luca Caracausi [1], Zaira Da Ros [1,*], Alice Premici [1], Enrico Gennari [2] and Emanuela Fanelli [1]

[1] Department of Life and Environmental Sciences, Polytechnic University of Marche, Via Brecce Bianche, 60131 Ancona, Italy; caracausi.luca@gmail.com (L.C.); alice.premici@gmail.com (A.P.); e.fanelli@univpm.it (E.F.)
[2] Oceans Research Institute, P.O. Box 1767, Mossel Bay 6500, South Africa; e.gennari@oceans-research.com
* Correspondence: z.daros@staff.univpm.it

Simple Summary: Sharks, as important members of marine food webs, are often targeted by commercial and sport fishing, and moreover they constitute a significant part of the bycatch along with marine mammals and turtles. Overfishing, especially of top-predator species, disrupts the entire food web through a process called "mesopredator release", where the removal of top predators leads to an increase in mid-level predators, altering the ecosystem balance. Despite their crucial ecological role, there is limited information on sharks' diets. Traditional methods like stomach content analysis and newer, less invasive techniques, such as stable isotope analysis (SIA) of muscle tissue biopsies, provide insights into their feeding habits. A study on the pyjama shark or striped catshark (*Poroderma africanum*), a species native to South Africa, used SIA to explore its trophic ecology. The study found that as pyjama sharks grow, their diet shifts from more planktonic species as juveniles to more benthic prey. Juvenile sharks have a more varied diet, while adults are more selective. Although the pyjama shark population is increasing, according to the IUCN, there is a lack of structured monitoring programmes and catch data. Understanding the trophic ecology of mesopredators like the pyjama shark is crucial for predicting the impacts of their predation on marine ecosystems.

Abstract: Sharks may occupy both intermediate and upper levels of marine food webs. They are overfished worldwide and constitute one of the largest portions of the bycatch. The removal of top-predator species has negative cascading effects on the entire food web, causing the "mesopredator release" phenomenon, which leads to an increase in mesopredators with consequent changes in the ecosystem's energetic balance. Despite their important ecological role, information on their trophic ecology is limited. This essential information can be obtained through the analysis of stomach contents and, more recently, by using less invasive techniques, such as the stable isotope analysis of muscle tissue, obtained through biopsies. Here, we analysed the trophic ecology of the pyjama shark or striped catshark *Poroderma africanum*, an endemic species of South Africa, by means of SIA. The results obtained from SIA were analysed using the R SIMMR and SIBER packages to estimate the contribution of potential food sources to the diet and to evaluate the extent of the trophic niches. The SIMMR outputs showed that adults select more benthic prey than juveniles, which consume more planktonic species, with juveniles being more generalist than adults, according to SIBER outputs. As assessed by IUCN, the population of *P. africanum* is increasing, and given its role as mesopredator, future monitoring efforts could be crucial to elucidating their potential effects in marine food webs.

Keywords: stable isotope analysis; catshark; mesopredator; South Africa

1. Introduction

Sharks are cartilaginous fishes that may occupy both intermediate and apical levels of marine food webs. They are often typified as opportunistic predators, with a wide trophic spectrum from plankton to marine mammals [1,2]. Their importance is explained by the likely effects they may have on particular prey species and their high connectivity in the food web [2–4]. Sharks have the potential to influence the community structure and the habitat use of prey organisms [5]. They are overfished worldwide as a commercial and sport fishing target, and they constitute one of the largest portions of the bycatch, together with marine mammals and turtles [6]. The removal of top-predator species has negative cascading effects on the entire food web, causing the so-called "mesopredator release" phenomenon, which leads to an increase in mesopredators with a consequent change in the ecosystem's energetic balance [5,7]. Despite their important ecological role, information on their trophic ecology is limited because of their intrinsic biological characteristics, such as naturally low population densities, generally highly mobile and elusive behaviour and large-distance migrations [8,9]. Information on the composition of their diet is essential and can be obtained through the analysis of stable isotopes (stable isotope analysis or SIA) [10]. SIA reflects assimilated food, but it cannot be used to identify specific prey [11]. Moreover, this method produces information on long-term diet, since the stable isotope values of predators reflect those of assimilated nutrients from ingested prey integrated over longer time periods [12,13]. SIA provides a better understanding of the trophic role of the studied species, which can in turn support appropriate management and conservation strategies [14]. The use of SIA in trophic studies is based on a biological–chemical concept: during normal metabolic functions, the heavier, rare isotope is retained, while the lighter, more common isotope is excreted [15]. The stable isotopes of carbon ($\delta^{13}C$) and nitrogen ($\delta^{15}N$) are most used and provide powerful tools for estimating carbon flow to consumers ($\delta^{13}C$) and related trophic positions of species within food webs ($\delta^{15}N$). The analysis of nitrogen isotopes provides information on trophic position, as $\delta^{15}N$ increases with increasing trophic level [11,16], while that of carbon isotopes provides insights on the foraging ecology and movement/migration patterns [17]. The type of tissue that will be analysed is recognised as a fundamental aspect of experimental design when applying SIA in ecological studies. Each tissue has its own characteristics, which include different metabolic turnover rates between different species and within the same species. When using muscle tissue, which integrates short- to medium-term information efficiently on assimilated diets [13], a small sample size is required, and it can be obtained through a small biopsy, thus not provoking any damage or stress to the animal. For this project, this approach was used for the pyjama shark or striped catshark, *Poroderma africanum*. This shark belongs to the Scyliorhinidae family and is endemic to the waters off South African coasts, especially Cape Province. It is a nocturnal species, and its main characteristic is the presence of seven distinct dark longitudinal broad stripes extending the entire length of the body, on the dorsal and lateral sides [18]. The species has no specific protection, but its retention in commercial line fisheries is prohibited. The International Union for Conservation of Nature (IUCN) has assessed its population as "Least Concern".

The main objectives of this study were to investigate the trophic ecology of *P. africanum*. Specifically, we aimed to i. evaluate its feeding habits with SIA; ii. assess any spatial differences in its diet in three sites near Mossel Bay harbour; and iii. discuss trends in its population status in the light of the "mesopredator release hypothesis".

2. Materials and Methods

2.1. Study Area and Sampling

The study area is in Mossel Bay, in the Western Cape province of South Africa, halfway between Cape Town and Port Elizabeth (each 400 km away). The main feature of this site is the occurrence of a rocky reef. One of the most important abiotic factors in the structure of South African rocky intertidal communities is the role of wave action, which includes implications for energy flow between the intertidal and adjacent ecosystem [19–21]. Along the bay, there is an important freshwater contribution deriving from three main rivers, which take their name from the respective cities that pass through: Hartenbos, Little Brakrivier and Groot Brakrivier. All three rivers, crossing agricultural areas and raised beaches, tend to funnel great amounts of nutrients into the sea, enhancing primary production [22]. For this project, three sampling sites were chosen inside the bay. These were, from the closest to the furthest from Mossel Bay harbour, "The Point" (34°10.983′ S 22°09.774′ E, hereafter TP), "Hartenbos" (34°07.775′ S 22°07.392′ E, hereafter HB) and "Groot Brakrivier" (34°05.469′ S 22°14.514′ E, hereafter GB) (Figure 1).

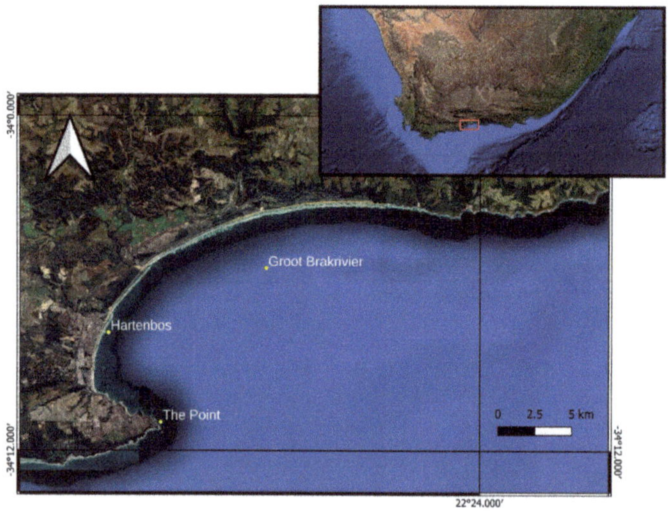

Figure 1. Study area in Mossel Bay, in the Western Cape province of South Africa, with the indication of the sampling sites (yellow symbols) of *P. africanum*. HB = Hartenbos, GB = Groot Brakrivier, TP = The Point (Downloaded on 2 May 2022 and modified from Google Earth Pro).

The collection of muscle samples took place during August 2021. Sharks were captured using a handline or a fishing pole and, once caught and pulled into the boat, they were placed in a container half-filled with salt water to reduce the stress during subsequent processes, like removing the hook from the mouth. Then, their total length (TL in cm), from the tip of the snout to the longest length of the tail, and pre-caudal length (PCL in cm), from the tip of the snout to the pre-caudal dimple, were measured [16]. For sample extraction, a 5 mm biopunch was inserted behind the first dorsal fin and a tiny piece of muscle tissue was taken. Subsequently, the samples were placed in a numbered test tube. At the end, the sharks were immediately released. Finally, all the samples were stored and maintained in a freezer at −20 °C until laboratory analysis.

Collection was authorised by the South African Department of Forestry Fisheries and Environment (RES2019-20) while the samples were exported under the TOPS permit 52015.

2.2. *Stable Isotope Analysis (SIA)*

Sample tissues of *P. africanum* were oven-dried at 60° for 24 h [23]. Each dried sample was grinded using a mortar and a small pestle, and ca. 1–1.2 mg was placed into tin capsules (Elemental Microanalysis Tin Capsules Pressed, Standard Weight 5 × 3.5 mm) using a rectangular tipped spatula. After weighing, each capsule containing the sample was closed using a rectangular-tipped spatula and tweezers and ordered in a plastic rack. Then, the rack was stored in a freezer at −20° until delivery to the specialised laboratory for subsequent analysis. Samples were analysed through an elemental analyser (Thermo Flash EA 1112, Thermo Scientific: Waltham, MA, USA) for the determination of total carbon and nitrogen, coupled with a continuous-flow isotope-ratio mass spectrometer (Thermo Delta Plus XP) for the determination of δ^{13}C and δ^{15}N, at the Laboratory of Stable Isotope Ecology of the University of Palermo (Italy). Lipid extraction is important as sharks have a high concentration of lipids because of the presence of large amounts of squalene. This could affect the δ^{13}C values and thus the results of the analysis. Here, lipids were not extracted from the samples because the number of samples was not sufficient. Thus, a correction equation was applied to $\delta^{13}C$ values, by using the relationship between C:N ratios and the $\delta^{13}C$ signatures according to [24]: $\delta^{13}C_{corrected} = \delta^{13}C_{untreated} - 3.32 + 0.99 \times C{:}N_{bulk}$.

2.3. Statistical Analysis

A three-fixed-factor design was used: 'TL', 'Sex' and 'Site'. The first two were two-levels factors ('AD'/'JUV' for TL, and 'Male'/'Female' for Sex). The median of the length frequency distribution of *P. africanum* separated juveniles (<82 cm) from adults (≥82 cm). The third factor (Site) has three levels ('HB'/'GB'/'TP'), according to the three sampling sites. Differences in $\delta^{15}N$ and $\delta^{13}C$ contents among the considered factors and their interactions were tested by univariate and multivariate PERMANOVA (Permutational Multivariate Analysis of Variance) and were carried out on the resemblance Euclidean matrix of untransformed $\delta^{15}N$ and $\delta^{13}C$ values, respectively. For all PERMANOVA tests, the significance value was set at $p < 0.05$. Both univariate and multivariate PERMANOVA tests were carried out under an unrestricted permutation of raw data, with 9999 permutations. The Monte Carlo test was applied because of the low sample size. In addition, if the main test of PERMANOVA highlighted significant differences, a pairwise test was performed to identify the source of the variation. Multivariate and univariate statistical analyses on the obtained results were conducted using PRIMER6 and PERMANOVA+ [25,26]. Correlations between $\delta^{15}N$ or $\delta^{13}C$ and the total length of *P. africanum* were calculated with PAST (version 4.0.9) [27].

2.4. Mixing Models

A Bayesian model SIMMR (Stable Isotope Mixing Models in R [28]) was run to estimate the contribution of the different food sources to the diet of *P. africanum* with the software R 4.0.5 (R Development Core Team 2009). The variability in the isotope values (mean and standard deviation) of prey species can be incorporated into the model [29]. Before running the model, the isotopic values of the sources and sharks were plotted, applying the correct trophic enrichment factors (TEFs) to potential sources to build a mixing polygon (Figure 2) [30].

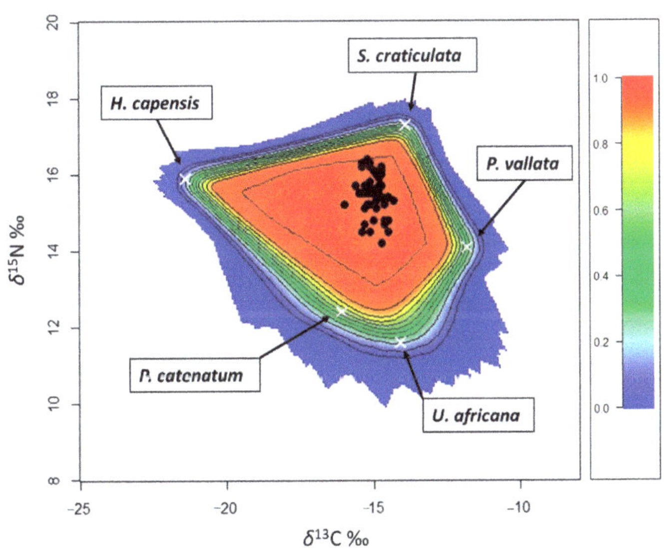

Figure 2. Mixing plot of the potential food sources for *P. africanum*: white crosses indicate mean isotopic values of the food sources, corrected with the TEF. Black dots represent the isotopic values of *P. africanum* samples. Probability contours are at the 5% level (outermost contour) and at every 10% level, following the colors shown in the legend in the right part of the graph. Full name of prey species as in Table 1.

TEFs used for $\delta^{15}N$ and $\delta^{13}C$ are 2.29 and 0.9, respectively [31]. The list of potential prey of *P. africanum* was taken from the literature, as there are no data from prey directly taken in South Africa (Table 1).

Table 1. List of potential prey of *P. africanum* and the related isotopic composition (plus standard deviations) as reported in the literature.

Sources	Mean $\delta^{13}C$ (‰)	Mean $\delta^{15}N$ (‰)	SD $\delta^{13}C$ (‰)	SD $\delta^{15}N$ (‰)
Perinereis nuntia vallata [32]	−12.7	11.8	0.3	0.1
Sesarma catenata [32]	−17.0	10.1	1	0.2
Upogebia africana [32]	−15.0	9.3	1	0.4
Hyporhamphus capensis [33]	−22.3	13.6	0.01	0.01
Phalium craticulatum [34]	−14.8	15.0	0.01	0.01

The SIBER package (Stable Isotope Bayesian Ellipses in R 3.5.3) [35] was used to compare isotopic niche widths among and within the communities. It was also used to calculate TA (Total Convex Hull Area) and SEA_C (Standard Ellipse Area corrected for low sample size, p interval = 0.40 to encompass 40% of our data) [36] for the different communities. Three models were run on *P. africanum* isotopic contents using different combinations. In the first, the three sites were used as groups and a unique community (value 1) was considered. In the second, the three sites were used as groups, but two communities were set, with each one corresponding to one of the two sexes. In the third model, the two levels of TL (juveniles below 81 cm TL and adults >82 cm TL) were used to determine two groups of the same community (Table 2).

Table 2. Groups and communities considered for running SIBER on SIA results.

Model	Groups	Communities
1	TP, HB and GB	1
2	TP, HB and GB	2 (M and F)
3	AD and JUV	1

3. Results

The length frequency distribution, considering all individuals, shows a different trend for males (n = 25) and females (n = 30) (Figure 3a). The results indicate that 18% of 25 sampled males of *P. africanum* have a TL between 81 and 90.1 cm. Most of the collected specimens have a TL between 91 and 100 cm. Females predominate in the three size classes below 90.1 cm. The results found that 5% have a TL above 91 cm. All size classes are represented at TP, from 61 to 100 cm. At GB, there are no individuals between 71 and 80.1 cm, while at HB, no individuals between 61 and 70.1 cm occurred (Figure 3b).

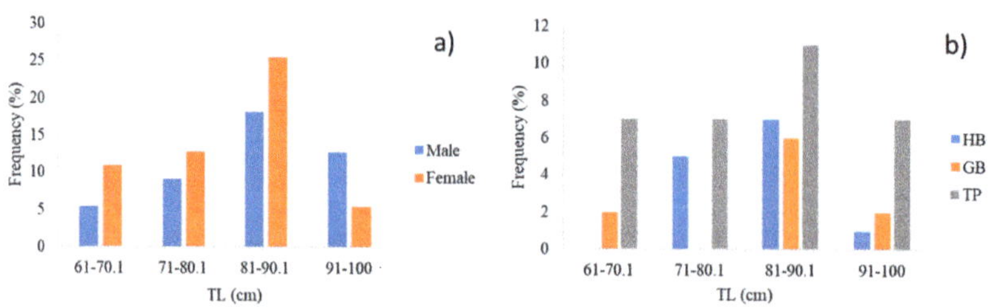

Figure 3. (**a**) Length frequency distribution of males and females of *P. africanum* and (**b**) length frequency distribution according to the site (HB = Hartenbos, GB = Groot Brakrivier and TP = The Point) and the total length (TL); n = 55.

3.1. Overall Isotopic Composition

Isotopic values for all considered levels within factors are shown in Table 3.

Table 3. $\delta^{13}C$ and $\delta^{15}N$ (mean and standard deviation = SD) for F (females), M (males), JUV (juveniles), AD (adults), HB (Hartenbos), GB (Groot Brakrivier) and TP (The Point).

	Mean $\delta^{13}C$ (‰)	SD $\delta^{13}C$ (‰)	Mean $\delta^{15}N$ (‰)	SD $\delta^{15}N$ (‰)
F	−15	0.3	15.5	0.4
M	−14.9	0.3	15.4	0.6
JUV	−15.1	0.3	15.4	0.4
AD	−14.9	0.3	15.5	0.6
HB	−15.0	0.2	16.0	0.3
GB	−14.9	0.3	15.7	0.3
TP	−15.0	0.4	15.2	0.4

$\delta^{13}C$ values are very similar in all three sites (ca. -15 ± 0.30‰), as are $\delta^{15}N$ values (ca. 15.7 ± 0.30‰). The correlation between TL and $\delta^{15}N$ ($R = 0.08$, $p > 0.05$, Figure 4a) was positive but not significant, while the correlation of TL with $\delta^{13}C$ ($R = 0.5$, $p < 0.001$; Figure 4b) was positive and significant.

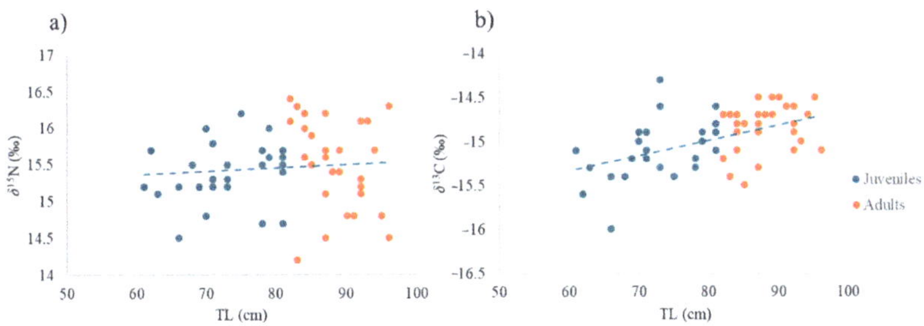

Figure 4. Scatterplot of (**a**) $\delta^{15}N$ (‰) values vs. total length (cm) and (**b**) $\delta^{13}C$ (‰) values vs. total length (cm).

The univariate PERMANOVA test carried on $\delta^{15}N$ values ($p < 0.05$) showed significant differences for the factor "Site" (Table 4a). Pairwise tests showed variations between the levels of the factors "Site" (between TP and the other two sites) and "Sex" (between males and females) (Table 4a). The univariate PERMANOVA test on $\delta^{13}C$ values showed significant differences only for factor "TL" (Table 4a).

Table 4. Results of the PERMANOVA main test (**a**) and of the pairwise comparisons (**b**) for $\delta^{15}N$ (left) and $\delta^{13}C$ (right). df = degrees of freedom; MS = mean square; Pseudo-F = statistic F; t = statistic t for pairwise comparisons; p(MC) = probability level with Monte Carlo test; TL = total length; HB = Hartenbos; GB = Groot Brakrivier; TP = The Point; F = female; M = male; * = $p < 0.05$; n.s. = not significant.

(a)		$\delta^{15}N$			$\delta^{13}C$		
Source	**df**	**MS**	**Pseudo-F**	**P(MC)**	**MS**	**Pseudo-F**	**P(MC)**
TL	1	0.22	1.60	n.s.	0.71	7.12	*
Sex	1	0.00	0.03	n.s.	0.00	0.04	n.s.
Site	2	3.18	23.38	*	0.04	0.35	n.s.
TL×Sex	1	0.00	0.00	n.s.	0.01	0.11	n.s.
TL×Site	2	0.19	1.39	n.s.	0.11	1.15	n.s.
Sex×Site	2	0.40	2.95	n.s.	0.14	1.46	n.s.
TL×Sex×Site	2	0.19	1.38	n.s.	0.06	0.57	n.s.
Residuals	43	0.14			0.10		
Total	54						

Table 4. *Cont.*

(b) Groups	t	P(MC)
HB, GB	0.94	n.s.
HB, TP	4.94	*
GB, TP	4.16	*
Groups	**t**	**P(MC)**
F, M	2.19	*

3.2. Mixing Models and Niche Width

Two Bayesian SIMMR models were used to estimate the potential food sources for *P. africanum*, considering the significance obtained with the univariate analysis. The first concerns the difference between sites: the gastropod *Phalium craticulatum*, in yellow, is the species that contributes the most, about 60 percent, to the diet in all areas, with some differences in the contribution of other species (Figure 5).

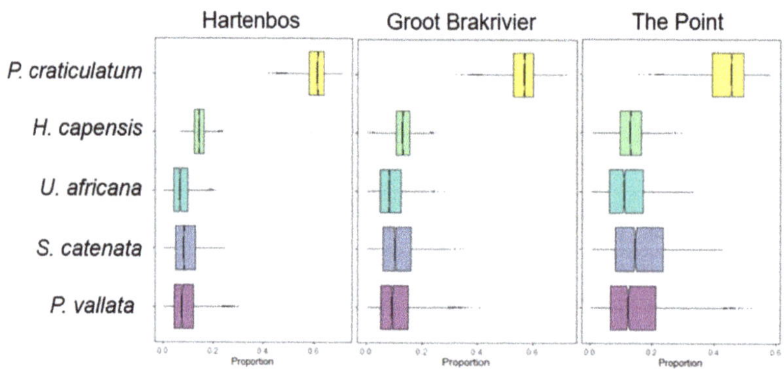

Figure 5. Proportions of each food source in the diet of the catshark collected in Hartenbos (HB), Groot Brakrivier (GB) and The Point (TP). Boxplots were obtained with stable isotope analysis mixing models. Each plot shows proportions for each food source in the specific site. Boxes indicate 50%, 75% and 95% Bayesian confidence intervals.

Standard ellipses obtained by SIBER show the isotopic niche widths of the groups. The niche width of P. *africanum* in TP (area near the harbour) is greater than in the other two sites (Figure 6, Table S1).

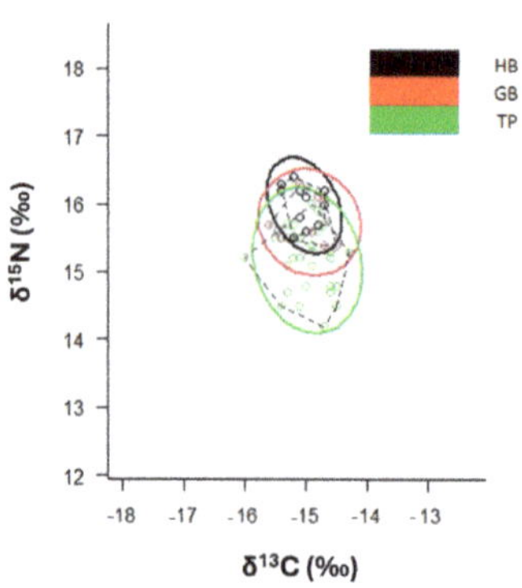

Figure 6. δ^{13}C-δ^{15}N scatterplot with standard ellipses (p interval = 40%) corrected for small sample size (SEA$_C$) for *P. africanum* collected in Mossel Bay, by site (Hartenbos = HB, Groot Brakrivier = GB and The Point = TP).

The second, however, concerns the difference between sizes: *Phalium craticulatum* has a greater contribution in adults, while *Hyphoramphus capensis* has a greater contribution in juveniles (Figure 7).

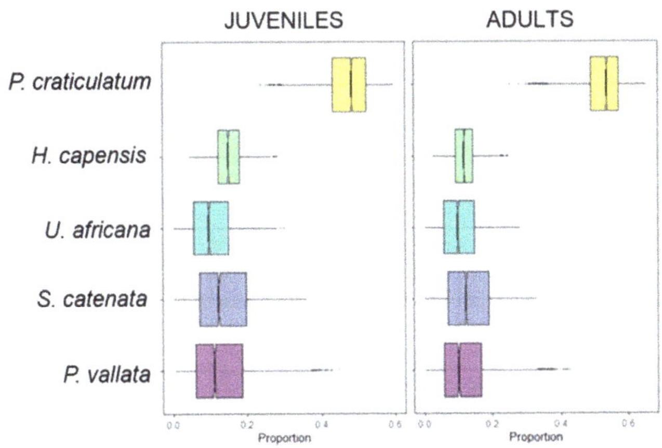

Figure 7. Proportions of each food source in the diet of the catshark collected for juveniles and adults. Boxplots were obtained with stable isotope analysis mixing models. Each plot shows proportions for each food source in the specific TL. Boxes indicate 50%, 75% and 95% Bayesian confidence intervals.

The SEA_C value of adults is stretched along $\delta^{15}N$-axis (Figure 8, Table S2).

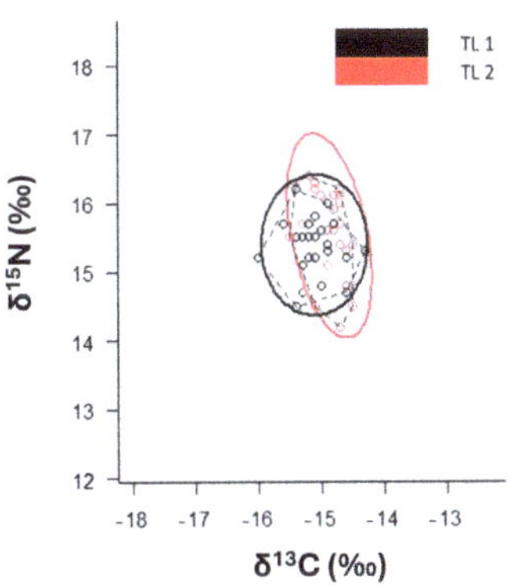

Figure 8. $\delta^{13}C$-$\delta^{15}N$ scatterplot with standard ellipses (p interval = 40%) corrected for small sample size (SEA_C) for *P. africanum* collected in Mossel Bay, by total length (TL 1 = juveniles smaller than 81 cm TL, TL 2 = adults larger than 82 cm TL).

4. Discussion

Usually, in catsharks, $\delta^{15}N$ content is greater in animals of a larger size [37,38]. As catsharks grow, they feed on prey positioned at higher trophic levels. They can select specimens of the same species, but of a larger size or located at a higher position within the food web, as also observed in other macropredator sharks, such as smooth-hound sharks [39]. However, this positive relationship between $\delta^{15}N$ and TL is not always valid [37]. In some cases, $\delta^{15}N$ values, and in turn the trophic level, remain fairly constant as the size increases. This happens for the specimens of *Poroderma africanum* analysed here. Conversely, the increasing $\delta^{13}C$ trend with increasing size suggests an ontogenetic dietary shift associated with changes in morphology, physiology or lifestyle [16]. The stable isotope value of carbon indicates whether the carbon source is more planktonic or more benthic [40], or marine vs. continental, thus providing indications of an inshore and offshore trend [41,42]. Higher $\delta^{13}C$ values (from -13 to -17‰) indicate that a species is relying on a benthic food web. Conversely, lower $\delta^{13}C$ values point to a greater dependence on a pelagic food web [40,41]. In this study, juveniles showed very low $\delta^{13}C$ values, while adults had a higher $\delta^{13}C$ value, with a significant increase in $\delta^{13}C$ with catsharks' size. Two hypotheses can be advanced for this finding: (i) juveniles eat more benthopelagic shrimps or other prey approaching the sea bottom, while adults feed on purely benthic items [43]; (ii) the shift in the isotopic signal could be due to a horizontal displacement with adults living more offshore than juveniles [44,45]. However, this catshark is not recognised as a migrator, so it is unlikely that there is an inshore–offshore movement for this species [46], particularly in this study area. Moreover, the output of the SIBER model carried out considering TL as a discriminant factor shows a higher selective diet of adults (that show a stretched SEA_C along the $\delta^{15}N$-axis) compared to the more generalist diet of juveniles. This is likely because most of the adults analysed in this study live near The Point (TP), which, probably due to input from organic discharges, has the most variable source of N among the three sampling sites. The TP site is close to the harbour, while the other two sites are characterised by the presence of two rivers. The output of the second SIMMR model, which considers the differences between juveniles and adults, confirms that the mean contribution of the gastropod *Phalium craticulatum* (a strictly benthic organism) to adults' diet is greater than for juveniles. Similarly, the pelagic fish *Hyporhamphus capensis* seems to give a greater contribution to juveniles' diet. However, since the isotopic data used on the model for prey were from the literature, the model results must be taken with caution and further studies are needed on *P. africanum's* diet to compare SIA to SCA results. Additionally, some environmental variables (downloaded at https://giovanni.gsfc.nasa.gov/giovanni/, accessed on 22 March 2024) may explain the observed pattern: river runoff, chlorophyll-a concentration, as a proxy of primary production, and particulate organic carbon (POC) concentration. Satellite data (MODIS-Aqua, Modisa, Level 3m) confirmed the difference in chlorophyll-a and POC concentrations among the three sites (data up 30th October 2021) (Figure 9a,b).

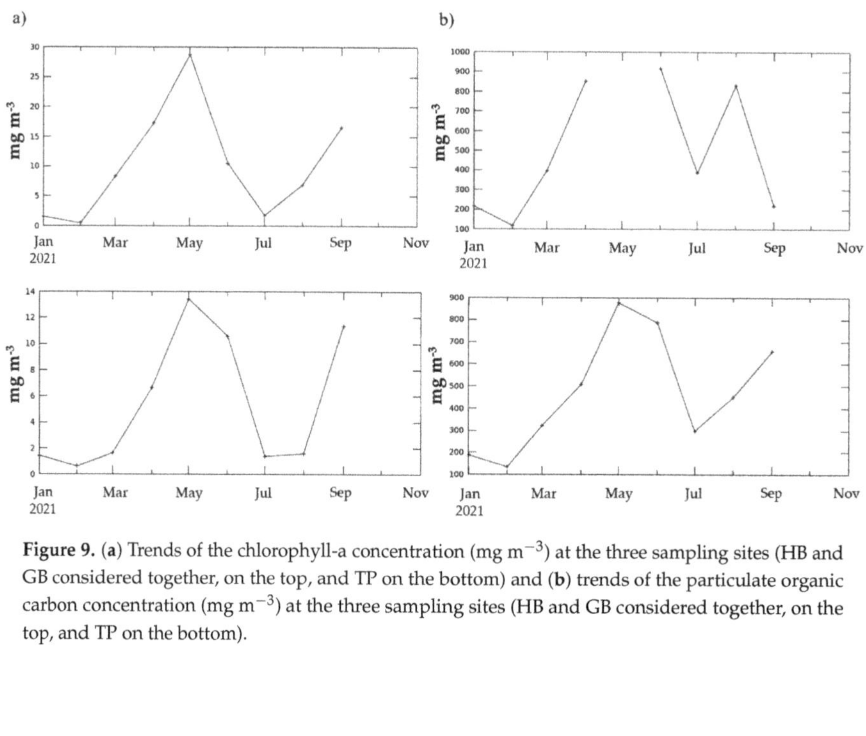

Figure 9. (**a**) Trends of the chlorophyll-a concentration (mg m^{-3}) at the three sampling sites (HB and GB considered together, on the top, and TP on the bottom) and (**b**) trends of the particulate organic carbon concentration (mg m^{-3}) at the three sampling sites (HB and GB considered together, on the top, and TP on the bottom).

Higher values of these two variables were recorded at HB and GB than at TP. No differences were registered among the runoff values in the three sites, in agreement with the similar $\delta^{13}C$ values observed in the three sites [47,48].

Although *Poroderma africanum* has an important ecological role in the local food web, as also highlighted by our results, catch data are scarce. This implies the lack of a well-defined management plan for the species that allows its adequate conservation or, at least, adequate monitoring of its population status in South African waters, as also reported by the IUCN assessment for this species.

5. Conclusions

The scyliorhinid *P. africanum* acts as a mesopredator, occupying the intermediate level of the food web it belongs to. If mesopredators increase, they could lead other species, which share the same resources, such as benthic prey, to modify their trophic niche and possibly also their habitats [5,49], thus avoiding competitive exclusion. Furthermore, an increase in mesopredators would even have the potential to cause the extinction of some prey, particularly the more sensitive species, who have low population growth rates or are more easily attacked by mesopredators [5]. Future studies, for example those conducted with stomach content analyses by recovering samples from bycatch, could confirm which are the specific prey ingested by *P. africanum*, shedding more light on its feeding ecology and trophic position, together with its ecological role.

Supplementary Materials: The following supporting information can be downloaded at https://www.mdpi.com/article/10.3390/ani14172559/s1: Table S1: Total Convex Hull Area (TA) and Standard Ellipse Areas corrected for small sample size (SEA_C) values of *P. africanum* between sites. Table S2: Total Convex Hull Area (TA) and Standard Ellipse Areas corrected for small sample size (SEA_C) values of *P. africanum* between TLs.

Author Contributions: Conceptualisation, E.F., E.G.; methodology, E.F., E.G.; software, L.C., E.F., Z.D.R.; validation, E.F., Z.D.R.; formal analysis, L.C., E.F., Z.D.R.; investigation, L.C., E.G.; resources, E.F., E.G.; data curation, A.P., L.C., E.F., Z.D.R.; writing—original draft preparation, L.C., E.F., Z.D.R.; writing—review and editing, all authors. All authors have read and agreed to the published version of the manuscript.

Funding: Stable isotope analyses conducted for this study were funded by the 2021 research project ELIS (tracking changes in habitat use and trophic ecology of ELasmobranchs, through stable ISotope analysis) of the Department of Life and Environmental Sciences of the Polytechnic University of Marche.

Institutional Review Board Statement: No Ethics Committee or Institutional Review Board approval was required for our manuscript, as all the specimens were released safe after capture according to the procedure carried out by the Oceans Research Institute. Collection was authorised by the South African Department of Forestry Fisheries and Environment (RES2019-20).

Informed Consent Statement: Not applicable.

Data Availability Statement: Data are made available upon request to the authors.

Acknowledgments: The authors wish to thank all the staff at the Oceans Research Institute, which supported the collection of *P. africanum* samples that were exported for research purposes under the TOPS permit 52015 enlarged by the Department of Forestry, Fishery, and the Environment of South African Government.

Conflicts of Interest: The authors declare no conflicts of interest.

References

1. Compagno, L.J.V. Alternative life-history styles of cartilaginous fishes in time and space. *Environ. Bio. Fishes* **1990**, *28*, 33–75. [CrossRef]
2. Ferretti, F.; Worm, B.; Britten, G.L.; Heithaus, M.R.; Lotze, H.K. Patterns and ecosystem consequences of shark declines in the ocean. *Ecol. Lett.* **2010**, *13*, 1055–1071. [CrossRef]
3. Bascompte, J.; Melián, C.J. Simple trophic modules for complex food webs. *Ecology* **2005**, *86*, 2868–2873. [CrossRef]

4. Ellis, J.K.; Musick, J.A. Ontogenetic changes in the diet of the sandbar shark, *Carcharhinus plumbeus*, in lower Chesapeake Bay and Virginia (USA) coastal waters. *Environ. Bio. Fishes* **2007**, *80*, 51–67. [CrossRef]
5. Ritchie, E.G.; Johnson, C.N. Predator interactions, mesopredator release and biodiversity conservation. *Ecol. Lett.* **2009**, *12*, 982–998. [CrossRef] [PubMed]
6. Oliver, S.; Braccini, M.; Newman, S.J.; Harvey, E.S. Global patterns in the bycatch of sharks and rays. *Mar. Pol.* **2015**, *54*, 86–97. [CrossRef]
7. Estes, J.; Crooks, K.; Holt, R.D. *Predators, Ecological Role of Encyclopedia of Biodiversity*, 2nd ed.; Elsevier: Amsterdam, The Netherlands, 2013. [CrossRef]
8. Heupel, M.R.; Knip, D.M.; Simpfendorfer, C.A.; Dulvy, N.K. Sizing up the ecological role of sharks as predators. *Mar. Ecol. Prog. Ser.* **2014**, *495*, 291–298. [CrossRef]
9. Matich, P.; Heithaus, M.R. Multi-tissue stable isotope analysis and acoustic telemetry reveal seasonal variability in the trophic interactions of juvenile bull sharks in a coastal estuary. *J. Anim. Ecol.* **2014**, *83*, 199–213. [CrossRef]
10. MacNeil, M.A.; Skomal, G.B.; Fisk, A.T. Stable isotopes from multiple tissues reveal diet switching in sharks. *Mar. Ecol. Prog. Ser.* **2005**, *302*, 199–206. [CrossRef]
11. Shiffman, D.S.; Gallagher, A.J.; Boyle, M.D.; Hammerschlag-Peyer, C.M.; Hammerschlag, N. Stable isotope analysis as a tool for elasmobranch conservation research: A primer for non-specialists. *Mar. Freshw. Res.* **2012**, *63*, 635–643. [CrossRef]
12. Domi, N.; Bouquegneau, J.M.; Krishna, D. Feeding ecology of five commercial shark species of the Celtic Sea through stable isotope and trace metal analysis. *Mar. Environ. Res.* **2005**, *60*, 551–569. [CrossRef]
13. MacNeil, M.A.; Drouillard, K.G.; Fisk, A.T. Variable uptake and elimination of stable nitrogen isotopes between tissues in fish. *Can. J. Fish. Aquat. Sci.* **2006**, *63*, 345–353. [CrossRef]
14. Hussey, N.E.; Chapman, D.D.; Donnelly, E.; Abercrombie, D.L.; Fisk, A.T. Fin-icky samples: An assessment of shark fin as a source material for stable isotope analysis. *Limnol. Oceanogr. Methods* **2011**, *9*, 524–532. [CrossRef]
15. Fry, B. *Stable Isotope Ecology*; Springer: New York, NY, USA, 2006; Volume 521. [CrossRef]
16. Van der Heever, G.M.; Van der Lingen, C.D.; Leslie, R.W.; Gibbons, M.J. Spatial and ontogenetic variability in the diet and trophic ecology of two co-occurring catsharks (*Scyliorhinidae*) off South Africa. *Afr. J. Mar. Sci.* **2020**, *42*, 423–438. [CrossRef]
17. Hobson, K.A. Tracing origins and migration of wildlife using stable isotopes: A review. *Oecologia* **1999**, *120*, 314–326. [CrossRef] [PubMed]
18. Compagno, L.J.V.; Ebert, D.A.; Smale, M.J. *Guide to the Sharks and Rays of Southern Africa*; Struik Publishers: Cape Town, South Africa, 1989.
19. McQuaid, C.D. The establishment and maintenance of vertical size gradients in populations of *Littorina africana knysnaensis* (Philippi) on an exposed rocky shore. *J. Exp. Mar. Biol. Ecol.* **1981**, *54*, 77–89. [CrossRef]
20. McQuaid, C.D.; Branch, G.M. Trophic structure of rocky intertidal communities: Response to wave action and implications for energy flow. *Mar. Ecol. Prog. Ser. Oldendorf* **1985**, *22*, 153–161. [CrossRef]
21. Emanuel, B.P.; Bustamante, R.H.; Branch, G.M.; Eekhout, S.; Odendaal, F.J. A zoogeographic and functional approach to the selection of marine reserves on the west coast of South Africa. *South Afr. J. Mar. Sci.* **1992**, *12*, 341–354. [CrossRef]
22. Watling, R.J.; Watling, H.R. Metal Surveys in South African Estuaries. III. Hartenbos, Little Brak and Great Brak Rivers (Mossel Bay). *Water SA* **1982**, *8*, 108–113.
23. Fanelli, E.; Da Ros, Z.; Martino, I.; Donato, F.; Azzurro, E.; Bargione, G.; Lucchetti, A. Crowding in the middle of marine food web: A focus on *Raja asterias* and other Mediterranean batoids. *Mar. Environ. Res.* **2023**, *183*, 105830. [CrossRef]
24. Post, D.M.; Layman, C.A.; Arrington, D.A.; Takimoto, G.; Quattrochi, J.; Montaña, C.G. Getting to the fat of the matter: Models, methods and assumptions for dealing with lipids in stable isotope analyses. *Oecologia* **2007**, *152*, 179–189. [CrossRef]
25. Clarke, K.R.; Warwick, R.M. Change in Marine Communities. *Approach Stat. Anal. Interpret.* **2001**, *2*, 1–168.
26. Anderson, M.; Gorley, R.N.; Clarke, K. *PERMANOVA+ for Primer: Guide to Software and Statistical Methods*; PRIMER-E Ltd.: Plymouth, UK, 2008.
27. Hammer, Ø.; Harper, D.A.; Ryan, P.D. PAST: Paleontological statistics software package for education and data analysis. *Palaeontol. Electr.* **2001**, *4*, 9.
28. Parnell, A. SIMMR: A Stable Isotope Mixing Model. R Package Version 0.4.5. 2021. Available online: https://CRAN.R-project.org/package=simmr (accessed on 30 June 2024).
29. Parnell, A.C.; Phillips, D.L.; Bearhop, S.; Semmens, B.X.; Ward, E.J.; Moore, J.W.; Jackson, A.L.; Grey, J.; Kelly, D.J.; Inger, R. Bayesian stable isotope mixing models. *Environmetrics* **2013**, *24*, 387–399. [CrossRef]
30. Phillips, D.L.; Inger, R.; Bearhop, S.; Jackson, A.L.; Moore, J.W.; Parnell, A.C.; Semmens, B.X.; Ward, E.J. Best practices for use of stable isotope mixing models in food-web studies. *Canad. J. L Zool.* **2014**, *92*, 823–835. [CrossRef]
31. Tilley, A.; López-Angarita, J.; Turner, J.R. Diet reconstruction and resource partitioning of a Caribbean marine mesopredator using stable isotope Bayesian modelling. *PLoS ONE* **2013**, *8*, e79560. [CrossRef]
32. Richoux, N.B.; Froneman, P.W. Assessment of spatial variation in carbon utilization by benthic and pelagic invertebrates in a temperate South African estuary using stable isotope signatures. *Estuar. Coast. Shelf Sci.* **2007**, *71*, 545–558. [CrossRef]
33. Bergamino, L.; Dalu, T.; Whitfield, A.K.; Carassou, L.; Richoux, N.B. Stable isotope evidence of food web connectivity by a top predatory fish (*Argyrosomus japonicus*: Sciaenidae: Teleostei) in the Kowie Estuary, South Africa. *Afr. J. Mar. Sci.* **2014**, *36*, 207–213. [CrossRef]

34. De Lecea, A.M.; Fennessy, S.T.; Smit, A.J. Processes controlling the benthic food web of a mesotrophic bight (KwaZulu-Natal, South Africa) revealed by stable isotope analysis. *Mar. Ecol. Prog. Ser.* **2013**, *484*, 97–114. [CrossRef]
35. Jackson, A.L.; Inger, R.; Parnell, A.C.; Bearhop, S. Comparing isotopic niche widths among and within communities: SIBER–Stable Isotope Bayesian Ellipses in R. *J. An. Ecol.* **2011**, *80*, 595–602. [CrossRef]
36. Layman, C.A.; Arrington, D.A.; Montaña, C.G.; Post, D.M. Can stable isotope ratios provide for community-wide measures of trophic structure? *Ecology* **2007**, *88*, 42–48. [CrossRef]
37. Kim, S.L.; Casper, D.R.; Galván-Magaña, F.; Ochoa-Díaz, R.; Hernández-Aguilar, S.B.; Koch, P.L. Carbon and nitrogen discrimination factors for elasmobranch soft tissues based on a long-term controlled feeding study. *Environ. Biol. Fishes* **2012**, *95*, 37–52. [CrossRef]
38. Caracausi, L.; Da Ros, Z.; Masia Lillo, P.; Ligas, A.; Fanelli, E. Trophic ecology of Scyliorhinus canicula in the Mediterranean Sea: Literature review and insights from a Tyrrhenian case-study. 2024; *Submitted.*
39. Di Lorenzo, M.; Vizzini, S.; Signa, G.; Andolina, C.; Palo, G.B.; Gristina, M.; Mazzoldi, C.; Colloca, F. Ontogenetic trophic segregation between two threatened smooth-hound sharks in the Central Mediterranean Sea. *Sci. Rep.* **2020**, *10*, 11011. [CrossRef]
40. Jennings, S.; Reñones, O.; Morales-Nin, B.; Polunin, N.V.; Moranta, J.; Coll, J. Spatial variation in the δ^{15}N and δ^{13}C stable isotope composition of plants, invertebrates and fishes on Mediterranean reefs: Implications for the study of trophic pathways. *Mar. Ecol. Prog. Ser.* **1997**, *146*, 109–116. [CrossRef]
41. Fanelli, E.; Menicucci, S.; Malavolti, S.; Da Ros, Z.; Biagiotti, I.; Canduci, G.; De Felice, A.; Leonori, I. The pelagic food web of the Western Adriatic Sea: A focus on small pelagics. *Sci. Rep.* **2023**, *13*, 14554. [CrossRef] [PubMed]
42. Rabehagasoa, N.; Lorrain, A.; Bach, P.; Potier, M.; Jaquemet, S.; Richard, P.; Ménard, F. Isotopic niches of the blue shark *Prionace glauca* and the silky shark *Carcharhinus falciformis* in the southwestern Indian Ocean. *Endang. Species Res.* **2012**, *17*, 83–92. [CrossRef]
43. Dainty, A.M. Biology and Ecology of Four Catshark Species in the Southwestern Cape, South Africa. MSc Thesis, University of Cape Town, Cape Town, South Africa, 2002.
44. Tanaka, H.; Ohshimo, S.; Takagi, N.; Ichimaru, T. Investigation of the geographical origin and migration of anchovy *Engraulis japonicus* in Tachibana Bay, Japan: A stable isotope approach. *Fish. Res.* **2010**, *102*, 217–220. [CrossRef]
45. Trueman, C.N.; MacKenzie, K.M.; Palmer, M.R. Identifying migrations in marine fishes through stable-isotope analysis. *J. Fish Biol.* **2012**, *81*, 826–847. [CrossRef]
46. Graham, B.S.; Koch, P.L.; Newsome, S.D.; McMahon, K.W.; Aurioles, D. Using isoscapes to trace the movements and foraging behavior of top predators in oceanic ecosystems. In *Isoscapes*; Springer: Dordrecht, The Netherlands, 2010; pp. 299–318. [CrossRef]
47. Lee, K.Y.; Graham, L.; Spooner, D.E.; Xenopoulos, M.A. Tracing anthropogenic inputs in stream foods webs with stable carbon and nitrogen isotope systematics along an agricultural gradient. *PLoS ONE* **2018**, *13*, e0200312. [CrossRef]
48. Davias, L.A.; Kornis, M.S.; Breitburg, D.L. Environmental factors influencing δ^{13}C and δ^{15}N in three Chesapeake Bay fishes. *ICES J. Mar. Sci.* **2014**, *71*, 689–702. [CrossRef]
49. Elmhagen, B.; Ludwig, G.; Rushton, S.P.; Helle, P.; Lindén, H. Top predators, mesopredators and their prey: Interference ecosystems along bioclimatic productivity gradients. *J. Anim. Ecol.* **2010**, *79*, 785–794. [CrossRef] [PubMed]

animals MDPI

Article

Diet of Three Cryptobenthic Clingfish Species and the Factors Influencing It

Domen Trkov [1,2,*], Danijel Ivajnšič [3,4], Marcelo Kovačić [5] and Lovrenc Lipej [1,2]

1 Marine Biology Station Piran, National Institute of Biology, Fornače 41, 6330 Piran, Slovenia; lovrenc.lipej@nib.si
2 Jožef Stefan Institute and Jožef Stefan International Postgraduate School, Jamova Cesta 39, 1000 Ljubljana, Slovenia
3 Faculty of Natural Sciences and Mathematics, University of Maribor, Koroška Cesta 160, 2000 Maribor, Slovenia; dani.ivajnsic@um.si
4 Faculty of Arts, University of Maribor, Koroška Cesta 160, 2000 Maribor, Slovenia
5 Natural History Museum Rijeka, Lorenzov Prolaz 1, 51000 Rijeka, Croatia; marcelo@prirodoslovni.com
* Correspondence: domen.trkov@nib.si

Simple Summary: Clingfish are small fish species that spend most of their time hiding in various shelters on the seabed. Due to this way of life, their ecology is little known, although they are important for the ecosystem. The aim of the research was to investigate the diet of three clingfish species (*Lepadogaster lepadogaster*, *L. candolii*, and *Apletodon incognitus*) using a method based on the analysis of prey from their faeces. The results show that crustaceans are the most important prey for all three species, although the composition of the diet also depends on various factors, such as the size of the fish and the prey, the behaviour of the fish, the home range of the fish, and the availability of food. These results provide us with important information about the participation of clingfish in the food web and deepen our knowledge of the fish's diet and the factors that influence it. The results also show that the occurrence of predatory fish depends on the presence of their prey.

Abstract: Cryptobenthic fish are small benthic fish species that normally live in various hiding places. Due to their large numbers, they are very important for energy transfer to higher trophic levels. However, due to their small size and hidden lifestyle, knowledge about them and their ecology, including their diet, is still limited. Using a non-destructive method based on faecal pellets, we investigated the diet of three clingfish species, *Lepadogaster lepadogaster*, *L. candolii*, and *Apletodon incognitus*, in the shallow northern Adriatic Sea. To better understand the results, we studied the fauna of potential prey in the habitats of the fish studied and also took fish specimens to observe their behaviour in the laboratory. The three species feed predominantly on crustaceans, particularly amphipods, copepods, and decapods. The proportion of the different taxa in the diet depends on the species of clingfish, the size of the specimens, and the size of the prey. In addition, the behaviour of the fish, the home range of the specimens, and the availability of food played an important role. The presence of certain crustacean groups in the environment also determines the occurrence of clingfish of different species and sizes.

Keywords: non-destructive method; fish; diet habits; faecal pellets; *Lepadogaster lepadogaster*; *Lepadogaster candolii*; *Apletodon incognitus*; crustaceans; northern Adriatic Sea

1. Introduction

A very important part of the fish communities in coastal waters is cryptobenthic fish fauna. A fish species or a life history stage of a fish species is cryptobenthic if individuals exclusively or predominantly spend their lifetime in cryptobenthic microhabitats, that is, in the restricted living spaces underneath the bottom surface of the substrate or biocover, with a physical barrier to open spaces [1]. Due to this lifestyle, they are rarely observed by divers and are usually not detected during conventional ichthyofauna surveys [2]. Therefore, knowledge on the ecology of cryptobenthic fishes remains very incomplete and their importance has been mostly overlooked in the past. Smith-Vaniz, Jelks, and Rocha [3] reported that about 64% of the fish fauna sampled with rotenone in their Caribbean research was not detected by visual observation. In the Mediterranean, Kovačić et al. [4], using three different methods, recorded 42 fish species in total. The two visual census methods recoded 31 species, while the anaesthetic method found 18 species, with an overlap of 7 species that were ambivalent in their occurrence. This is probably one of the reasons why cryptobenthic fish species were considered rare in the past [5], while more recent studies have shown that they are quite common and abundant in the Mediterranean [6] (and references therein [7]). Studies on cryptobenthic fishes from tropical reefs have even shown that they have the potential to influence the ecosystem due to their high abundance and therefore represent an important part of biodiversity in coastal areas that can significantly influence ecosystem functions [8]. While many authors report on the importance (e.g., energy transfer) of cryptobenthic fish species in tropical seas [8–10], there is still a lack of knowledge on cryptobenthic fish species from the Mediterranean and their ecology [7,11,12].

One of the least known taxa of cryptobenthic fish species in the Mediterranean Sea is clingfishes (family Gobiesocidae) [7,13]. Clingfishes have a flattened, scaleless body and are mainly smaller than 10 cm [14,15]. They are characterised by a suction disc consisting of pelvic fins, a reduced swim bladder [14,16], and the absence of a stomach [17]. Three species have been recorded in Slovenian waters so far, namely *Lepadogaster lepadogaster* (Bonnaterre, 1788), *L. candolii* (Risso, 1810), and *Apletodon incognitus* Hofrichter and Patzner, 1997 [18–20]. *L. lepadogaster* and *L. candolii* are the most widespread European clingfish species, distributed throughout the Mediterranean and also along the eastern Atlantic coast from England to northwest Africa, the Canary Islands, and Madeira [2]. *A. incognitus* has been only recently described [21] and little is known about its ecology. The species is known to occur in the northern Mediterranean Sea, including the Adriatic Sea, and in the eastern Atlantic Ocean near the Azores [21,22]. Although all three species differ in their choice of habitats and especially depth distribution, there are also small overlaps in their habitat use that could be influenced by food supply [20]. Understanding feeding relationships is very important as it could reveal different strategies to reduce competition, such as resource partitioning [23]. In addition, knowledge of foraging habits provides information about ecological processes at the individual, population, and community levels [24]. Feeding and foraging habits also determine the position (trophic level) of animals in food webs and define their ecological role [25]. However, most studies on the fish diet are based on the examination of stomach contents. The main disadvantage of this method is that the studied animals must be killed and dissected, so for many species studied, large numbers of animals must be sacrificed [24,26], which can lead to significant changes in the population structure of fish species in some areas [27,28]. The potential negative effects of over-extraction on *L. lepadogaster* populations have already been highlighted by King [29]. At the same time, such methods are prohibited in marine protected areas (MPAs), and it is unlikely to obtain a permit to study legally protected species by using lethal methods [20]. Since more than half of the natural Slovenian coastline is protected by a 200 m wide marine belt and belongs to MPAs [30], the use of non-destructive sampling methods is essential. With this in mind, we applied a non-destructive method developed for the study of small benthic species, which has been shown to be very effective for *L. lepadogaster* [31].

The aims of the study are: (1) to apply a non-destructive method to three clingfish species, *L. lepadogaster*, *L. candolii*, and *A. incognitus*, and evaluate it; (2) to study the diets of these species and their position within the food webs; and (3) to determine factors influencing their diet composition. Within each species, we studied the diet of the specimens of the different sexes, size classes, and from different bottom depths. In the case of *L. lepadogaster*, seasonal differences in diet were also studied.

2. Materials and Methods

2.1. Study Area

The study was conducted in the Slovenian part of the Gulf of Trieste, the northernmost part of both the Adriatic Sea and the Mediterranean Sea. The Gulf of Trieste is a shallow gulf (20–25 m) with an area of about 500 km^2 [32]. The coastline of the Slovenian part of the Gulf of Trieste is about 46 km long [33]. The coastal bottom consists of boulder fields and gravel banks that extend to a depth of about 10 m, from where it is replaced by sediment bottom [20]. Soft sedimentary bottom of fluvial origin can also be found in the shallow waters of the bays [32]. Seagrass meadows (mostly *Cymodocea nodosa*) are found on the shallow sediment bottom at depths between 1 and 11 m [34]. Most of the flysch cliffs have retained their natural state, while the coastal plains are heavily exposed to anthropogenic influences [33].

The salinity in the Gulf of Trieste is typically marine but is influenced by freshwater inflows, ranging from 33 to 38.5‰. The water temperature normally fluctuates from 8 °C in winter to 24 °C in summer [32]. The average tidal range is high compared to the rest of the Adriatic Sea, with the low water level up to 80 cm below the mean sea level [35–38]. The Isonzo River is the most important freshwater source, which has a significant influence on seasonal plankton dynamics (autotrophic plankton and zooplankton). In addition, the development of the autotrophic plankton biomass determines the patterns of the consumer community [39].

2.2. Field Work

2.2.1. Fish Sampling

The sampling of cryptobenthic fish was carried out from January 2017 to March 2019 at various locations in the Slovenian part of the Gulf of Trieste. All sampling was carried out in coastal waters. The specimens were collected by snorkelling and by SCUBA diving in the infralittoral zone. In addition, some specimens were also collected in the mediolittoral and tidal pools at low tide. The main habitat types sampled were boulder fields, gravel banks, seagrass meadows with *Cymodocea nodosa* (Ucria) Ascherson, 1870 and *Posidonia oceanica* (Linnaeus) Delile, 1813, as well as sandy-muddy bottoms [20]. A total of 206 sampling sessions of approximately 1 h each were conducted at 72 randomly selected sites (Figure 1).

The random search for fishes was performed in various hiding places, e.g., under stones, rocks, shells or in natural cavities such as caves, holes, crevices, etc. To facilitate the collection of the fish, the narcotic Quinaldine (Sigma-Aldrich St. Louis, MO, USA) was used. Quinaldine was diluted at 1:15 with alcohol [40,41]. The anaesthetic was sprayed into the hiding place using a laboratory wash bottle. The anaesthetised fish were then caught with a hand net. Specimens of *L. lepadogaster* were also collected by lifting stones and capturing specimens with a hand net (d = 40 cm). Specimens attached to the underside of the stone were then dropped directly into the net. Specimens of *A. incognitus* inhabiting oyster shells were captured by placing a plastic bag over the oyster and then chasing the fish into the bag. The captured specimens were kept in a 100 mL plastic chamber with a lid, which had small holes for the inflow of oxygenated water which were small enough to prevent faecal pellets from passing through.

At each collecting site, the following data were recorded: date, location, macrohabitat (e.g., seagrass meadow, rocky area, sediment bottom), microhabitat (e.g., rock pile, noble pen shell with dead oysters), hiding place (e.g., under a rock, in an oyster shell), and bottom depth. For sampling in the mediolittoral, the depth was calculated using the depth of the sampling site and the zero point from the tidal amplitude calendar [36–38].

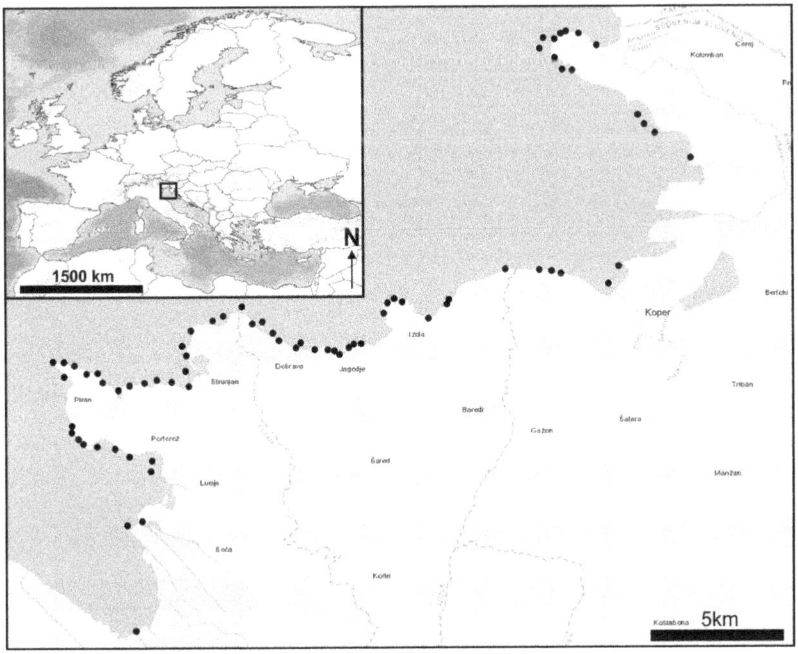

Figure 1. Sampling sites (black dots) along the Slovenian coast where clingfish were searched for.

2.2.2. Invertebrate Sampling

To investigate the structure of the fauna in the clingfish habitat, samples of benthic invertebrates were taken in May 2018 (warm season) and January 2019 (cold season). We sampled the shelters most frequently used by the clingfish species. Under stones inhabited by *L. lepadogaster*, 10 sample squares (25 × 25 cm, 625 cm^2) were sampled in the cold season and 10 sample squares in the warm season, while under stones inhabited by *L. candolii*, 10 sample squares were collected only in the warm season, as they were most frequently caught at this time. In the shelters with clingfishes, macroinvertebrates were collected with the substrate (e.g., pebbles, sand), using a hand net (20 × 20 cm, mesh size 500 μm). The square size of 25 × 25 cm was chosen because a frame of 20 × 20 cm (400 cm^2) is considered the minimum area for sampling Mediterranean infralittoral communities [42]. Substrate and macroinvertebrates were stored in plastic bags and brought to the Marine Biology Station Piran, National Institute of Biology (MBS-NIB). Since *A. incognitus* is associated with *Pinna nobilis* Linnaeus, 1758 (Trkov et al. [18], an endangered species whose populations are declining [43], no faunal samples were taken in those shelters.

2.3. *Laboratory Work*

2.3.1. Fish

After sampling, the fish were brought to the MBS-NIB as quickly as possible i.e., in less than one hour, where they were photographed with the Olympus TG-4, measured to the nearest 0.01 mm with a calliper (total length) and weighed to the nearest 0.01 g with a Sartorius TE 1502S balance (wet weight). Particular attention was paid to the condition of the clingfish, which were always kept moist during the measurements. The fish were identified at the species level using scientific literature with species diagnoses, descriptions, and identification keys [2,14,21,44,45]. The sex of specimens was determined based on the length and size of the urogenital papillae, which are larger and more elongated in males than in females [11,15]. For the size-based analyses, the specimens of each species were, preliminarily tested for size differences using Levene's test for equality of variances and the corresponding ANOVA analysis with Tukey contrasts post hoc test in the R statistical environment and then divided into different size classes. The specimens of *L. lepadogaster* and *L. candolii* were divided by Hofrichter [11] into five size classes, while those of *A. incognitus* were divided into four size classes. For ease of interpretation, the size classes correspond to the age-size classes proposed by Hofrichter ([11]; Table 1). However, the age estimate based on size could contain a considerable error, as the age ranges of the fish cohorts usually overlap, so the age estimate is only a rough approximation, therefore we have not used it in our analyses, except in Table 1. Due to the small sample size in some size classes, size classes in this research were merged as follows: *L. lepadogaster*: ≤3, 4, and 5; *L. candolii:* 1, 2, 3, and ≥4; *A. incognitus:* ≤2, 3, and ≥4.

Table 1. Size and age classes of three clingfish species as proposed by Hofrichter [11].

Size Class	Age Class (years)	*L. candolii*	*L. lepadogaster*	*A. incognitus*
1	0+	<25 mm	<22 mm	<16 mm
2	1+	26–39 mm	23–37 mm	17–24 mm
3	2+	40–51 mm	38–50 mm	25–34 mm
4	3+	52–70 mm	51–63 mm	>35 mm
5	4+	>71 mm	>64 mm	-

2.3.2. Invertebrate Fauna

The substrate collected in the fish shelters was examined for invertebrates as available food for clingfish. The invertebrates were identified using the scientific literature with species diagnoses, descriptions, and identification keys [46–49] and counted. The main prey taxa were also measured and weighed (Olympus SZx16 stereomicroscope with an Olympus DP74 camera; Sartorius CP 225D balance; wet weight from 70% ethanol) to obtain length–weight correlation curves for application on ingested prey.

2.3.3. Fish Diet

The non-destructive method based on faecal pellets [31] was used to determine fish diet. After measuring and weighing, each freshly caught clingfish was placed in an 11 × 13 × 14 cm chamber with filtered seawater (125 µm) and an aerator above the bottom to prevent the defragmentation of the faecal pellets. Freshly caught clingfish were left in the chambers for 24 h. Every three to four hours (during the day), the animals were checked to see if they had produced faecal pellets. The pellets were carefully removed from the bottom of the chambers using a modified pipette and fixed in 70% EtOH. Pellets consist of a peritrophic membrane and the inner part with undigested prey pieces. In some cases, the pellets were broken, and the contents of the entire chamber were filtered through a 125 µm mesh plankton net and checked. After defecation, the specimens were released unharmed at the site where they had been collected. After each use, the chambers were disinfected with 70% EtOH to prevent the transmission of diseases and parasites. Based on a small sample of freshly captured *A. incognitus* examined using a non-destructive method, we also include eight specimens from the MBS-NIB collection (seven captured in 2007 and one in 2019) whose gut contents were examined by dissection.

The contents of the faecal pellets and gut were analysed under an Olympus stereomicroscope and photographed with a camera. Some faecal pellets were also weighed using a Sartorius CP 225D balance (wet weight of 70% EtOH). Prey were identified to the lowest taxa level possible to be identified and then counted. A large proportion of the digested prey was whole or almost whole, while digested prey, which was broken down into smaller pieces, was recognised by typical body parts such as the carapace. The prey that was most digested, such as polychaetes and amphipods, were identified by their jaws [31].

The wet weight of the most important prey (e.g., decapods) was calculated using correlation curves between the length and weight of the available food collected in the sampling area (from samples preserved in 70% EtOH) (see Section 2.3.2). Undigested prey and prey that only occasionally appeared in the faeces or prey whose weight could not be calculated from the correlation curves, were weighed directly from the faecal pellets, or the average weight (of the animals collected in the sampling area) was used.

2.3.4. Observations in Aquaria

Three specimens of each species, i.e., *L. lepadogaster*, *L. candolii*, and *A. incognitus* were housed in three aquaria at MBS-NIB, where they were observed for better understanding of their feeding habits and behavioural patterns. The aquaria measured 150 × 50 × 50 cm and were set up to mimic the natural habitat of the clingfish. For *L. lepadogaster* and *L. candolii*, the aquaria were set up to mimic boulder fields with pebbles and larger stones, while *A. incognitus* was housed in an aquarium that mimicked a sediment bottom with a *Pinna nobilis* shell and oysters on top. The behaviour of each fish was observed for three days, 10 min in the morning and 10 min in the afternoon, for a total of 60 min per species. Occasionally, some observations were also made by night out of the scheduled observations.

2.3.5. Data Analysis

Three different quantitative methods were used to capture the complexity of the studied fish diet: frequency of occurrence (F%; [50]), numerical abundance (N%; [51]), and gravimetric composition (B%; [51]). In addition, the modified index of relative importance (IRI) [52,53] was calculated using the following formula: IRI = F% (N% + B%), where F% is the frequency of occurrence, N% is the numerical abundance, and B% is the gravimetric composition. To facilitate comparisons between different foods, Cortés [54] recommended expressing the IRI as a percentage. The IRI% for a given food category f is IRI % = (IRI/$\sum$IRI) × 100.

The dietary diversity of prey was expressed by the index of trophic diversity (ITD), which is a modified Shannon–Wiener diversity index (H′; [55]): ITD = 1 − H′. ITD values range from 0 (no diversity) to 1 (full diversity). The ITD was calculated for all three species and for different size groups.

To investigate the trophic level of the species, the TROPH value was calculated for all three species and for different size groups. The trophic level was calculated as follows [25,56,57]: TROPHi = 1 + $\sum$DCij × TROPHj, where DCij represents the proportion of prey j in the diet of consumer species i; TROPHj represents the proportion of prey j in the trophic level. The program TrophLab (a stand-alone Microsoft Access routine for estimating trophic levels; June 2000 version) was downloaded from www.fishbase.org (accessed on 5 March 2019) [56] and used to calculate the TROPH index of the species studied. For means, standard deviation (SD) was used to calculate.

A nonparametric permutative multivariate analysis of variance (PERMANOVA) based on the abundance data of the recorded prey taxa in the faecal pellets and the Bray–Curtis dissimilarity measure matrix was used to evaluate the diet differences in the numerical abundance of prey among studied species and to check for the diet differences by sex, size, bottom depth, and seasons for each species. In addition, a linear regression test (t-test) was used to test the linear dependence between the size of the prey and the size of the fish.

To compare diet among species, we used the Bray–Curtis dissimilarity index, which provides information on ecological or dietary niche overlap [58,59]: BCij = 1 − (2Cij/(Si + Sj)), where Cij is the sum of the lower values only for the prey taxa shared by both clingfishes. Si and Sj are the total number of prey taxa counted in the diet of each species. Values range from 0 (no overlap) to 1 (complete overlap). To show the importance of each prey taxa at different depths, the SIMPER function was used [60]: d[ijk] = abs(x[ij] − x[ik])/sum(x[ij] + x[ik]), where x is the abundance of prey taxa i in sampling units j and k. The total index is the sum of the individual contributions across all S prey taxa d[jk] = sum(i = 1…S) d[ijk]. The SIMPER function performs pairwise comparisons of groups of sampling units and determines the average contributions of the individual prey taxa to the average overall dissimilarity according to Bray–Curtis (https://www.rdocumentation.org/packages/vegan/versions/2.6-2/topics/simper; accessed on 14 May 2020). When using the summary function (ordered = TRUE), the results include the cumulative contribution and are ordered by the contribution of prey taxa to the potential difference in prey composition and abundance.

PERMANOVA and all other tests were calculated using the R statistical environment [61] and the Vegan package (version 2.6-2) [62], if not noted differently at the particular test above.

For the invertebrate fauna collected in the habitat of *L. lepadogaster* and *L. candolii*, the relative frequency of occurrence (F%) and the relative abundance of individuals of the different taxa (N%) per species habitat were calculated.

3. Results

3.1. Diet Study of Three Clingfish Species

A total of 363 specimens of *L. lepadogaster*, *L. candolii*, and *A. incognitus* were included in the study (Table S1). The non-destructive method was performed on 356 specimens (Table 2), all of which survived until their release into the wild. Faecal pellets were obtained from 96.9% of the specimens. Due to the small sample size of *A. incognitus*, we included eight specimens from the MBS-NIB collection whose gut contents were analysed.

Table 2. Diet attributes of three clingfish species obtained with a non-destructive method.

	L. lepadogaster	*L. candolii*	*A. incognitus*
Proportion of defecated specimens (%)	96.9	96.8	97.4
Number of defecated specimens	188	120	37
Number of faecal pellets	295	161	56
Average number of faecal pellets per fish	1.6	1.3	1.5
Number of prey items per species	1334	1779	749
Average number of prey items per fish	7.1	14.8	16.6
Maximum number of prey items per fish	39	143	131
Minimum number of prey items per fish	1	1	1
Average number of prey items per faecal pellet	4.5	11.0	12.7
Maximum number of prey items per faecal pellets	39	143	70
Minimum number of prey items per faecal pellets	1	1	1

3.1.1. *Lepadogaster lepadogaster*

A total of 194 specimens were collected. Their total length ranged from 28.06 mm to 78.02 mm (on average, 57.60 ± 8.54 mm), while their weight ranged from 0.23 g to 5.85 g (2.17 ± 0.96 g). Among the 182 sexually identifiable specimens, 78 were females and 104 were males (sex ratio 1:1.3). The measurements were performed on 80 faecal pellets. The average length of the pellets was 4.64 mm, the average width was 2.63 mm, and the average weight was 16.64 mg.

A total of 47 taxa, which were divided into 18 main taxonomic groups, were identified as prey. In addition to prey, the faecal pellets often contained various sediment particles such as sand grains, invertebrate shell fragments, and algal particles. Sediment particles could account for up to 16.5% of the weight of the faecal pellets. Copepods (N% = 30.1), mostly represented by harpacticoids, were by far the most numerous prey group, followed by amphipods (N% = 20.3) and decapods (N% = 12.6). The latter two prey categories were also the most abundant prey groups in terms of the frequency of occurrence. Amphipods were ingested by 67.6% of the specimens, while decapods were preyed on by 56.4% of the fish. The most important prey group in terms of biomass were decapods (B% = 60.9), followed by amphipods (B% = 18.3), which together accounted for the majority (79.2%) of the biomass. Decapods were the most important prey group in terms of relative importance (IRI% = 43.6; Figure 2) and were mainly represented by the crab *Pisidia* sp. (82.7% of the numerical abundance of decapods). According to the diagnostic characteristics, the crabs of the genus *Pisidia* mainly belong to *P. bluteli* (Risso, 1816) and less frequently to

P. longimana (Risso, 1816). The results showed that the size of *Pisidia* sp. was not linearly related to the size of the fish (*t*-test for linear regression; $p = 0.3216$, $\alpha = 0.05$). Alternative prey groups were amphipods (IRI% = 27.5), such as species of the family Gammaridae (97.4% of amphipods), and copepods (IRI% = 13.0), mostly harpacticoids. Gastropods, such as juveniles of *Bittium reticulatum*, *Alvania* sp., and *Rissoa* sp., accounted for 8.3% of the IRI. Based on the diet composition, we calculated the TROPHs index, which was 3.18 ± 0.44, and the ITD index, which was 0.83.

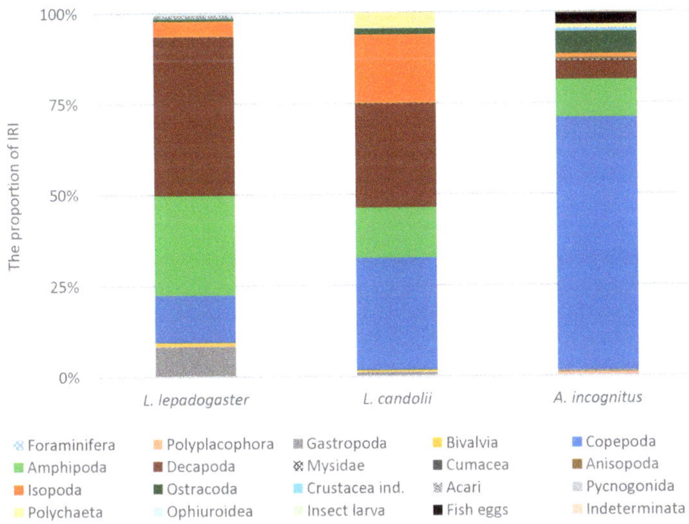

Figure 2. Index of the relative importance (%) of different prey groups for three clingfish species.

The quantitative composition of the diet of *L. lepadogaster* was not significantly different between sexes (PERMANOVA; $p = 0.648$, $\alpha = 0.05$) and also no significant difference was found in diet composition between the different size classes (PERMANOVA; $p = 0.135$, $\alpha = 0.05$). However, copepods dominated the diet of the size group ≤ 3 years and their abundance and frequency of occurrence decreased with the size of the group. This trend was also observed in Acari and ostracods. In contrast, the abundance and frequency of occurrence of decapods, isopods, and gastropods increased with the size of the group. The TROPHs index ranged from 3.18 ± 0.36 for the smallest size group and increased to 3.35 ± 0.52 for the largest size group, while the ITD ranged from 0.75 in size group ≤3 to 0.82 for size group 5 (Table 3).

Table 3. ITD and TROPHs indexes for three clingfish species. Some classes were grouped together due to the small sample size.

Size Class	ITD			TROPHs		
	L. lepadogaster	*L. candolii*	*A. incognitus*	*L. lepadogaster*	*L. candolii*	*A. incognitus*
1	-	0.63	-	-	3.03	-
2	-	0.69	0.78 *	-	3.14	3.26 *
3	0.75 *	0.76	0.31	3.18 *	3.19	3.05
4	0.84	0.81 **	0.68 **	3.21	3.22 **	2.98 **
5	0.82	-	-	3.35	-	-
Per species	0.83	0.73	0.51	3.18 ± 0.44	3.12 ± 0.36	3.13 ± 0.26

* The group also contains specimens of a smaller class; ** the group also contains specimens of a larger class.

The differences in the diet composition among individuals at the different bottom depths were statistically significant (PERMANOVA; $p = 0.001$, $\alpha = 0.05$). The most significant prey group in shallow water was copepods, while in deeper water it was decapods and gastropods. In addition, the SIMPER function showed that decapods were the most important prey group at the maximum depth of *L. lepadogaster* occurrence and their importance decreased with decreasing bottom depth. It also showed that the importance of copepods decreased with increasing depth. In addition, the numerical abundance of prey showed that copepods and bivalves were more abundant in shallower waters. In contrast, the abundance and frequency of occurrence of decapods, ophiuroids, and gastropods increased with increasing depth.

The seasonal differences in the diet composition were statistically significant (PERMANOVA; $p = 0.001$, $\alpha = 0.05$). The presence of amphipods in the diet was most significant in winter, copepods in spring, and decapods in summer and autumn. In winter, the most abundant groups were amphipods and gastropods, which were also the most abundant prey groups in the diet. In spring, copepods were by far the most abundant prey group, accounting for over 52% of the numerical abundance. In summer, the numerical abundance of copepods decreased, while the abundance of decapods increased and became the most abundant and frequent prey group in the diet. The abundance of isopods, acari, and amphipods also increased in summer. In the autumn, amphipods were the most abundant prey group, followed by decapods, whose abundance decreased slightly compared to summer. However, the frequency of decapods in the diet was highest in the autumn (79.3%). A comparison of the size of *Pisidia* sp. crabs consumed shows that fish consumed the largest specimens of *Pisidia* sp. in winter, while their size decreased over the seasons and was smallest in autumn. In addition, seasonal differences in the diet of males were separately tested and were statistically significant (PERMANOVA; $p = 0.001$, $\alpha = 0.05$), with the presence of copepods in the diet being most intensive in spring.

3.1.2. *Lepadogaster candolii*

A total of 120 specimens were included in the dietary study. The total length of the 113 specimens measured ranged from 13.97 mm to 86.94 mm (average 41.48 ± 16.00 mm), while their weight ranged from 0.03 g to 6.05 g (1.08 ± 1.10 g). Of the 57 sexually identifiable specimens, 17 were male and 40 were female (sex ratio 1:2.4). A total of 119 faecal pellets were measured. The average weight of the faecal pellets was 10.42 mg and they were on average 3.53 mm long and 2.10 mm wide.

A total of 42 different prey taxa were found and recognised in the diet. The prey was categorised into 18 major taxonomic groups. In addition to prey remains, the faecal pellets sometimes contained various sediment particles such as sand grains, old invertebrate shell fragments, and algal particles, which could account for up to 64.1% of the weight of the faecal pellets. The most frequently preyed groups were copepods (F% = 74.2), followed by decapods (F% = 60.0), isopods (F% = 59.7), and amphipods (F% = 56.7). Copepods were by far the most numerous prey (N% = 47.2), followed by isopods (N% = 11.3). In terms of biomass, decapods were the most important (B% = 46.3), followed by isopods (B% = 25.4) and amphipods (B% = 20.5). Copepods (IRI% = 30.9) and decapods (IRI% = 28.5) were the most important prey in terms of IRI. Copepods were almost exclusively represented by harpacticoids. In addition, seven parasitic copepods (family Caligidae) were found in the diet. Decapods were mainly represented by the crabs of the genus *Pisidia* (55.0% of all decapods) and *Athanas nitescens* (38.3% of all decapods). They belonged mainly to *P. bluteli* and less frequently to *P. longimana*. Other prey groups were isopods (IRI% = 18.9) and amphipods (IRI% = 13.8). Among the isopods, many parasitic larvae of Gnathiidae were found in the diet, and in one case, fish scales were also found in the faecal pellet (probably ingested together with the parasite). The TROPHs index was 3.12 ± 0.36 and the ITD was 0.73.

The quantitative composition of diets showed no statistical difference between the sexes (PERMANOVA; $p = 0.308$, $\alpha = 0.05$). Copepods were the most significant and most abundant prey group for both sexes. However, copepods, gastropods, and bivalves were more abundant in the diet of males, while decapods, isopods, and polychaetes were more abundant in the diet of females. In females, decapods (82.7%) were the most frequent prey group, followed by copepods (71.2%), while in males, copepods and amphipods (both 62.5%) were the most frequent prey groups, followed by decapods (56.3%). The difference in diet composition between specimens from different size classes was statistically significant (PERMANOVA; $p = 0.002$, $\alpha = 0.05$). The most important prey groups for smaller juvenile specimens were copepods and polychaetes, while adult specimens preferred decapods. Copepods were most abundant and frequent in the diet of smaller clingfishes, and their proportion decreased with increasing group size, while the abundance and frequency of occurrence of decapods and isopods increased with increasing group size. The ITD index ranged from 0.63 in size group 1 to 0.81 in size group ≥4, while the TROPHs index ranged from 3.03 ± 0.27 in size group 1 to 3.22 ± 0.40 in the largest size group ≥4.

The differences in the diet composition among individuals from the different bottom depths were statistically significant (PERMANOVA; $p = 0.001$, $\alpha = 0.05$). The abundance and frequency of occurrence of copepods and bivalves in the diet increased with increasing bottom depth, while the number of gastropods and decapods decreased with increasing depth. In addition, the SIMPER function showed that decapods were the most important prey group in the 2 to 3 m depth range and their importance decreased with increasing depth. The importance of copepods did not change with increasing depth, as shown by the SIMPER function. Due to the uneven distribution of samples across all four seasons, seasonal differences in diet were not analysed for *L. candolii*.

3.1.3. *Apletodon incognitus*

A total of 45 specimens were included in the dietary analysis, of which 37 specimens were examined using the faecal pellet method, while the gut contents were examined in eight specimens from the MBS-NIB collection. The total lengths of the 38 specimens measured ranged from 15.06 mm to 45.21 mm (average 32.38 ± 9.27 mm), while their weights ranged from 0.03 g to 1.29 g (0.48 ± 0.37 g). Of the 34 sexually identifiable specimens, 9 were female and 25 were male (sex ratio 1:2.8). A total of 41 faecal pellets were measured. The pellets weighed an average of 1.95 mg and were 2.34 mm long and 1.58 mm wide.

In the diet, 23 different taxa, which were classified into 18 main taxonomic groups, were recognised. In some cases, in addition to the prey remains, various sediment particles such as sand grains, invertebrate shell fragments, and algal particles accounted for up to 35.1% of the faecal pellet weight. Most of the prey in the diet of *A. incognitus* were copepods (N% = 69.0), which were also the most frequent prey (F% = 77.8), followed by ostracods (F% = 55.6) and amphipods (F% = 46.7). The most important prey groups in terms of biomass were decapods (B% = 22.3), amphipods (B% = 17.5), and copepods (B% = 14.6). The decapod crustaceans were mainly represented by various shrimp species (e.g., *A. nitescens*) and crabs of the genus *Pisidia*. By far the most important prey based on the IRI were copepods (IRI% = 69.4), mainly represented by harpacticoids, followed by amphipods (IRI% = 10.4), which were mainly represented by Gammaridea (88.0% of amphipods) and less by species of the family Caprellidae. Alternative prey groups were ostracods (IRI% = 6.2) and decapods (IRI% = 5.1). Based on the diet composition, a TROPHs index of 3.13 ± 0.26 and an ITD of 0.51 were calculated.

The quantitative composition of diet showed a statistical difference between males and females (PERMANOVA; $p = 0.003$, $\alpha = 0.05$). However, it should be noted that the sample of females was very small. Copepods, fish eggs, and ostracods were most significant in the male diet. Fish eggs were only found in the diet of the males. In both species, however, copepods predominated in terms of number and the frequency of occurrence. In females, ostracods, copepods, and decapods were the most frequent prey groups.

The difference in diet composition between the different size groups was statistically significant (PERMANOVA; $p = 0.009$, $\alpha = 0.05$). In all three size groups, copepods were the most significant and numerous prey group. The smallest specimens had the most diverse diet, with a higher abundance and frequency of occurrence of amphipods, decapods, anisopods, and isopods compared to the two larger groups. Only in the largest group, which consisted only of males, fish eggs were found. In addition, there was a significant difference in diet composition between nesting and non-nesting males (PERMANOVA; $p = 0.022$, $\alpha = 0.05$), with fish eggs being the most significant prey of nesting males. The ITD and TROPHs indices were also calculated for each size group. The TROPHs index ranged from a minimum of 2.98 $\pm$ 0.36 in the largest size group $\geq$4 to 3.26 $\pm$ 0.44 in size group $\leq$2. The ITD index was lowest in size group 3 (0.31), while it was highest in size group $\leq$2 (0.78).

Based on the bottom depth range, specimens were divided into three depth groups (2–3 m, 3–4 m, > 4 m), but no statistical significance was found in diet composition between specimens in the different groups (PERMANOVA; $p = 0.247$, $\alpha = 0.05$). Copepods were the most abundant and frequent prey group in all three depth classes. Due to the uneven distribution of samples across all four seasons, seasonal differences in diet were not analysed.

3.1.4. Comparison between Species

Crustaceans were the most important prey group for all clingfish species. However, there were significant differences in the diet composition among species (PERMANOVA; $p = 0.001$, $\alpha = 0.05$) that significantly differed ($p < \alpha$; $\alpha = 0.05$) in size. Copepods were the most significant prey group in the diet of *A. incognitus* and to a slightly lesser extent of *L. candolii*, for which amphipods, decapods, and isopods were the most significant prey group. For *L. lepadogaster*, amphipods and decapods were the most significant prey groups, which is also confirmed by the IRI (%), which is highest for these two prey groups. The Bray–Curtis dissimilarity index showed that the greatest overlap in diet was between *L. candolii* and *L. lepadogaster* and the least overlap was between *A. incognitus* and *L. lepadogaster* (Table 4), which is confirmed by the IRI (%) of the different prey groups for all three species. In contrast to the diet of *L. lepadogaster*, where decapods were mainly represented only by crabs of the genus *Pisidia*, the diet of *L. candolii* included a large number of *A. nitescens* (38.3% of decapods) in addition to *Pisidia* sp. (55.0% of decapods). This was also observed in the fauna samples under stones, where *A. nitescens* made up 7.8% of the decapods in the lower mediolittoral and 26.5% in the infralittoral, where most of the *L. candolii* were found. The TROPHs index per species was highest in *L. lepadogaster* (3.18 $\pm$ 0.44), followed by *A. incognitus* (3.13 $\pm$ 0.26) and *L. candolii* (3.12 $\pm$ 0.36), while the ITD index was highest in *L. lepadogaster* (0.83), followed by *L. candolii* (0.73) and was lowest in *A. incognitus* (0.51).

Table 4. Bray–Curtis dissimilarity index of three clingfish species.

	L. candolii	*A. incognitus*
A. incognitus	0.445	-
L. lepadogaster	0.391	0.458

3.2. *Fauna in the Clingfish Habitat*

The sampling of prey groups under stones inhabited by *L. lepadogaster* showed that amphipods were one of the most numerous and frequent prey groups in the area, followed by decapods (Table 5). A total of 1364 specimens of different prey groups were found in 20 sampling quadrants (25 $\times$ 25 cm^2) in the lower mediolittoral and upper infralittoral.

Table 5. Occurrence of different prey groups under stones in the habitat of *L. lepadogaster* and *L. candolii*.

		L. lepadogaster		*L. candolii*	
Higher taxon	Lower taxon	N(%)	F(%)	N(%)	F(%)
Mollusca	Gastropoda	3.7	55.0	0.4	30
	Bivalvia	1.4	30.0	2.0	70
Crustacea	Decapoda	11.5	75.0	16.5	100
	Amphipoda	63.1	100.0	9.6	100
	Anisopoda	-	-	0.2	20
	Isopoda	3.8	50.0	1.3	60
	Mysida	2.6	65.0	0.2	10
Pantopoda	Pycnogonida	0.1	5.0	0.1	10
Annelida	Polychaeta	10.9	90.0	56.5	100
Echinodermata	Ophiuroida	2.9	70.0	13.0	100

The sampling of prey groups under stones inhabited by *L. candolii* showed that amphipods, decapods, polychaets and ophiuroids were the most numerous and frequent prey groups in this area. A total of 913 specimens of different prey groups were found in 10 sampling quadrants (25 × 25 cm) in the 2–3 m depth range. Due to the method used to sample the fauna (mesh size of the net), no information could be provided on the abundance and occurrence of copepods and other meiofaunal organisms in the area.

3.3. Fish Observations in the Aquaria

The movements of *L. lepadogaster* were mainly restricted to the underside of the rock. Most of the time they wait in ambush until the prey comes close enough to grab it, or they approach the prey with small jumps, when necessary, but they were never observed leaving the shelter.

L. candolii proved to be much more mobile compared to *L. lepadogaster*. The specimens usually hid by clinging to the underside of a rock, but they were also observed moving around the shelter. *L. candolii* not only waited in ambush for prey but actively searched for it everywhere in the aquarium. Small specimens were observed to hide less, especially at night (observed outside the scheduled observation periods) when they came out of the shelters and picked up copepods from the aquarium walls.

The specimens of *A. incognitus* in the aquarium behaved similarly to *L. candolii*. Most of the time, they hid in oyster shells, but they were also observed to move around. The specimens of *A. incognitus* waited in ambush for prey. When they spotted a potential prey, they approached with small jumps and then snatched it, like *L. candolii*. It was also observed that the clingfish did not attack when the prey was not moving. This behaviour was observed when they were feeding on specimens of the shrimp *Hippolyte* sp.

4. Discussion

4.1. Feeding Habits of Three Clingfish Species

Since all fish survived the faecal collection method, it can be considered non-destructive and suitable for use in MPAs and for protected fish species, as previously stated by Trkov and Lipej [31]. The faecal pellet-based method proved to be very useful and efficient for use with clingfish, as they were obtained from 96.9% of all specimens examined. The specimens of all three species often produce more than one faecal pellet, indicating a high degree of prey filling given their very short digestive tract. This is consistent with Depczynski and Bellwood [10], who found that cryptobenthic fish feed continuously or on energy-rich food. A high proportion (93%) of *L. lepadogaster* specimens that had prey in their gut was also reported by King [29], confirming the observations of Hofrichter [11] that *L. lepadogaster* fed throughout the day. This is also confirmed by observations on specimens of *A. incognitus*, which showed that the animals produce an average of 3.3 faecal pellets per day (one faecal pellet every 7–8 h; D.T. personal observation). This means that the species feed quite frequently and the prey obtained from the food samples accounts for only about one-third of their daily consumption. Various sediment particles (e.g., sand grains) frequently observed in the faecal pellets were also reported by Hofrichter (1993). However, such particles were probably consumed incidentally with the prey and play no role in digestion to grind the food (as in birds, for example; [63]), as the digested prey was not crushed.

Observations in the aquarium show that all studied clingfish species rely mainly on vision to hunt, which is characteristic of many fish that sit in wait for prey [64,65]. Due to the wide range of prey taxa found in the diet of all three clingfishes, they could be considered carnivorous opportunists. This is consistent with the studies of other authors [23,29,66,67] and has also been observed in other clingfish species [68]. Indeed, highly opportunistic foraging for a wide range of small prey is a common characteristic of many cryptobenthic fish species [10,69].

In all three clingfish species, crustaceans were the most important prey group, which is consistent with many published studies [11,15,23,66,67,70–72]. The importance of crustaceans, especially amphipods, copepods, and decapods, in the diet of opportunistic cryptobenthic fish was also emphasised by Brandl et al. [69]. In all three clingfish species, harpacticoid copepods were the most abundant prey group, which has been observed in many fish species that feed on meiofaunal organisms (benthic organisms that pass through a 0.5 mm sieve and are retained on a 45 µm sieve; [73]). A high proportion of relatively small harpacticoid copepods in the diet could be a consequence of the fact that they have a 35% higher caloric value than amphipods and the proportion of successful prey capture is much higher for harpacticoid copepods than for amphipods [74]. However, it is important to consider which species of copepods and amphipods were used in the study. The high caloric content and low catch costs are probably the main reasons for the large proportion of copepods in the diet of clingfish, although their jaw anatomy allows clingfishes to eat much larger prey. The jaw shape studied by Hofrichter [11] seems to be closely related to the choice of food. The jaw is longest in *L. lepadogaster* and least elongated in *A. incognitus*. Since clingfish cannot generate suction pressure, this jaw shape and the ability to open the jaw up to an angle of 60° [11,15] allow *L. lepadogaster* to grasp large prey such as decapod crustaceans. This is also reflected in their relatively high trophic position in relation to their small body size, which is characteristic of cryptobenthic fish [12]. Furthermore, the teeth of clingfish are not designed to crush their prey but only to hold it [11,15]. Consequently, the prey is eaten whole, and due to the rapid digestion, it is often relatively undigested and hard parts remain uncrushed. This is probably the reason why the faecal pellets contain a relatively thick peritrophic membrane that surrounds the prey remains and protects the intestinal walls from damage that could be caused by sharp prey remains [75]. Since the peritrophic membrane is formed in the midgut (D.T. personal observation), this suggests that all digestion takes place there.

The importance of crustaceans as prey for the clingfish is also reflected in the red courtship colouration, which is reflected in *L. lepadogaster* in bright red dorsal, anal, and caudal fins, in *L. candolii* in three to four oblique, bright red stripes on the gill cover and three to five oblique, bright red spots extending from the middle of the back to the middle of the basal part of the dorsal fin, and in *A. incognitus* the head of the male was intensely red [76]. Pigments or their precursors, especially carotenoids (red, orange, and yellow colours), cannot be synthesised but must be ingested with food, which means that colour expression depends on individual feeding success and physiological performance [77]. As carotenoids are limited in availability and one of their main sources is crustaceans (e.g., crabs; [78]), this could suggest that the red colouration of male clingfish is related to their feeding success, which depends on their ability to defend a territory (hiding place) that provides a sufficient number of crustaceans as prey. This is probably related to the habitat choice of mating males of *L. lepadogaster* and *L. candolii*, which are most often found near the intertidal border in the upper infralittoral [20], which is known to be one of the most productive areas [79]. This shows how important crustaceans are as a prey group for the studied clingfish species.

4.1.1. *Lepadogaster lepadogaster*

The diet is consistent with the findings of other researchers. A high proportion of crustaceans, represented by copepods and amphipods, has been observed in the diet of many other tidewater fish species [23]. Amphipods and copepods have also been recognised as one of the most abundant prey groups in the diet of *Gobius paganellus* [66, 80,81], a species frequently observed in the same habitat as *L. lepadogaster* (D.T. personal observation). The great importance of decapods and amphipods in the diet of *L. lepadogaster* was also observed by Wilson [82]. Thus, the high number and abundance of amphipods and decapods in the diet correspond to a high number and abundance of them among stones, indicating the opportunistic feeding habits of *L. lepadogaster*. Despite the high abundance of copepods, their importance (IRI) in the diet was rather low compared to decapods and amphipods, which is due to their negligible biomass. In fact, the average weight of decapods in the diet of *L. lepadogaster* was 1026 times greater than the average weight of copepods. This is consistent with the observations of King [29], who noted that copepods, despite their high abundance, represent only a small total volume of prey, while conversely, decapods, despite their low abundance, represent a large total volume of prey. However, the weights of prey in this study should be interpreted with caution as they were calculated using correlation curves (e.g., Decapoda) or average weights (e.g., Copepoda), so their weight may slightly differ from the actual values.

4.1.2. *Lepadogaster candolii*

Despite their negligible biomass, copepods are the most important prey in terms of IRI due to their high proportion in the diet of *L. candolii*, followed by decapods, isopods, and amphipods. The importance of copepods and amphipods in the diet has also been observed by other authors [23,71,72]. The high proportion of decapods and amphipods in the diet is consistent with their high abundance and frequency of occurrence under the rocks inhabited by *L. candolii*. However, despite the high number of polychaets and ophiuroids under rocks, they were not as abundant in the diet of *L. candolii*. The reason for this could be that polychaets hide in the sediment or, in the case of serpulid polychaets, in tubes, making them less accessible to clingfishes, while ophiuroids do not seem to be a preferred prey, as they were rarely ingested.

The presence of parasitic copepods (family Caligidae), parasitic larvae of Gnathiidae, and fish scales in the diet suggests that *L. candolii* also feeds on fish parasites, which is consistent with the observations of Mazé [67]. These results support the finding that *L. candolii* is an occasional cleaner fish whose cleaning activity in the Mediterranean was first observed by Weitzmann and Mercader [83]. Among the Mediterranean clingfish species, cleaning behaviour was also observed in *Diplecogaster bimaculata* [84]. In a few cases, parasitic larvae of Gnathiidae were also found in the diet of *L. lepadogaster*. In addition, they were observed as parasites on specimens of *L. lepadogaster* and *L. candolii*, while parasitic copepods were also observed in the latter species (D.T. personal observation) Therefore, it is possible that there is an intraspecific cleaning behaviour in clingfish.

4.1.3. *Apletodon incognitus*

This species has only recently been described [11,21] and this is the first detailed insight into the diet of this species. It is a small species, mostly associated with *P. nobilis* shells [20], where it feeds mainly on small crustaceans, of which harpacticoid copepods are by far the most important. Harpacticoid copepods have also been recognised as the most important food for the similar small stenotopic clingfish *Opeatogenys gracilis* (Canestrini, 1864), living on seagrass leaves [85], where we did not find any specimen of *A. incognitus*, but such observations have been reported by Hofrichter and Patzner [7].

4.2. Factors Affecting the Diet of Clingfishes

Various factors influence the diet of clingfish. Roughly, the factors can be divided into external environmental factors (e.g., food availability, season, habitat) and internal factors related to the fish itself (e.g., size of the fish, home range, behaviour). However, the factors are strongly intertwined and it is difficult to separate them completely. For example, the diet of clingfish can be influenced by the season, which in turn is related to the seasonal change in the fish behaviour (e.g., nesting) and the availability of prey (prey settlement and seasonal variability in abundance). In addition, the available food also depends on the depth and the habitat.

4.2.1. Fish and Prey Size

Fish size is one of the most important factors in food selection [73]. A comparison of the average size of fish species shows that larger species have a larger food niche. This was observed in *L. lepadogaster* as the average largest of these three species with 47 taxa found in its diet, while 42 different prey taxa were recognised in the diet of the slightly smaller *L. candolii*, and only 23 different taxa were observed in *A. incognitus* as the smallest species. This was reflected also in the different ITD values among species (Table 3). These results are in line with previous findings of other authors [15,73,86], who emphasised that a larger body size allows fish to feed on a wider range of prey. However, the composition of the diet depends not only on the size of the fish but also on the energetic benefits derived from hunting certain prey [73]. The differences in prey composition and size with increasing fish size are related to the optimisation of energy intake for growth [87]. Therefore, larger predatory fish eat larger prey [73]. However, if they eat larger prey, the fish have to expend less for the same amount of food. Consequently, eating larger prey also leads to a lower average number of prey ingested, as observed in *L. lepadogaster*, which ingested the lowest average number of prey per fish (7.1) compared to *L. candolii* (14.8) and *A. incognitus* (16.6). In general, smaller specimens (smaller size classes) of *L. lepadogaster* and *L. candolii* feed mainly on small prey such as copepods, and with increasing size (larger size classes), the specimens move on to larger prey such as decapod crustaceans. This is also reflected in the TROPHs index, which increases with the size of the fish species and with the size of the specimens (size classes) within the species, which is consistent with the observations of Hayden et al. [12] in other cryptobenthic species. Small animals, such as copepods, are most likely to be eaten by young fish stages or those of small adult size [73]. Meiofauna (especially harpacticoid copepods) are an important prey group for fish in the size range from 30 to 60 mm, depending on jaw structure [73]. As has been observed, the size of these three clingfishes allows them to still feed efficiently on copepods, which enables them to meet their energy requirements in the absence of larger prey, especially during the breeding season. In the smaller *A. incognitus*, copepods were the most important prey even at the adult stage, and a switch to larger prey was not observed. This is probably due to the small size of the adults, which allows efficient utilisation of smaller prey, while it could be also connected to the nesting behaviour.

4.2.2. Home Range and Behaviour

The limited distribution range of cryptobenthic fish species, which naturally restricts access to prey [10], could explain their opportunistic diet. Species with a smaller home range are therefore more likely to feed opportunistically on a wide range of prey than species with a large distribution range. This is confirmed by the ITD value, since the dietary diversity indicates the degree of opportunistic feeding, which was highest in *L. lepadogaster*. This species lives the most cryptically of all three species, has the smallest home range, and is highly territorial [18]. Based on the underside of the stone [20], *L. candolii* had a larger home range and also fed outside the shelter [88,89], which is reflected in a less opportunistic diet (ITD = 0.73). *A. incognitus* was the species with the least cryptobenthic lifestyle (most often found outside the shelter [20]), so it is considered to have the largest home range of all three species, which is reflected in the least opportunistic diet of all three species

(ITD = 0.51). However, it is not clear which has a stronger effect on the composition of dietary diversity in these three species, the home range or the fish size, as both explain the results well.

The behaviour of the fish also proved to be an important factor influencing the composition of the fish diet. The aquarium observations showed that *L. lepadogaster* moved mainly inside the space below the stone, while *L. candolii* and *A. incognitus* were much more mobile and moved outside the shelter (e.g., on the tops of the stones). This feeding behaviour was also reflected in the composition of the diet, as the specimens of *L. candolii* and *A. incognitus* had more prey in their diet that normally occur on the top of stones on algae (e.g., Cumacea and Caprellidae) than *L. lepadogaster*.

Nesting behaviour, which is related to the time of year, also affects the diet of the fish. During the breeding season, the males take care of the nest, so their movements are limited to the nest [15]. Consequently, they do not move around in search of food but are mainly dependent on prey that comes or is carried into the immediate vicinity of the nest by the water current. The high proportion of copepods in the males' diet could therefore be related to their territorial behaviour and nest guarding. The higher proportion of copepods in the diet of males was observed in all three species, although they were larger than the females and would be expected to feed more on larger prey such as decapods. The importance of copepods in the diet of *L. lepadogaster* males was observed especially during the mating season (spring), when copepods were the most significant prey, while outside the mating season, copepods were much less important. It should be noted that infralittoral harpacticoid copepods are most abundant in the Gulf of Trieste in summer [90], but there are no data on copepods from the mediolittoral. Furthermore, copepods were found to play an important role in the diet of nesting males of *A. incognitus*, which is probably related to their hiding mode, as larger individuals (which were mostly nesting males) usually hide in oyster shells attached to the shells of *P. nobilis* and therefore rely mainly on prey that happens to swim/come into the oyster shells (e.g., copepods). This is also reflected in the less diverse diet and lower TROPHs index of larger specimens (mostly nesting males) of *A. incognitus* compared to smaller ones. This type of diet is possible due to the extensive diurnal migration of harpacticoid copepods between sediment and grass blades [91,92], which thus enter empty oysters on *P. nobilis* in seagrass beds. In addition, *A. incognitus* nests in summer, when infralittoral harpacticoid copepods are most abundant in the Gulf of Trieste [90]. However, feeding the males with copepods may result in larger prey (e.g., decapods) being available to the females during the mating season, as these require more energy for egg production.

4.2.3. Food Availability

Depth and Habitat

The differences in the diet of clingfish depend on their selectivity and on the availability of different prey. The availability of different prey depends on the bottom depth and the associated occurrence of different habitats. The differences in diet composition as a function of depth were most evident in *L. lepadogaster* and *L. candolii*. In both species, the largest individuals were observed to be most abundant just below the tide line [20]. This rocky upper infralittoral is also known as a habitat for brown algal forests of the genus *Cystoseira* [93], which is among the most productive habitats [94] and provides niches for various species of invertebrates such as molluscs and crustaceans [93,95,96], which are potential food for clingfish. However, the presence of large specimens of both species in this area is probably related to the availability of decapods, which were their main prey. Based on the identification characteristics of the prey [49], the decapods belonged to the genus *Pisidia*, more specifically to *P. bluteli* or *P. longimana*. These two species, together with *A. nitescens*, make up most of the decapods in the diet of *L. lepadogaster and L. candolii*. In addition, crabs of the genus *Pisidia* are also known to be important food for many other fish species [97]. *P. bluteli* and *P. longimana* are mainly found in the upper infralittoral near the tide line, where they hide under stones [46,98], while *A. nitescens* occurs on a mixture of sediment, phytal, and hard bottom [46], being most abundant on pebbles in the upper infralittoral [99]. This is consistent with our results on fauna sampling under stones, which show that *A. nitescens* is more abundant in the habitat of *L. candolii*, which is also reflected in the composition of the diet. In fact, *L. candolii* had a higher proportion of *A. nitescens* in its diet than *L. lepadogaster*. Thus, the high proportion of decapods in the diet of *L. lepadogaster* and *L. candolii* in the lower mediolittoral and upper infralittoral is consistent with the presence of decapods in this habitat, while the high proportion of copepods in the diet of both species away from the tide line is related to the presence of smaller specimens and alternative prey at these depths. In addition, the large number of copepods in the diet of *L. candolii* from deeper waters may be related to the fact that the sand deeper on the seabed replaces the pebbles. Consequently, the proportion of decapods that prefer pebbles as substrate in the diet of *L. candolii* decreases with depth, while the proportion of harpacticoid copepods in the diet that live in/on sand [91] increases with depth. These results suggest that the preferred prey determines the presence and abundance of clingfish species, as well as the specimens of a certain size. This is not surprising, as cryptobenthic fish appear to have a higher mass-specific metabolic rate than larger fish species due to their small body size [69]. Consequently, energy consumption in small fish is very high and they barely tolerate periods of starvation or reduced food intake [100]. Small fish are therefore dependent on constant feeding with high-quality food (e.g., crustaceans) [69], the exploitation of which is limited by the size of the prey that the fish can catch and ingest [101].

Season

Seasonal differences were found in the occurrence of certain prey in the diet of *L. lepadogaster*, whereas this factor was not tested in *L. candolii* and *A. incognitus*, due to the uneven distribution of the fish sample over the seasons. The summer peak of decapods in the diet of *L. lepadogaster* coincides with a high abundance of decapod larvae in summer in temperate latitudes [102]. No reproductive data are available for *P. bluteli* and *P. longimana*, whereas the closely related species *P. longicornis*, which lives in somewhat deeper waters, is known to have a summer larval colonisation. In addition, the highest mortality was observed in this species in summer [97]. The newly settled *Pisidia* are most numerous at the end of summer and the beginning of fall [97], which explains their high abundance and frequency in the diet of clingfish at this time. This is also confirmed by the results showing that specimens of *Pisidia* sp. were on average the smallest in the diet of *L. lepadogaster* in summer and especially in autumn. Since *L. lepadogaster* and *Pisidia* sp. are low-mobility species [15,97], this probably means that when the fish eat most of the *Pisidia* sp. in their hiding places, the abundance and frequency of *Pisidia* sp. in the diet consequently decreases, which was observed in winter and spring. In times of shortages of suitable decapods, *L. lepadogaster* increasingly feeds on other prey groups (e.g., snails and copepods). Such a shift in diet from decapods and isopods in the warmer season to molluscs in the colder season was also observed by Compaire et al. [103]. Furthermore, this is consistent with Zander and Hagemann [104], who reported that fish feed on alternative prey in spring when there is a shortage of benthic crustaceans of suitable size. However, the decline of decapods in the diet of *L. lepadogaster* may also be due to decapods reaching a size at which they are difficult or impossible for fish to eat; however, this is less likely for *P. bluteli* and *P. longimana* with a relatively small size at which they can still be eaten by adult *L. lepadogaster* specimens.

Fish eggs' presence in the diet is also related to the season and nesting behaviour. The feeding on fish eggs by *L. candolii* and *L. lepadogaster* was previously observed by Almada et al. [88] and Hofrichter [11]. Fish eggs in the diet could be the result of predation on other fish nests, the removal of the remains of their own hatched eggs, or, most likely, cannibalism. Cannibalism can occur when individuals prey on the nests of other males or when nesting males prey on their own eggs due to the shortage of food during the nesting season (small amount of potential prey in the immediate vicinity of the nest). Guarding males therefore sometimes eat some or all of their own eggs to keep themselves alive during the nesting season [105]. This cannibalism was particularly observed in *A. incognitus*, as fish eggs were usually found in the faecal pellets of the nesting males. Such consumption of eggs from their own spawn was also observed in guarding males of *Gobius auratus* Risso, 1810 [104]. Another interesting prey item ingested by nesting males of *A. incognitus* was serpulid polychaets, which were frequently observed in occupied oyster shells. This could indicate that serpulid polychaets are also an alternative prey during the mating season when the males take care of the nest and their movements are restricted to the oyster shell. Fish feeding with polychaets at a time when there is a lack of more suitable food (in spring) has been observed previously by Zander and Hagemann [104].

5. Conclusions

The faecal pellet-based method proved to be efficient in its application to clingfish, as all specimens survived the method and faecal pellets were obtained from 96.9% of all specimens examined.

Based on their diet, all three clingfishes *Lepadogaster lepadogaster*, *L. candolii*, and *Apletodon incognitus* can be considered carnivorous opportunists that feed mainly on crustaceans, with amphipods, copepods, and decapods being the most important taxa. The greatest overlap in diet occurred between *L. candolii* and *L. lepadogaster*, and the least between *A. incognitus* and *L. lepadogaster*.

In all three species, differences were found in the diet of specimens of different size classes, while in *L. lepadogaster* and *L. candolii*, differences were also observed in the diet of specimens from different bottom depths. Differences in diet between the sexes were only observed in *A. incognitus*, while seasonal differences in diet were confirmed in *L. lepadogaster*. Diet composition depends on the species of clingfish, the size of the specimens and the size of the prey they can eat, the behaviour of the fish and the associated home range of the specimens, and also on the availability of food, which depends on habitat, depth, and season.

The presence of certain crustacean taxa in the environment determines the occurrence of clingfish of different species and sizes.

The study of fauna as potential prey and observation in the aquarium have both proven to be very useful methods and considerably complement the results of diet composition in order to obtain a comprehensive understanding of the dietary habits of the species.

Supplementary Materials: The following supporting information can be downloaded at: https://www.mdpi.com/article/10.3390/ani14192835/s1, Table S1: Diet composition of *Lepadogaster lepadogaster*, *L. candolii*, and *Apletodon incognitus*.

Author Contributions: Conceptualisation, D.T. and L.L.; methodology, D.T. and L.L.; software, D.I.; validation, D.T., D.I., M.K. and L.L.; formal analysis, D.T. and D.I.; investigation, D.T. and L.L.; resources, D.T. and L.L.; data curation, D.T. and D.I.; writing—original draft preparation, D.T.; writing—review and editing, D.T., D.I., M.K. and L.L.; visualisation, D.T.; supervision, L.L.; project administration, D.T.; funding acquisition, L.L. All authors have read and agreed to the published version of the manuscript.

Funding: This research was funded by the Slovenian Research Agency, grant number P1-0237.

Institutional Review Board Statement: This study was conducted at the Marine Biology Station Piran of the National Institute of Biology of Slovenia, where researchers have a state permit to study fish. In addition, the sucker fish species *Apletodon incognitus*, *Lepadogaster lepadogaster* and *Lepadogaster candolii* are not protected according to national legislation in Slovenia (Decree on protected wild animal species; Official Journal of RS, no. 46/04, 109/04, 84/05, 115/07, 32/08—odl. US, 96/08, 36/09, 102/11, 15/14, 64/16 in 62/19) and therefore no special permits are required to handle them for research purposes. We confirm that the ethical costs of the research are outweighed by the scientific value of the research. Knowledge of the feeding habits of *A. incognitus*, *L. lepadogaster* and *L. candolii* is important for understanding the dynamics of local ecosystems and contributes to the general increase in knowledge. The authors assert that the present study conforms to the generally accepted "3Rs":

replacing animals with alternatives wherever possible, reducing the number of animals used, and refining experimental conditions and procedures to minimize harm to the animals.

Informed Consent Statement: Not applicable.

Data Availability Statement: Data are contained within the article and Supplementary Materials.

Acknowledgments: We would like to thank Tihomir Makovec and Borut Mavrič for their help and support during the field and laboratory work.

Conflicts of Interest: The authors declare no conflicts of interest.

References

1. Kovačić, M.; Patzner, R.A.; Schliewen, U. A First Quantitative Assessment of the Ecology of Cryptobenthic Fishes in the Mediterranean Sea. *Mar. Biol.* **2012**, *159*, 2731–2742. [CrossRef]
2. Henriques, M.; Lourenco, R.; Almada, F.; Calado, G.; Goncalves, D.; Guillemaud, T.; Cancela, M.L.; Almada, V.C. A Revision of the Status of *Lepadogaster lepadogaster* (Teleostei: Gobiesocidae): Sympatric Subspecies or a Long Misunderstood Blend of Species? *Biol. J. Linn. Soc.* **2002**, *76*, 327–338. [CrossRef]
3. Smith-Vaniz, W.F.; Jelks, H.L.; Rocha, L.A. Relevance of Crypticfishes in the Biodiversity Assessments: A Case Study at Buck Island Reef National Monument, St. Croix. *Bull. Mar. Sci.* **2006**, *79*, 17–48.
4. Kovačić, M.; Glavičić, I.; Paliska, D.; Soldo, A.; Valić, Z. A Comparison of Methods of Visual Census and Cryptobenthic Fish Collecting, an Integrative Approach to the Qualitative and Quantitative Composition of the Mediterranean Temperate Reef Fish Assemblages. *J. Mar. Sci. Eng.* **2024**, *12*, 644. [CrossRef]
5. Miller, P.J. Gobiidae. In *Fishes of the North-Eastern Atlantic and the Mediterranean*; Whitehead, P.J.P., Bauchot, M.-L.H., Nielsen, J.-C., Tortonese, J.E., Eds.; UNESCO: Paris, France, 1986; pp. 1019–1085.
6. Kovačić, M.; Glavičić, I.; Paliska, D.; Valić, Z. A First Qualitative and Quantitative Study of Marine Cave Fish Assemblages of Intracave Cavities. *Estuar. Coast. Shelf Sci.* **2021**, *263*, 107624. [CrossRef]
7. Hofrichter, R.; Patzner, R.A. Habitat and Microhabitat of Mediterranean Clingfishes (Teleostei: Gobiesociformes: Gobiesocidae). *Mar. Ecol.* **2000**, *21*, 41–53. [CrossRef]
8. Ackerman, J.L.; Bellwood, D.R. Reef Fish Assemblages: A Re-Evaluation Using Enclosed Rotenone Stations. *Mar. Ecol. Prog. Ser.* **2000**, *206*, 227–237. [CrossRef]
9. Ackerman, J.L.; Bellwood, D.R. Comparative Efficiency of Clove Oil and Rotenone for Sampling Tropical Reef Fish Assemblages. *J. Fish. Biol.* **2002**, *60*, 893–901. [CrossRef]
10. Depczynski, M.; Bellwood, D.R. The Role of Cryptobenthic Reef Fishes in Coral Reef Trophodynamics. *Mar. Ecol. Prog. Ser.* **2003**, *256*, 183–191. [CrossRef]
11. Hofrichter, R. Taxonomie, Verbreitung Und Ökologie von Schildfischen Der Unterfamilie Lepadogastrinae (*Gobiesocidae, Teleostei*). Ph.D. Thesis, Naturwissenschaftliche Fakultät, Paris Lodron Universität Salzburg, Salzburg, Austria, 1995.
12. Hayden, B.; Kovačić, M.; Kirinčić, M.; Marčić, Z. Comparative Trophic Ecology of Microhabitat-associated Guilds of Reef Fishes in the Adriatic Sea. *J. Fish. Biol.* **2022**, 1–15. [CrossRef]
13. Patzner, R.A. Habitat Utilization and Depth Distribution of Small Cryptobenthic Fishes (Blenniidae, Gobiesocidae, Gobiidae, Tripterygiidae) in Ibiza (Western Mediterranean Sea). *Environ. Biol. Fishes* **1999**, *55*, 207–214. [CrossRef]
14. Brandl, S.J.; Wagner, M.; Hofrichter, R.; Patzner, R.A. First Record of the Clingfish *Apletodon dentatus* (*Gobiesocidae*) in the Adriatic Sea and a Description of a Simple Method to Collect Clingfishes. *Bull. Fish. Biol.* **2011**, *13*, 65–69.
15. Hofrichter, R. *Beitrag Zur Kenntnis Der Mediterranen Schildfische (Teleostei, Gobiesocidae) Mit Besonderer Berücksichtigung Der Fortpflanzung von Lepadogaster lepadogaster*; Naturwissenschaftlichen Fakultät der Universität Salzburg: Salzburg, Austria, 1993.
16. Briggs, J.C. A Monograph of the Clingfishes (Order Xenopterygii). *Stanf. Lchth. Bull.* **1955**, *6*, 1–244.
17. Harder, W. *Anatomy of Fishes. Part 2: Figures and Plates*; E. Schweizerbart'sche Verlagsbuchhandlung (Nagele u. Obermiller): Stuttgart, Germany, 1975.
18. Trkov, D.; Mavrič, B.; Orlando-Bonaca, M.; Lipej, L. Marine Cryptobenthic Fish Fauna of Slovenia (Northern Adriatic Sea). *Ann. Ser. Hist. Nat.* **2019**, *29*, 59–72. [CrossRef]
19. Lipej, L.; Orlando-Bonaca, M.; Richter, M. New Contributions to the Marine Coastal Fish Fauna of Slovenia. *Ann. Ser. Hist. Nat.* **2005**, *15*, 165–172.
20. Trkov, D.; Ivajnšič, D.; Kovačić, M.; Lipej, L. Factors Influencing Habitat Selection of Three Cryptobenthic Clingfish Species in the Shallow North Adriatic Sea. *J. Mar. Sci. Eng.* **2021**, *9*, 789. [CrossRef]
21. Hofrichter, R.; Patzner, R.A. A New Species of *Apletodon* from the Mediterranean Sea and the Eastern Atlantic with Notes on the Differentiation between *Apletodon* and *Diplecogaster* Species (Pisces: Teleostei: Gobiesociformes: Gobiesocidae). *Senckenb Biol.* **1997**, *77*, 15–22.
22. Bilecenoglu, M.; Kaya, M. The Occurrence of *Apletodon incognitus* Hofrichter & Patzner, 1997 (*Gobiesocidae*) in the Eastern Mediterranean Sea. *Acta Ichthyol. Piscat.* **2006**, *36*, 143–145.
23. Velasco, E.M.; Gómez-Cama, M.C.; Hernando, J.A.; Soriguer, M.C. Trophic Relationships in an Intertidal Rockpool Fish Assemblage in the Gulf of Cádiz (NE Atlantic). *J. Mar. Syst.* **2010**, *80*, 248–252. [CrossRef]

24. Sánchez-Hernández, J.; Servia, M.J.; Vieira-Lanero, R.; Cobo, F. Ontogenetic Dietary Shifts in a Predatory Freshwater Fish Species: The Brown Trout as an Example of a Dynamic Fish Species. In *New Advances and Contributions to Fish Biology*; Türker, H., Ed.; InTech: Rijeka, Croatia, 2013; pp. 271–298.
25. Pauly, D.; Froese, R. Trophic Levels of Fishes. In *FishBase 2000: Concepts, Design and Data Sources*; Pauly, D., Froese, R., Eds.; ICLARM: Manila, Philippines, 2000; p. 127.
26. Kamler, J.F.; Pope, K.L. Nonlethal Methods of Examining Fish Stomach Contents. *Rev. Fish. Sci.* **2001**, *9*, 1–11. [CrossRef]
27. Hartleb, C.F.; Moring, J.R. An Improved Gastric Lavage Device for Removing Stomach Contents from Live Fish. *Fish. Res.* **1995**, *24*, 261–265. [CrossRef]
28. Light, R.W.; Adler, P.H.; Arnold, D.E. Evaluation of Gastric Lavage for Stomach Analyses. *N. Am. J. Fish. Manag.* **1983**, *3*, 81–85. [CrossRef]
29. King, P.A. Littoral and Benthic Investigations on the West Coast of Ireland: XXII, The Biology of a Population of Shore Clingfish *Lepadogaster lepadogaster* (Bonnaterre, 1788) at Inishbofin, Co. Galway. *Proc. R Ir. Acad. B* **1989**, *89*, 47–58.
30. Vidmar, B.; Turk, R. Marine Protected Areas in Slovenia: How Far Are We from the 2012/2020 Target? *Varst. Narave* **2011**, *1*, 159–170.
31. Trkov, D.; Lipej, L. A Non-Destructive Method for Assessing the Feeding Habits of Coastal Fish. *Mediterr. Mar. Sci.* **2019**, *20*, 453–459. [CrossRef]
32. Ogorelec, B.; Mišič, M.; Faganeli, J. Marine Geology of the Gulf of Trieste (Northern Adriatic): Sedimentological Aspects. *Mar. Geol.* **1991**, *99*, 79–92. [CrossRef]
33. Lipej, L.; Turk, R.; Makovec, T. *Ogrožene Vrste in Habitatni Tipi v Slovenskem Morju*; Zavod Republike Slovenije za Varstvo Narave: Ljubljana, Slovenia, 2006.
34. Lipej, L.; Ivajnšič, D.; Makovec, T.; Mavrič, B.; Šiško, M.; Trkov, D.; Orlando-Bonaca, M. *Raziskava z Oceno Stanja Morskih Travnikov v Krajinskem Parku Strunjan*; Morska Biološka Postaja, Nacionalni Inštitut za Biologijo: Piran, Slovenia, 2018.
35. Malačič, V.; Viezzoli, D.; Cushman-Roisin, B. Tidal Dynamics in the Northern Adriatic Sea. *J. Geophys. Res. Oceans* **2000**, *105*, 26265–26280. [CrossRef]
36. Ličer, M.; Fettich, A.; Jeromel, M. *Prognozirano Plimovanje Morja Tide Tables 2018*; Ministry of the Environment and Spatial Planning, Slovenian Environment Agency: Ljubljana, Slovenia, 2018.
37. Ličer, M.; Fettich, A.; Jeromel, M. *Prognozirano Plimovanje Morja Tide Tables 2017*; Ministry of the Environment and Spatial Planning, Slovenian Environment Agency: Ljubljana, Slovenia, 2017.
38. Ličer, M.; Fettich, A.; Jeromel, M. *Prognozirano Plimovanje Morja Tide Tables 2019*; Ministry of the Environment and Spatial Planning, Slovenian Environment Agency: Ljubljana, Slovenia, 2019.
39. Mozetič, P. Seasonal and Inter-Annual Plankton Variability in the Gulf of Trieste (Northern Adriatic). *ICES J. Mar. Sci.* **1998**, *55*, 711–722. [CrossRef]
40. Gibson, R.N. The Use of the Anaesthetic Quinaldine in Fish Ecology. *J. Anim. Ecol.* **1967**, *36*, 301. [CrossRef]
41. Patzner, R.A. Sea Urchins as a Hiding Place for Juvenile Benthic Teleosts (*Gobiidae* and *Gobiesocidae*) in the Mediterranean Sea. *Cybium* **1999**, *23*, 93–97.
42. Montesanto, B.; Panayotidis, P. The *Cystoseira* Spp. Communities from the Aegean Sea (NE Mediterranean). *Mediterr. Mar. Sci.* **2001**, *2*, 57–68. [CrossRef]
43. Kersting, D.; Mouloud, B.; Cizmek, H.; Grau, A.; Jimenez, C.; Katsanevakis, S.; Oztürk, B.; Tuncer, S.; Tunesi, L.; Vázquez-Luis, M.; et al. *Pinna nobilis*. The IUCN Red List of Threatened Species 2019. *IUCN Red List. Threat. Species* **2019**, e.T160075998A160081499. [CrossRef]
44. Jardas, I. *Jadranska Ihtiofauna*; Školska knjiga: Zagreb, Croatia, 1996.
45. Wagner, M.; Bračun, S.; Kovačić, M.; Iglésias, S.P.; Sellos, D.Y.; Zogaris, S.; Koblmüller, S. *Lepadogaster purpurea* (Actinopterygii: Gobiesociformes: Gobiesocidae) from the Eastern Mediterranean Sea: Significantly Extended Distribution Range. *Acta Ichthyol. Piscat.* **2017**, *47*, 417–421. [CrossRef]
46. Melzer, R.; Ceseña, F.; Buršić, M.; Lehmann, T.; Mayer, R.; Mavrič, B.; Makovec, T.; Pfannkuchen, M.; McHenry, J.; Heß, M. *Knights, Ballerinas and Invisibles: The Decapod Crustaceans of the Brijuni Marine Protected Area = Vitezovi, Balerine i Nevidljivi: Rakovi Deseteronošci Zaštićenog Morskog Područja Nacionalnog Parka Brijuni*; Javna ustanova Nacionalni park Brijuni: Pula, Croatia, 2019.
47. Falciai, L.; Minervini, R. *Guida Ai Crostacei Decapodi d'Europa*; Franco Muzzio Editore: Rome, Italy, 1992.
48. Riedel, R. *Fauna e Flora Del Mediterraneo*; Franco Muz-zio Editore: Rome, Italy, 2010.
49. Koukouras, A.; Mavidis, M.; Noël, P.Y. The Genus *Pisidia* Leach (*Decapoda, Anomura*) in the Northeastern Atlantic Ocean and the Mediterranean Sea. *Crustaceana* **2002**, *75*, 451–463. [CrossRef]
50. Hynes, H.B.N. The Food of Fresh-Water Sticklebacks (*Gasterosteus aculeatus* and *Pygosteus pungitius*), with a Review of Methods Used in Studies of the Food of Fishes. *J. Anim. Ecol.* **1950**, *19*, 36. [CrossRef]
51. Macdonald, J.S.; Green, R.H. Redundancy of Variables Used to Describe Importance of Prey Species in Fish Diets (Bay of Fundy). *Can. J. Fish. Aquat. Sci.* **1983**, *40*, 635–637. [CrossRef]
52. Pinkas, L.M.; Oliphant, S.; Iverson, I.L.K. Food Habits of Albacore, Bluefin Tuna and Bonito in Californian Waters. *Calif. Fish. Game* **1971**, *152*, 1–105.
53. Simenstad, C.A. Fish Food Habits Analysis. *Principal Investigators Report. Environ. Assess. Alaskan Cont. Shelf* **1979**, *4*, 441–450.

54. Cortés, E. A Critical Review of Methods of Studying Fish Feeding Based on Analysis of Stomach Contents: Application to Elasmobranch Fishes. *Can. J. Fish. Aquat. Sci.* **1997**, *54*, 726–738. [CrossRef]
55. Shannon, C.E.; Weaver, W. *The Mathematical Theory of Communication*; University of Illinois Press: Chicago, IL, USA, 1963.
56. Pauly, D.; Froese, R.; Sa-a, P.S.; Palomares, M.L.; Christensen, V.; Rius, J. *TrophLab Manual 2000*; ICLARM: Manila, Philippines, 2000.
57. Pauly, D.; Palomares, M.L. Approaches for Dealing with Three Sources of Bias When Studying the Fishing down Marine Food Web Phenomenon. In *Fishing down the Mediterranean Food Webs?* Briand, F., Ed.; CIESM Workshop Series: Kerkyra, Greece, 2000; pp. 61–66.
58. Bray, J.R.; Curtis, J.T. An Ordination of the Upland Forest Communities of Southern Wisconsin. *Ecol. Monogr.* **1957**, *27*, 325–349. [CrossRef]
59. Ricotta, C.; Podani, J. On Some Properties of the Bray-Curtis Dissimilarity and Their Ecological Meaning. *Ecol. Complex.* **2017**, *31*, 201–205. [CrossRef]
60. Clarke, K.R. Non-Parametric Multivariate Analyses of Changes in Community Structure. *Austral Ecol.* **1993**, *18*, 117–143. [CrossRef]
61. R Core Team. *R: A Language and Environment for Statistical Computing*; R Foundation for Statistical Computing: Vienna, Austria, 2016; Available online: https://www.R-project.org/ (accessed on 25 September 2024).
62. Oksanen, J.; Blanchet, F.G.; Kindt, R.; Legendre, P.; Minchin, P.R.; O'hara, R.B.; Simpson, G.L.; Solymos, P.; Stevens, M.H.H.; Wagner, H. *Vegan: Community Ecology Package*; R Package Version 2.6-2; R Foundation for Statistical Computing: Vienna, Austria, 2022. Available online: https://cran.r-project.org/web/packages/vegan/index.html (accessed on 25 September 2024).
63. Gionfriddo, J.P.; Best, L.B. Grit Use by Birds. In *Current Ornithology*; Springer US: Boston, MA, USA, 1999; pp. 89–148.
64. Hogue, E.W.; Carey, A.G. Feeding Ecology of 0-Age Flatfish at a Nursery Ground on the Oregon Coast. *Fish. Bull.* **1982**, *80*, 555–564.
65. Magnhagen, C.; Wiederholm, A.-M. Parasitic Spawning and Paternity Assurance in Gobiidae View Project The FiRe Research Programme (1994–1998) on Reproductive Disturbances in Baltic Sea Fish View Project Oecologia (Berl) Food Selectivity Versus Prey Availability: A Study Using the Marine Fish *Pomatoschistus microps*. *Oecologia* **1982**, *55*, 311–315. [CrossRef] [PubMed]
66. Gibson, R.N. The Vertical Distribution and Feeding Relationships of Intertidal Fish on the Atlantic Coast of France. *J. Anim. Ecol.* **1972**, *41*, 189–207. [CrossRef]
67. Mazé, R.A. Estudio de La Dieta de Las Poblaciones Intermareales de Dos Especies de *Lepadogaster* (*Teleostei, Gobiesocidae*) En La Costa Cantábrica, España. *Boletín Real. Soc. Española Hist. Nat. Sección Biológica* **2007**, *102*, 85–92.
68. Paine, R.T.; Palmer, A.R. *Syciases sanguineus*: A Unique Trophic Generalist from the Chilean Intertidal Zone. *Copeia* **1978**, *1*, 75–81. [CrossRef]
69. Brandl, S.J.; Goatley, C.H.R.; Bellwood, D.R.; Tornabene, L. The Hidden Half: Ecology and Evolution of Cryptobenthic Fishes on Coral Reefs. *Biol. Rev.* **2018**, *93*, 1846–1873. [CrossRef]
70. Day, F. *The Fishes of Britain and Ireland*; Williams and Norgate: London, UK, 1884; Volume 2.
71. Dunne, J. Gobiesocidae Occuring in the Coastal Waters of Connemara. *Irish Fish. Investig. (Ser. A)* **1983**, *23*, 32–36.
72. Gibson, R.N. The Food and Feeding Relationships of Littoral Fish in the Banyuls Region. *Via Milieu A* **1968**, *19*, 447–456.
73. Gee, J.M. An Ecological and Economic Review of Meiofauna as Food for Fish. *Zool. J. Linn. Soc.* **1989**, *96*, 243–261. [CrossRef]
74. Volk, E.C.; Wissmar, R.C.; Simenstad, C.A.; Eggers, D.M. Relationship between Otolith Microstructure and the Growth of Juvenile Chum Salmon (*Oncorhynchus keta*) under Different Prey Rations. *Can. J. Fish. Aquat. Sci.* **1984**, *41*, 126–133. [CrossRef]
75. Stobbs, R.E. Feeding Habits of the Giant Clingfish *Chorisochismus dentex* (Pisces: Gobiesocidae). *South Afr. J. Zool.* **1980**, *15*, 146–149. [CrossRef]
76. Trkov, D. *Ecology of Mediterranean Cryptobenthic Fish Fauna: Lessons Learnt from Clingfishes (*Gobiesocidae*)*; Jožef Stefan International Postgraduate School: Ljubljana, Slovenia, 2020.
77. Erisman, B.E.; Allen, L.G. Color Patterns and Associated Behaviors in the Kelp Bass, *Paralabrax clathratus* (Teleostei: Serranidae). *Bull. South Calif. Acad. Sci.* **2005**, *104*, 45–62. [CrossRef]
78. García-Chavarría, M.; Lara-Flores, M. The Use of Carotenoid in Aquaculture. *Res. J. Fish. Hydrobiol.* **2013**, *8*, 38–49.
79. Cabanellas-Reboredo, M.; Vázquez-Luis, M.; Mourre, B.; Álvarez, E.; Deudero, S.; Amores, Á.; Addis, P.; Ballesteros, E.; Barrajón, A.; Coppa, S.; et al. Tracking a Mass Mortality Outbreak of Pen Shell *Pinna nobilis* Populations: A Collaborative Effort of Scientists and Citizens. *Sci. Rep.* **2019**, *9*, 1–11. [CrossRef]
80. Azevedo, J.M.N.; Simas, A.M.V. Age and Growth, Reproduction and Diet of a Sublittoral Population of the Rock Goby *Gobius paganellus* (Teleostei, Gobiidae). *Hydrobiologia* **2000**, *440*, 129–135. [CrossRef]
81. Dunne, J. Littoral and Benthic Investigations on the West Coast of Ireland: IX. Section A (Faunistic and Ecological Studies). The Biology of the Rock Goby, *Gobius paganellus* L., at Carna. *Proc. R. Ir. Acad. B* **1978**, *78*, 179–191.
82. Wilson, J.P.F. A Note on the Biology of the Cornish Clingfish *Lepadogaster lepadogaster* (Bonnaterre). *Ir. Nat. J.* **1981**, *20*, 209–210.
83. Weitzmann, B.; Mercader, L. First Report of Cleaning Activity of *Lepadogaster candolii* (*Gobiesocidae*) in the Mediterranean Sea. *Cybium* **2012**, *36*, 487–488.
84. Patzner, R.A.; Debelius, H. *Partnerschaft Im Meer*; Engelbert Pfriem Verlag: Wuppertal, Germany, 1984.
85. Vizzini, S.; Mazzola, A.; Scilipoti, D. Notes on the Biology and Ecology of *Opeatogenys gracilis* (Canestrini 1864) (Pisces: *Gobiesocidae*) from Coastal Environments in Sicily (Mediterranean). In *Mediterranean Ecosystems*; Springer: Milan, Italy, 2001; pp. 221–224.

86. Zander, C.D.; Berg, J. Feeding Ecology of Littoral Gobiid and Blennioid Fishes of the Banyuls Area (Mediterranean Sea). II: Prey Selection and Size Preference. *Vie Milieu* **1984**, *34*, 149–157.
87. Stoner, A.W.; Livingston, R.J. Ontogenetic Patterns in Diet and Feeding Morphology in Sympatric Sparid Fishes from Seagrass Meadows. *Copeia* **1984**, *1984*, 174. [CrossRef]
88. Almada, V.C.; Garcia, G.J.M.; Santos, R.S. Padrões de Actividade e Estrutura Dos Territórios Dos Machos Parentais de *Parablennius pilicornis* Cuvier (Pisces: Blenniidae) Da Costa Portuguesa. *Análise Psicológica* **1987**, *2*, 261–280.
89. Gonçalves, D.M.; Gonçalves, E.J.; Almada, V.C.; Almeida, S.P. Comparative Behaviour of Two Species of *Lepadogaster* (Pisces: *Gobiesocidae*) Living at Different Depth. *J. Fish. Biol.* **1998**, *53*, 447–450. [CrossRef]
90. Vrišer, B. Sezonska Dinamika in Variabilnost Harpaktikoidov (Copepoda-Harpacticoida) v Tržaškem Zalivu: Triletna Raziskava. *Ann. Ser. Hist. Nat.* **1996**, *9*, 53–60.
91. Hicks, G.R.F. Distribution and Behaviour of Meiofaunal Copepods Inside and Outside Seagrass Beds. *Mar. Ecol. Prog. Ser.* **1986**, *31*, 159–170. [CrossRef]
92. Walters, K.; Bell, S.S. Diel Patterns of Active Vertical Migration in Seagrass Meiolauna. *Mar. Ecol. Prog. Ser.* **1986**, *34*, 95–103. [CrossRef]
93. Orlando-Bonaca, M.; Trkov, D.; Klun, K.; Pitacco, V. Diversity of Molluscan Assemblage in Relation to Biotic and Abiotic Variables in Brown Algal Forests. *Plants* **2022**, *11*, 2131. [CrossRef]
94. Ballesteros, E. Production of Seaweeds in Northwestern Mediterranean Marine Communities: Its Relation with Environmental Factors. *Sci. Mar.* **1989**, *2–3*, 357–364.
95. Pitacco, V.; Orlando-Bonaca, M.; Mavrič, B.; Popović, A.; Lipej, L. Mollusc Fauna Associated with the *Cystoseira* Algal Associations in the Gulf of Trieste (Northern Adriatic Sea). *Mediterr. Mar. Sci.* **2014**, *15*, 225. [CrossRef]
96. Piazzi, L.; Bonaviri, C.; Castelli, A.; Ceccherelli, G.; Costa, G.; Curini-Galletti, M.; Langeneck, J.; Manconi, R.; Montefalcone, M.; Pipitone, C.; et al. Biodiversity in Canopy Forming Algae: Structure and Spatial Variability of the Mediterranean *Cystoseira* Assemblages. *Estuar. Coast. Shelf Sci.* **2018**, *207*, 132–141. [CrossRef]
97. Robinson, M.; Tully, O. Dynamics of a Subtidal Population of the Porcellanid Crab *Pisidia longicornis* (Decapoda: *Crustacea*). *J. Mar. Biol. Assoc. United Kingd.* **2000**, *80*, 75–83. [CrossRef]
98. Holthuis, L.B. Report on a Collection of Crustacea Decapoda and Stomatopoda from Turkey and the Balkans. *Zool. Verh. Leiden.* **1961**, *47*, 1–67.
99. Pallas, A.; Garcia-Calvo, B.; Corgos, A.; Bernardez, C.; Freire, J. Distribution and Habitat Use Patterns of Benthic Decapod Crustaceans in Shallow Waters: A Comparative Approach. *Mar. Ecol. Prog. Ser.* **2006**, *324*, 173–184. [CrossRef]
100. Miller, P.J. Miniature Vertebrates. In *The Implications of Small Body Size*; Oxford University Press: New York, NY, USA, 1996.
101. Mihalitsis, M.; Bellwood, D.R. A Morphological and Functional Basis for Maximum Prey Size in Piscivorous Fishes. *PLoS ONE* **2017**, *12*, e0184679. [CrossRef]
102. Pfister, C. Demographic Consequences of Within Year Variation in Recruitment. *Mar. Ecol. Prog. Ser.* **1997**, *153*, 229–238. [CrossRef]
103. Compaire, J.C.; Cabrera, R.; Gómez-Cama, C.; Soriguer, M.C. Trophic Relationships, Feeding Habits and Seasonal Dietary Changes in an Intertidal Rockpool Fish Assemblage in the Gulf of Cadiz (NE Atlantic). *J. Mar. Syst.* **2016**, *158*, 165–172. [CrossRef]
104. Zander, C.D.; Hagemann, T. Feeding Ecology of Littoral Gobiid and Blennioid Fish of the Banylus Area (Mediterranean Sea). III. Seasonal Variations. *Sci. Mar.* **1989**, *53*, 441–450.
105. Gross, R.M.; Sargent, R.C. The Evolution of Male and Female Parental Care in Fishes. *Am. Zool.* **1985**, *25*, 807–822. [CrossRef]

animals

MDPI

Article

Influence of Contaminants Mercury and PAHs on Somatic Indexes of the European Hake (*Merluccius merluccius*, L. 1758)

Monica Panfili [1], Stefano Guicciardi o Guizzardi [1], Emanuela Frapiccini [1,*], Cristina Truzzi [2,*], Federico Girolametti [2], Mauro Marini [1], Alberto Santojanni [1], Anna Annibaldi [2], Silvia Illuminati [2] and Sabrina Colella [1]

1 Institute for Marine Biological Resources and Biotechnologies, National Research Council (IRBIM-CNR), 60125 Ancona, Italy; monica.panfili@cnr.it (M.P.); stefano.guicciardioguizzardi@cnr.it (S.G.o.G.); mauro.marini@cnr.it (M.M.); alberto.santojanni@cnr.it (A.S.); sabrina.colella@cnr.it (S.C.)

2 Department of Life and Environmental Sciences, Università Politecnica delle Marche, 60131 Ancona, Italy; f.girolametti@staff.univpm.it (F.G.); a.annibaldi@staff.univpm.it (A.A.); s.illuminati@staff.univpm.it (S.I.)

* Correspondence: emanuela.frapiccini@cnr.it (E.F.); c.truzzi@staff.univpm.it (C.T.)

Simple Summary: Recent awareness highlights the significant impact of contaminants on the Mediterranean marine ecosystem and fishery resources. Monitoring these pollutants is crucial due to their accumulation in marine organisms and the health risks they pose through consumption. This study examines the levels of total mercury and PAHs in the muscle tissue of European hake from an important fishing ground in the Adriatic Sea. Seasonal and gender patterns as well as correlations with somatic indexes were explored to provide cost-effective bioindicators for pollution monitoring and mitigation.

Abstract: This research investigates the dynamics of contaminant exposure in European hake (*Merluccius merluccius*, L. 1758) from the Adriatic Sea (Central Mediterranean Sea) by examining the levels of total mercury (THg) and polycyclic aromatic hydrocarbons (PAHs) in the muscle fish tissues. The study explores the correlations between these pollutants and somatic indexes to identify the early warning signals of pollution and ecological effects. The levels of pollutants are influenced by season and sex. Lipids appear to have a minimal effect on the PAH levels, whereas they exhibit a positive correlation with mercury levels in the muscle. No significant relationships between the pollutants and condition indexes were observed, except for a positive correlation between THg and the gonadosomatic index, indicating a potential impact on the reproductive health of fish. In contrast, PAHs showed no meaningful correlation with condition indexes. Differences in contaminant accumulations and lipid levels between sexes reflect variations in metabolic activity, reproductive costs, and adaptive strategies to seasonal changes and energy demands. This study highlights the importance of long-term monitoring to improve pollution management, environmental conservation, and the protection of marine organisms' health.

Keywords: total mercury; PAHs; *Merluccius merluccius*; lipid content; seasonal variability; condition indexes; Adriatic Sea

1. Introduction

In recent years, there has been increasing awareness of the significant influence of pollutants on the marine ecosystems within the Mediterranean basin, particularly affecting demersal and pelagic fishery resources. This partially enclosed sea, known for its limited water circulation, is one of the most heavily polluted bodies of water globally [1]. These resources are already under significant pressure due to overfishing, habitat degradation, and not uniformly efficient management practices [2–7].

This concern has prompted extensive research into the diverse effects of pollutants, focusing on the long-term preservation of marine resources [8–10].

Aquatic organisms, including fish, accumulate pollutants both directly from contaminated water and indirectly through the food web [7,11,12]. The effects of pollutants on aquatic organisms vary based on the toxicity and concentration of the contaminants [11], with significant variability among individuals, species, and life stages [13–15]. These effects can influence organ functions, reproductive status, and population size, determined by the species' tolerance and survival capabilities [16].

Fish are notable for their ability to metabolize, concentrate, and retain pollutants, making them important bioindicators for evaluating environmental stress and detecting toxicant-induced alterations and degradations over time and space [11,12,17–20]. They also offer insights into human exposure risks through the aquatic food chain [21–26]. Consequently, monitoring pollutant levels in marine organisms is crucial to ensure environmental safety by establishing legislative limits.

The Adriatic Sea, a productive fishing ground and integral part of the Mediterranean, is vulnerable to environmental pollution from human activities like resource exploitation, agricultural runoff, coastal urban development, heavy shipping traffic, and substantial inputs from industrialized Italian rivers [7,27–30]. Its unique oceanographic and geographical characteristics enhance contaminant biomagnification in organisms [31–34]. For these reasons, it is urgent to elucidate the pollution status in the Adriatic Sea, focusing on the accumulation and persistence of contaminants like trace elements and PAHs in the environment and marine organisms, especially in edible fishes.

However, measuring xenobiotic concentrations and implementing monitoring programs can be costly and may not fully reflect their impact on biological systems [35,36]. Moreover, chronic exposure to even low levels of toxic elements can have adverse effects on ecosystems which is often not fully investigated [37,38].

Recent advances in ecotoxicology have extensively employed bioindicators and condition indexes in fish populations as efficient early warning tools for assessing chemical exposure impacts on environmental quality [39–45]. These indexes can help initiate bioremediation strategies before irreversible environmental damage occurs.

Somatic indexes, such as the gonadosomatic index (GSI), hepatosomatic index (HSI), and the Le Cren condition factor (Kn), can be indicative of fish health and describe physiological changes due to pollutant exposure and environmental stressors [46–48]. The GSI provides information about gonadal health and maturation, with significant evidence showing that pollutant exposure can reduce GSI and cause morphological changes in gonads [49,50]. The HSI is a well-known bioindicator of contaminant exposure, as the liver plays a crucial role in detoxification, with pollutants causing an increase in liver size due to hypertrophy (enlargement) or hyperplasia (increased number of liver cells), or both [51–53]. The Kn [54]) is a quantitative indicator of individual well-being, reflecting recent food availability and energy reserves [55,56].

This study investigates GSI, HSI, and Kn in European hake (*Merluccius merluccius*, L. 1758) from the Adriatic Sea, relating them to the levels of mercury (Hg) and polycyclic aromatic hydrocarbon (PAHs) levels recorded in the fish muscle. This approach aims to gain a deeper understanding of the environmental and biological factors influencing pollutant accumulation and provide insights into the physiological responses of European hake to environmental stressors, potentially revealing simple, cost-effective bioindicators suitable for routine pollution monitoring and mitigation for this important species in the Adriatic Sea.

This species is widely distributed across the Atlantic and the Mediterranean and Black Sea and supports important fisheries in several Mediterranean regions, making it a key species for both economic and ecological reasons. However, it has faced significant overfishing, although in recent decades, its critical overexploitation condition has slowly improved, with the landings in the Mediterranean decreasing from 52,394 tonnes in 1994 to 17,824 tonnes in 2021 [57]. In 2022, hake landings in the Mediterranean Sea totaled 18,388 tons with 14% from the northern and central Adriatic [58]. The risk of stock collapse has been highlighted by several scientists [6,59–61], underscoring the need for comprehensive studies on the impacts of pollutants on this species and the broader ecosystem.

Mercury and PAHs exhibit distinct physicochemical properties and interaction patterns that influence their toxicity. Both contaminants partially metabolize and accumulate in lipid-rich tissues like fish muscle, posing potential risks to human health through consumption [62]. Recent research shows that PAH mixtures induce cellular stress and metabolic disruptions in fish, leading to oxidative stress, endocrine issues, and development problems, which affect their survival, growth, and reproduction [9,63–65].

Mercury, even in low concentrations, can damage cells, tissues, proteins, and genes; disrupt physiological processes; and impair behavior and reproductive health [66–70]. It can also trigger energy-consuming detoxification processes, reducing the energy available for growth [70].

The data on Hg levels have already been previously analyzed [71], but in this study, we have expanded on those analyses by examining the relationship between the Hg and PAH levels observed in the muscle tissue of *Merluccius merluccius* and various somatic indexes, such as GSI, HSI, and Kn, to identify the early warning signals of pollution and ecological effects.

2. Materials and Methods

2.1. Study Area and Fish Sampling

European hake (*Merluccius merluccius*) individuals (n = 163) were seasonally sampled by commercial bottom trawlers from May 2018 to January 2019 in a productive offshore fishing ground spanning the Northern and Central Adriatic Sea (FAO-GFCM geographical sub-area 17, Figure 1).

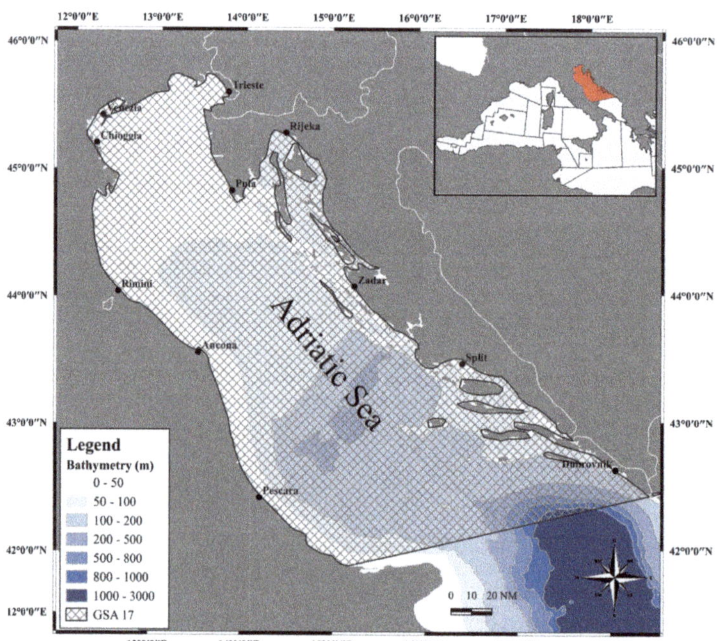

Figure 1. Map of the study area GSA 17, elaborated by I. Costantini (software: QGIS Development Team, 2024. QGIS Geographic Information System. Open Source Geospatial Foundation Project. http://qgis.osgeo.org, accessed on 30 May 2024).

In the laboratory, each individual's body length was measured with a precision of 0.5 cm, total weight was recorded to 0.1 g (Radwag WLC–6F1/K precision scale, Radwang Wagi Elektroniczne, Radom, Poland), and sex was determined [72].

The sampling procedures did not include any animal experimentation and animal ethics approval was, therefore, not necessary under the Italian legislation (D.L. 4 March 2014, n. 26, art. 2). The methodology and sampling of this study were based on dead specimens from professional fishing.

The gonads and liver were weighed to the nearest 0.001 g (Mettler Toledo XP204, analytical precision scale) in order to calculate the somatic indexes, GSI, HSI, and Kn, using the following equations:

$$\text{GSI (\%)} = \text{gonad weight/GW body} \times 100$$

$$\text{HSI (\%)} = \text{liver weight/GW body} \times 100$$

$$\text{Kn} = \text{W}/a\ \text{TL}^{b}$$

where a and b were the regression parameters of the length-weight relationship, GW was the gutted weight, and TL was the total length.

2.2. Laboratory Analyses and Total Mercury Determination

All the analytical procedures were conducted in a clean room laboratory conforming to the ISO 14644-1 Class 6 standards, with certain areas meeting ISO Class 5 requirements under laminar flow. Prior to use, all the laboratory equipment underwent a rigorous acid-cleaning procedure [73]. The specimens were weighed using an AT261 analytical balance (Mettler Toledo, AG Laboratory & Weighing Technologies, Greifensee, Switzerland), readability 0.01 mg, repeatability standard deviation 0.015 mg). Different fish tissues, such as the gonads, liver, and muscle, were dissected using a scalpel that had been previously cleaned with acetone to avoid contaminating the sample.

Variable volume micropipettes and neutral tips from Brand (Transferpette, Wertheim, Germany) and scalpels with sterile stainless-steel blades from Granton (Mod. 91021, Sheffield, UK) were employed. Ultrapure water was obtained from a Milli–Q water system (Merck Millipore in Darmstadt, Germany). The acetone and petroleum ether utilized for lipid extraction were RS grade for pesticide analysis (Carlo Erba, Milano, Italy). Certified reference material for Hg content, specifically dogfish muscle DORM-2 (NRCC, Ottawa, ON, Canada), was utilized.

For Hg determination in fish muscle, approximately 0.05 g of each sample, which was minced and homogenized (homogenizer MZ 4110, DCG Eltronic, Monza, Italy), was directly analyzed via thermal decomposition amalgamation atomic absorption spectrometry (TDA AAS) utilizing a Direct Mercury Analyzer (DMA-1, FKV, Milestone, Sorisole, Italy) [74–77]. The sequential steps followed the protocol indicated by Annibaldi et al., [78]. Detection cell ranges were set between 0.03–200 ng and 200–1500 ng.

Mercury concentrations were quantified using the calibration curve technique. For each specimen, the analysis was performed in triplicate and the THg concentration of a blank was subtracted from the sample measurements to account for potential mercury contamination during the analysis. The use of DORM–2 certified reference material ensured analytical quality control. The mean experimental THg value of 4.47 $\pm$ 0.10 mg kg^{-1} wet weight (ww), closely matching ($p > 0.05$) the certified values (4.43 $\pm$ 0.05 mg kg^{-1} ww), indicated method accuracy and repeatability.

2.3. Polycyclic Aromatic Hydrocarbon (PAH) Extraction and Analysis

The Quick Easy Cheap Effective Rugged and Safe (QuEChER) method was employed to extract and purify PAHs from the muscle of *M. merluccius*. This method represents a simple, rapid, and cost-effective alternative to the conventional extraction methods for multi-residue analysis [79]. By requiring only a few steps (extraction and clean-up), it reduces both time and solvent consumption. The QuEChERS method, originally designed for pesticides, underwent modifications to other persistent organic pollutants, such as PAHs [80].

For PAH analysis, a portion of approximately 5 g of muscle was taken dorsally, directly under the anterior dorsal fin and well above the lateral line, using solvent-cleaned scalpels and scissors. Each sample was wrapped in aluminum foil and stored in a freezer at −18 °C.

The extraction of PAHs using the QuEChERS method was conducted using some kits for the partition of the compounds from the aqueous solution and acetonitrile as solvent extraction. This partitioning process involved the addition of $MgSO_4$ and NaCl. Subsequently, the samples underwent purification through dispersive solid-phase extraction (dSPE) clean-up using C18, $MgSO_4$, and primary secondary amine (PSA). The purified extracts were then evaporated under a gentle flow of N_2 and finally recovered with acetonitrile for chemical analysis. The analysis was conducted using UHPLC (Ultimate 3000, Thermo Scientific, Waltham, MA, USA), which was equipped with a fluorescence detector (RF2000, Thermo Scientific). For separation, a Hypersil Green PAH column (2.1 × 150 mm, 1.8 µm, 120 Å) in a reversed phase was used.

Total PAHs include the sum of the sixteen most environmentally relevant PAHs, which are naphthalene, acenaphthylene, acenaphthene, fluorene, phenanthrene, anthracene, fluoranthene, pyrene, benz(a)anthracene, chrysene, benzo(b)fluoranthene, benzo(k)fluoranthene, benz(a)pyrene, dibenz(a,h)anthracene, benzo(ghi)perylene, and indeno(1,2,3–c,d)pyrene, listed in the US Environmental Protection Agency (EPA) priority pollutant list.

The analysis for quality control assessment involved both the use of an external standard multipoint calibration technique and procedural blanks (n = 6). Calibration curves were performed via serial dilutions ranging from 1:400 to 1:3200 v/v of a standard PAH solution of EPA 610 PAH Mix (Supelco, Bellafonte, PA, USA). The mean recovery rate was calculated at 87% (±5%) and no correction for surrogate recoveries was applied. The limits of detection (LOD) and quantification (LOQ) were calculated following the ICH Q2B guidelines [81] (ICH, 2005), resulting in LODs ranging from 0.001 to 0.01 ppb and LOQs ranging from 0.004 to 0.4 ppb. The concentrations of all the PAH compounds in the blank extract samples remained below the LOQ.

2.4. Lipid Content in Fish Muscle

The muscle tissue from each sample specimen underwent mincing, homogenization, precise weighing, and freeze-drying using an Edwards EF4 Modulyo freeze dryer in Crawley, Sussex, UK, until reaching a constant weight (±0.2 mg). The total lipid content was determined in triplicate aliquots from each specimen and the average moisture percentage was calculated.

Microwave-Assisted Extraction (MAE) involved placing 0.5 g of each portion into a Teflon extraction vessel with 10 mL petroleum ether and 5 mL acetone in a Microwave–Accelerated Reaction System (MARS-5, 1500 W; CEM, Mathews, NC, USA; [82–84].

The resulting extract underwent filtration through Whatman GF/C filter papers (Ø 90 mm, GE Healthcare Life Sciences, Buckinghamshire, UK) containing anhydrous sodium sulfate (Carlo Erba) and was rinsed twice with an additional 2 mL of a petroleum ether-acetone mixture (2:1 v/v). Subsequently, the filtrate was evaporated under laminar flow inert gas (N_2) until reaching a constant weight, after which the mass of the extracted lipids was determined.

2.5. Statistical Analysis

The dataset consisted of 163 observations and eight variables. Among these, two were factor variables: season (with four levels: spring, summer, autumn, and winter) and sex (with two levels: male (M) and female (F)). The remaining six were numeric variables: total mercury, total PAHs, lipids, HSI, GSI, and CF.

Due to the limited availability of muscle samples, it was not feasible to test for both contaminants (PAHs and Hg) in all individuals. Out of 163 samples, Hg was analyzed in 74 individuals, PAHs in 151 individuals, and both contaminants were analyzed together in only 51 individuals.

The Season levels were categorized according to Artegiani et al., (1997a,b) [32,33] as follows: spring: April–June; summer: July–September; autumn: October–December; and winter: January–March.

There were some randomly distributed missing values in the dataset: 89 for total mercury, 12 for total PAHs, 21 for lipids, 25 for GSI, 34 for HSI, and 3 for Kn. However, in each of the statistical analyses, only the individuals with a full record (no missing value) were considered.

A 4 × 2 Analysis of Variance (ANOVA) design with interactions was employed to explore the effects of the two independent variables (season and sex) on the dependent variables (total mercury, total PAHs, lipids, HSI, GSI, and Kn). However, the limited number of observations and the stratification for the factor levels reduced the amount of data per cell, sometimes less than 20, posing challenges for assessing the normality distribution of the data [85]. In order to overcome the problems of normality distribution and homogeneity of variances, we directly applied a nonparametric ANOVA, based on the Aligned Rank Transform (ART), with the corresponding post hoc comparisons [86,87]. For the nonparametric two-way ANOVA, we used the function art from the R package ARTool. The function art.con from the same package was used for the post hoc comparisons. A reference p value of 0.05 was considered for significance. For the post hoc tests, the default Tukey HSD correction for multiple comparisons was applied [88].

For the post hoc comparisons, in the discussion of the results we focused on the main results, which include the differences among the seasons for each sex and the differences between the sexes in each season. As described above, the omnibus ANOVA and post hoc comparisons are based on the Aligned Rank Transform (ART). However, when discussing the results from a biological perspective, medians were referenced for the variables of interest since the median is more meaningful than the mean in the case of skewed distributions, as observed in our data.

Due to the limited number of data, which prevented assessing the normality assumption (see above), the correlation coefficients among the numeric variables were calculated using the nonparametric Kendall's tau, which has fewer assumptions than the Pearson coefficient [89]. Initially, this analysis was conducted without stratification for season and sex; subsequently, the data were stratified for season and sex and Kendall's tau was recalculated. Due to the numerous pairwise tests, the usual p value for significance (0.05) was adjusted using Bonferroni's correction [90] for the number of tests (15), resulting in an adjusted significance level of 0.003 (=0.05/15).

All the statistical analyses were performed using the free statistical software R ver 4.4.0 [91].

3. Results

3.1. Total Mercury Levels in Fish Muscle

The study estimated the total mercury level in the muscle tissue of 74 European hake individuals (38 females and 36 males) considering seasonal variations and sex differences.

The total mercury concentration (THg) ranged from 0.03 to 0.37 mg kg^{-1} ww (Table 1). Detailed data on the THg levels and body size of *Merluccius merluccius* are presented in Table S1 in Supplementary Materials. The omnibus ANOVA results and the pairwise post hoc comparisons are reported in Supplementary Materials (Table S2).

The results of the omnibus two-way ANOVA with interaction indicated that there is a marked effect of season ($F(3, 66) = 5.625$, p value = 0.002), no effect of sex ($F(1, 66) = 0.095$, $p = 0.759$), and a weak effect of interaction ($F(3, 66) = 3.124$, $p = 0.032$; Table S2).

Table 1. Total Hg concentrations (mg kg^{-1} ww) detected in the muscle of *M. merluccius* as a function of sex and season. # = number of valid values; median = median of the values; mean = mean of the values; sd = standard deviation; min = minimum value; max = maximum value.

Sex	Season	#	Median	Mean	sd	Min	Max
F	Spring	5	0.137	0.158	0.081	0.073	0.287
F	Summer	10	0.182	0.196	0.095	0.073	0.374
F	Autumn	10	0.103	0.103	0.033	0.045	0.152
F	Winter	13	0.096	0.091	0.02	0.059	0.122
M	Spring	10	0.155	0.155	0.032	0.111	0.217
M	Summer	7	0.125	0.131	0.074	0.071	0.289
M	Autumn	10	0.117	0.130	0.061	0.072	0.267
M	Winter	9	0.112	0.127	0.040	0.085	0.202

In the following, we will discuss only the main post hoc results. In females, the Hg concentration exhibited a seasonal trend from a peak in summer to a minimum in winter (Table 1 and Figure 2a). The value in summer is almost double that in winter and this difference is strongly significant (t.ratio(66) = 4.465, p value < 0.001). In males, a seasonal trend is not evident, and the data are compatible with the null hypothesis of no season influence. In each season, the data are compatible with the null hypothesis of no sex influence, see Table S2.

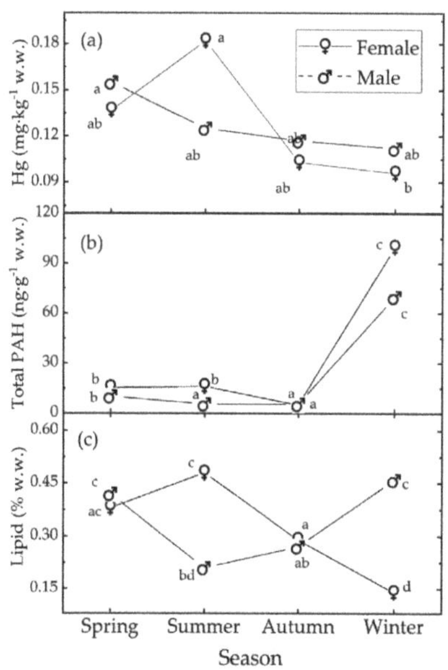

Figure 2. Median levels of (**a**) Hg, (**b**) total PAH, and (**c**) % lipids observed in the muscle tissue of *M. merluccius* as a function of sex and season. The medians with different letters differ at the p value = 0.05. The letters refer to the post hoc ART results; see text and Supplementary Materials.

3.2. Total PAH Levels in Fish Muscle

The study also examined the total levels of polycyclic aromatic hydrocarbons (PAHs) in 151 individuals (75 females and 76 males), with concentrations ranging from 2.70 to 185.79 ng g^{-1} wet weight (Table 2). Detailed PAH concentrations are provided in Table S3 in Supplementary Materials.

Table 2. Total PAH concentrations (ng g^{-1} ww) detected in the muscle of *M. merluccius* as a function of sex and season. # = number of valid values; median = median of the values; mean = mean of the values; sd = standard deviation; min = minimum value; max = maximum value.

Sex	Season	#	Median	Mean	sd	Min	Max
F	Spring	15	15.317	16.152	6.885	5.890	30.680
F	Summer	17	16.238	17.078	6.863	8.462	36.833
F	Autumn	9	4.990	5.998	2.287	3.236	8.959
F	Winter	34	99.680	89.221	41.464	23.725	185.788
M	Spring	33	10.211	16.931	22.421	3.713	117.229
M	Summer	14	5.593	5.835	1.443	3.881	8.206
M	Autumn	19	5.495	6.084	1.929	2.704	9.450
M	Winter	10	69.775	70.844	41.542	20.380	129.278

The results of the omnibus two-way ANOVA with interaction indicated that there is a strong effect of season (F (3, 143) = 90.573, $p < 0.001$), sex (F (1, 143) = 50.798, $p < 0.001$), and their interaction (F (3, 143) = 6.336, $p < 0.001$). These results, and the pairwise post hoc comparisons, are reported in Table S4 in Supplementary Materials.

In females, the total PAH levels exhibited a clear seasonal trend: they remained almost stable during spring and summer, declined in autumn to the lowest value, and peaked in winter (Table 2 and Figure 2b). Males showed a similar pattern, with the lowest PAH levels in autumn and the highest in winter; though in this case, the summer levels were closer to those in autumn than in spring. During spring, autumn, and winter, our data are compatible with the null hypothesis of no sex influence. In contrast, in summer, the data suggest a sex influence, with the female samples showing higher pollution levels than the male samples (t.ratio (143) = 7.093, p value < 0.001; Table S4).

3.3. Lipid Levels in Fish Muscle

Considering the propensity for PAHs and Hg to accumulate in lipids, the study estimated the lipid levels in the muscle of 142 European hake individuals (66 females and 76 males), taking into account seasonal variations and sex differences. The total lipid levels, expressed as % wet weight (w.w.), ranged from 0.14 to 0.52 (Figure 2c).

The results of the omnibus two-way ANOVA with interaction indicated that there is a strong effect of season (F (3, 134) = 56.796, $p < 0.001$), sex (F (1, 134) = 105.164, $p < 0.001$), and their interaction (F (3, 134) = 113.647, $p < 0.001$). These results, and the pairwise post hoc comparisons, are reported in Table S5 in Supplementary Materials.

In females, lipids exhibited a clear seasonal trend, increasing from spring, reaching the maximum in summer, and decreasing to a minimum value in winter, indicating the lowest lipid reserves during this season (Figure 2c). Males also showed a seasonal pattern, but their peak levels were observed in winter, with the minimum in summer. In spring and autumn, our data are compatible with the null hypothesis of no sex influence on lipid levels (Table S5). However, in summer and winter, our data suggest a sex influence: in summer, the female samples had higher lipid levels than the male samples (t.ratio (134) = 8.197, $p < 0.001$), while the opposite was observed in winter (t.ratio (134) = −10.832, $p < 0.001$).

3.4. Somatic Indexes: GSI, HSI, and Le Cren Kn

3.4.1. Gonadosomatic Index (GSI)

The study estimated the GSI in 138 European hake individuals (69 females and 69 males) considering seasonal variations and sex differences. The GSI ranged from 0.05 to 11.23 (Table 3).

The results of the omnibus two-way ANOVA with interaction indicate a strong effect of season, (F (3, 130) = 10.599, $p < 0.001$), sex (F (1, 130) = 31.193, $p < 0.001$), and their interaction (F (3, 130) = 13.369, $p < 0.001$). Detailed results and the pairwise post hoc comparisons are available in Table S6 in Supplementary Materials.

Table 3. GSI as a function of sex and season. # = number of valid values; median = median of the values; mean = mean of the values; sd = standard deviation; min = minimum value; max = maximum value.

Sex	Season	#	Median	Mean	sd	Min	Max
F	Spring	15	0.684	1.157	1.286	0.247	4.534
F	Summer	17	2.126	3.545	3.295	0.345	11.232
F	Autumn	19	0.685	1.510	2.355	0.053	8.536
F	Winter	18	0.421	0.393	0.117	0.136	0.562
M	Spring	32	0.809	0.792	0.394	0.196	2.009
M	Summer	14	0.728	0.850	0.419	0.353	1.615
M	Autumn	17	0.182	0.252	0.155	0.062	0.583
M	Winter	6	0.244	0.239	0.080	0.128	0.323

In females, the GSI reached its highest value in the summer, corresponding to the maximum gonadal development during the spawning peak (Figure 3a). The GSI then gradually decreased in autumn (October–December), returning to the value observed in spring, and reached its lowest in winter, indicating that more individuals were recorded in the post-reproductive gonadal maturation condition, although European hake is a protracted reproductive season species [92].

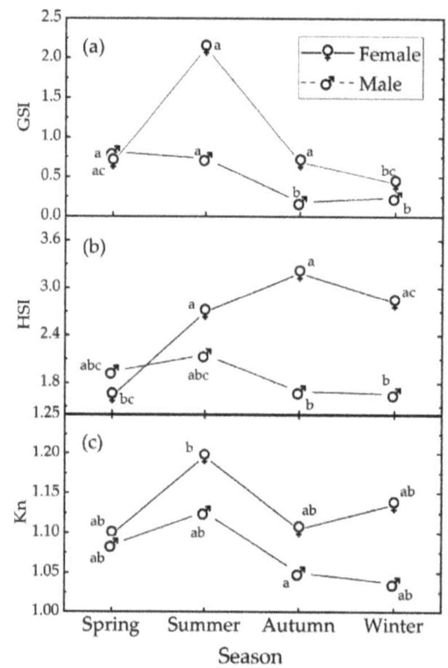

Figure 3. Median values of (**a**) GSI, (**b**) HSI, and (**c**) Kn Le Cren indexes as a function of sex and season. Medians with different letters differ at the p value = 0.05. The letters refer to the post hoc ART results; see text and Supplementary Materials.

Males also exhibited a seasonal pattern, but the peak value was observed in spring/summer, with the minimum in autumn/winter [93].

In winter, spring, and summer, our data are compatible with the null hypothesis of no sex influence on GSI levels. However, in autumn, our data suggest a sex influence, with the female samples showing higher GSI levels than the male samples (t.ratio (130) = 5.095; $p < 0.001$).

3.4.2. Hepatosomatic Index (HSI)

HSI was estimated in 129 individuals of European hake (59 females and 70 males), taking into account seasonal variations and sex differences. The HSI level ranged from 0.86 to 5.69. Detailed data are reported in Table 4.

Table 4. HSI as a function of sex and season. # = number of valid values; median = median of the values; mean = mean of the values; sd = standard deviation; min = minimum value; max = maximum value.

Sex	Season	#	Median	Mean	sd	Min	Max
F	Spring	6	1.636	1.6500	0.389	1.036	2.126
F	Summer	16	2.690	3.059	1.111	1.720	5.422
F	Autumn	19	3.187	3.087	1.189	1.242	5.688
F	Winter	18	2.820	2.879	0.867	1.082	4.272
M	Spring	32	1.941	2.398	1.103	0.856	4.797
M	Summer	14	2.152	2.481	0.852	1.393	3.939
M	Autumn	18	1.693	2.038	0.959	1.132	4.724
M	Winter	6	1.663	1.653	0.296	1.250	2.072

The results of the omnibus two-way ANOVA with interaction indicated that there is no effect of the season (F (3, 121) = 1.041, $p = 0.377$), and a strong effect of sex (F (1, 121) = 10.776, $p = 0.001$) and their interaction (F (3, 121) = 5.505, $p = 0.001$). These results and the pairwise post hoc comparisons are reported in Table S7 in Supplementary Materials.

In females, the HSI showed a mild seasonal trend, gradually increasing from its lowest value in spring to its highest in autumn, followed by a slight decrease in winter. Males exhibited no statistical seasonal pattern, with the highest HSI value occurring in summer and the lowest in winter (Figure 3b). The data indicated a sex influence only in autumn with females having a higher HSI level than males (t.ratio (121) = 3.497, $p = 0.015$).

3.4.3. Le Cren's Kn

Le Cren's Kn was calculated in 160 individuals (85 females and 75 males) considering seasonal variations and sex differences. This index ranged from 0.87 to 1.48 (Table 5).

Table 5. Le Cren's Kn index as a function of sex and season. # = number of valid values; median = median of the values; mean = mean of the values; sd = standard deviation; min = minimum value; max = maximum value.

Sex	Season	#	Median	Mean	sd	Min	Max
F	Spring	15	1.098	1.084	0.085	0.921	1.233
F	Summer	17	1.196	1.159	0.079	1.011	1.262
F	Autumn	19	1.105	1.098	0.109	0.874	1.328
F	Winter	34	1.136	1.116	0.092	0.918	1.242
M	Spring	33	1.085	1.097	0.080	0.969	1.324
M	Summer	14	1.126	1.124	0.119	0.933	1.356
M	Autumn	17	1.050	1.065	0.091	0.912	1.284
M	Winter	11	1.037	1.097	0.149	0.985	1.485

The omnibus two-way ANOVA with interaction showed no season effect (F (3, 152) = 2.528, $p = 0.059$), a weak sex effect (F (1, 152) = 5.760, $p = 0.018$), and no interaction effect (F (3, 152) = 1.121, p value = 0.342). Detailed results are available in Table S8 in Supplementary Materials.

Regarding Le Cren's Kn index, similar values were observed across all the seasons for both sexes (Figure 3c). The post hoc comparisons indicate that the Kn index data are substantially compatible with the null hypothesis of no influence of season or sex; i.e., despite the significant sex effect in the omnibus ANOVA test, none of the relative post hoc comparisons were significant (Table S8). This is due to the higher power of the omnibus ANOVA test according to Freund et al. (2010) [94].

3.5. Correlations between Pollutants, Lipids, and Somatic Indexes

A comprehensive statistical analysis was conducted to examine the relationships between the pollutants, lipids recorded in the muscle of European hake, and various somatic indexes. Despite the thorough investigation, only a few of these relationships were found to be statistically significant. The detailed results are summarized in Table 6, which provides a summary of the pairwise Kendall's tau correlations for the numerical variables. Kendall's tau is a measure of correlation that assesses the strength and direction of the association between two variables. This nonparametric statistic is particularly useful for understanding the relationships in ordinal data or when the data do not meet the assumptions of parametric tests.

Table 6. Kendall's tau correlation matrix for the six numerical variables. The bold figures are statistically significant values.

	Lipids	Tot PAHs	GSI	HSI	Le Cren CF	THg
Lipids	1.000	−0.106	**0.254**	−0.072	−0.078	**0.447**
Tot PAHs		1.000	−0.079	0.071	−0.004	−0.175
GSI			1.000	**0.255**	0.154	**0.288**
HSI				1.000	**0.268**	−0.169
Le Cren CF					1.000	−0.046
THg						1.000

Initially, this analysis was conducted on the entire dataset without stratification for season and sex; subsequently, the data were stratified for season and sex and Kendall's tau was recalculated (see Tables S9–S16 in Supplementary Materials).

Throughout the year, a significant positive relationship between the total Hg and the GSI was observed for both sexes combined, with no noticeable seasonal variations. In contrast, no significant relationship was detected between the total PAHs and GSI.

Additionally, a positive relationship between the Hg and lipid content was noted, although this correlation was not significant within individual seasons. For PAHs, there was no significant relationship observed.

A generally positive correlation between the lipid content and GSI was observed for both sexes combined throughout the year. This trend was not so evident when examining individual seasons and single-sex groups. This correlation is crucial, as lipid accumulation is vital for the accumulation of reserves in preparation for gamete development and spawning (see Tables S9–S16 in Supplementary Materials and values for detailed data).

Furthermore, a positive correlation was noted between GSI and HSI, as well as between HSI and Le Cren's Kn. Interestingly, a negative relationship between PAHs and Hg was observed throughout the year for both sexes combined, although this was not statistically significant. This result suggests that as the PAH levels increase, the Hg levels tend to decrease, and vice versa. Overall, no significant seasonal patterns were evident.

4. Discussion

The accumulations of contaminants in muscle tissue are influenced by various abiotic (water, sediment, and geographic location), biotic (size, sex, age, and reproduction stage), and ecological factors like growth rate, feeding sources, and trophic level [13,14,18,95–97].

These factors, individually or collectively, can cause significant stress, leading to reduced growth, impaired reproduction, increased disease susceptibility, and decreased ability to tolerate further stress. At the population level, stress effects can result in reduced recruitment and compensatory reserve.

Toxicants can affect organisms at various biological levels, making it inadequate to rely on a single stress response for a comprehensive assessment. The bioindicator approach, which evaluates multiple responses, effectively measures sublethal stress effects on fish. This approach serves as an early warning system and could help clarify the relationships between stressors and their broader biological impacts.

Bioindicators and condition indexes provide a comprehensive view of fish health and environmental stress, reducing misinterpretation and offering cost-effective pollution monitoring tools essential for evaluating long-term contaminant impacts.

This study applies the bioindicator approach to assess stress responses in relation to European hake (*Merluccius merluccius*), one of the most important commercial species fished in the Adriatic Sea. The fish were analyzed specifically for Hg and PAH levels in their muscle tissue and condition indexes.

This species, characterized by indeterminate fecundity [98–100], a protracted spawning season [101], and a complex reproductive strategy [102], is particularly vulnerable to the negative individual- and population-level impacts of chronic contaminant exposure.

In the Mediterranean, European hake typically spawns in spring-summer, peaking from April to July, though this timing can vary depending on the location and environmental conditions [103–108].

Both sexes experience metabolic and hormonal changes [109] during reproductive stages, affecting pollutant accumulation in muscle tissue.

This study found Hg levels in muscle tissue (0.03 to 0.37 mg kg^{-1} dw) generally consistent with those reported in the Adriatic Sea [110,111] and other parts of the Mediterranean Sea [112–114]. On the contrary, Jureša and Blanuša (2003) [115] and Perugini et al. (2014) [116] observed values well higher than these in the Adriatic Sea.

No significant differences in the THg levels were observed between the sexes, but females exhibited important seasonal variations, with peak levels in summer and lower levels in winter. In contrast, males maintained stable THg levels throughout the year (Figure 2a).

Several studies have shown that the Hg levels in the edible tissues of females decrease after spawning [117–120]. This reduction in Hg levels may be due to a sort of "cleaning process" where contaminants are eliminated during oocyte release [121–123].

This pattern is less evident in males, whose lipid content and energy reserves did not change greatly during the year. This is likely due to continuous spermatogenesis in adult specimens without a significant resting phase [93,124].

In most of the muscle samples analyzed, the total PAH levels recorded in this study fell within the "minimally polluted" category (10–99 ng/g), with a mean concentration of 33.1 $\pm$ 40.6 ng/g ww [125]. The total PAH concentrations detected in this study are comparable to those reported in the study by [126] on European hake caught in the Gulf of Taranto during the period from February to July 2010. However, the values found in the *M. merluccius* specimens from the Adriatic Sea examined in this study show lower concentrations compared to those sampled in the Gulf of Taranto (40–350 ng/g ww). Conversely, the PAH levels in the present study are higher than those found in the hake sampled in the Gulf of Naples in the study by [127] which reported the average total PAH values of 6 ng/g ww, and those reported in the study by [128], whose total PAH concentrations in the muscle of *M. merluccius* ranged from 3 to 7 ng/g ww. Meanwhile, the levels detected in the present study are very similar to those reported by [127] for samples from the Adriatic Sea, where the average total PAH concentrations in *M. merluccius* were about 44 ng/g.

Higher levels of total PAH concentration were recorded during the winter season compared to the summer, for both the male and female individuals (Table S2, Figure 2b). This can be attributed to both anthropogenic and natural factors. In the colder months, there is indeed a higher release of these substances into the atmosphere compared to the summer months due to the increased use of domestic heating systems as well as greater vehicular traffic during winter. In the marine environment, the winter months are associated with increased riverine discharge due to elevated levels of atmospheric precipitation. Additionally, lower winter temperatures inhibit all the PAH biodegradation processes performed by microorganisms, affecting the solubility, dispersion, and bioavailability of these contaminants. During the winter period, the reduced solar radiation intensity could decrease the efficiency of PAH photodegradation, leading to the increased persistence of these compounds in the marine environment, particularly within marine sediments. Additionally, marine sediments are subjected to remixing and bioturbation phenomena during the winter period, which can disrupt the equilibrium at the water-sediment interface and facilitate the release of previously deposited PAHs.

However, the samples were derived from commercial fishing operations and did not include the full range of upper size and age classes. As a result, the study did not investigate the relationship between contaminant levels and fish size, potentially missing important variations in contaminant distribution within the population. Future research with a larger sample size could provide more detailed insight and improve the reliability of the contamination assessment. Lipids, while serving as energy reserves, are considered crucial for the accumulation of pollutants like polycyclic aromatic hydrocarbons (PAHs) and mercury (Hg). Due to their hydrophobic nature, these compounds tend to associate with lipid-rich tissues in organisms. These pollutants can exert toxic effects and be transferred across trophic levels, affecting ecosystems [11,18,129–131].

In our study, lipids seem to exert a limited influence on the PAH levels in the muscle of European hake while showing a strong positive correlation with Hg. Lipid distributions in muscle tissue exhibited distinct seasonal trends and differences between sexes, displaying notable changes during summer and winter.

In females, the lipid levels were highest from spring to summer, likely due to energy storage for reproduction and oocyte development, and then decreased to their lowest reserves in winter after reproductive effort and egg release.

In contrast, males exhibited less variation in lipid content throughout the year, although they showed distinct seasonal trends, possibly due to continuous spermatogenesis, with the "resting phase" almost absent [93,124]. Males' lipid levels were highest in winter, aligning with pre-spawning energy reserves.

These differences in lipid level reflect variations in metabolic activity, costs of reproduction, and adaptive strategies of both sexes to seasonal environmental changes and energy demands.

Previous studies have shown extreme variability in correlations between lipid content and organic contaminants. Some reported positive correlations between PAH accumulation and lipids [132–134], while others found weak [131], negative [135], or no correlations [136,137].

Similarly, the importance of lipids in relation to Hg levels is quite contradictory [120, 138,139].

This inconsistency suggests that lipid content may not be the key factor for tissue-specific pollutant accumulation in fish, indicating the variability of this relationship among different fish species, environmental conditions, and especially trophic habits [140–144].

Furthermore, high variation in lipid reserves between individuals of the same species is common in wild fish [145] and may reflect significant differences in nutritional status and reproductive potential within a population. This strong inter-population variability likely explains the low correlation values observed between the different variables studied.

Somatic indexes like the GSI, HSI, and Kn were valuable for assessing pollutant impacts on aquatic ecosystems. These indexes are sensitive indicators of changes in fish physiology, allowing researchers to monitor the subtle effects of pollutants on fish populations. Extensive research supports their use as reliable proxies for assessing fish responses to environmental stressors [46–48].

A comprehensive analysis conducted throughout the year revealed no relationships between the two pollutants and condition indexes, except for a positive correlation between the total Hg and the GSI, without significant seasonal and sex variations. In contrast, PAHs showed no meaningful correlation with condition indexes. This could be due to the fact that the study focused on wild fish rather than those exposed to controlled levels of contamination, where the effects of the contaminants on these indexes would likely be more directly measurable.

A negative but not statistically significant relationship between PAHs and Hg was observed, suggesting that as PAH levels increase, Hg levels tend to decrease, and vice versa with no significant seasonal patterns.

Conversely, significant relationships were found among the condition indexes, highlighting energy intake utilization. Specifically, a positive correlation was noted between GSI and HSI, as well as between HSI and Kn.

To understand the reproductive strategy in terms of energy investment, assessing the seasonal variability of somatic and gonadic conditions is essential. Fish alternated their energy use between body growth and reserves (fat, muscle, and liver) and gonadic growth throughout the year.

In spring and summer, females allocated energy to reproduction both from concurrent feeding [93,107,146] and from lipid reserves to support a long reproductive cycle; in these seasons, the GSI increased progressively from April to August before sharply decreasing in autumn. This pattern indicates energy allocation towards reproductive efforts.

Males exhibited low GSI with minimal seasonal variation, showing less energy investment in gonad development. In autumn, the GSI for both sexes decreased, reaching its lowest point in winter, signifying post-spawning recovery [92].

The HSI was generally higher in females and remained constant in males throughout the year. Le Cren's Kn showed similar trends for both sexes, with females consistently having slightly higher values. Seasonal variability in body condition, in terms of weight and length, was more pronounced in females, suggesting an increase in body size from spring to summer, followed by a decline in the condition thereafter.

In spring and summer, during the preparation for spawning, females' lipid content and Kn values increased due to the accumulation of muscle and lipid reserves. Although gonad production increased during this period, the maintenance of body condition and liver mass growth suggests that external energy sources, such as food, supported both gonadic growth and body mass maintenance. This indicates that energy intake during the spawning season plays a crucial role in growth capacity, rather than relying solely on energy reserves accumulated several months prior. Similar to females, males used muscle lipids as energy reserves during this period.

In autumn, during the post-spawning period, the female hake exhibited an increase in HSI and a decrease in the lipid content and Kn, indicating the use of muscle lipids and proteins as energy reserves. Despite the increase in gonad production, stable body condition levels and liver mass growth suggest ongoing external energy support.

Males showed a decrease in Kn, but an increase in the lipid content, indicating reserve accumulation.

In winter, females showed a decrease in HSI and an increase in Kn, with a decline in the lipid content, reflecting the use of lipids as fuel during food shortages, almost depleting muscle lipids. Males, however, exhibited a decrease in Kn and an increase in the lipid content, continuing to accumulate reserves.

In Adriatic waters, high Kn and HSI levels from summer to winter indicate ample feeding resources for female somatic and gonadic growth. After the main spawning peak, a decrease in GSI and an increase in HSI suggest that adults do not end the spawning season exhausted. Hake continue feeding during the breeding season, unlike in cold waters where spawning occurs in winter with scarce food, and vitellogenesis relies on stored energy from previous months [100].

These observations highlight seasonal differences in energy use between sexes, influenced by environmental conditions and food availability.

5. Conclusions

The study revealed that the contaminant levels in European hake (*Merluccius merluccius*) from the Adriatic are below international safety limits. However, regular monitoring is essential for maintaining the health of marine ecosystems. The research also emphasizes the importance of understanding and mitigating the impact of pollutants on marine organisms' reproductive health to protect marine ecosystems worldwide.

A complex interaction between the PAHs and Hg levels in muscle tissues is mainly influenced by both sex and seasonal factors. PAH levels are more affected by seasonal changes with higher concentrations in the winter season, while Hg levels in females are consistently associated with lipid content throughout the year.

No significant correlation was found between PAHs and gonadosomatic or morphometric indexes, likely because this study focused on wild fish rather than those exposed to controlled contamination, where contaminant effects would be more measurable. Seasonal fluctuations in contaminant levels seem to be chiefly related to metabolic activity, the reproductive cycle, and the energy costs in terms of the energy of reproduction, growth, and physiological processes being different for the sexes, affecting tissue composition in terms of lipid content. This variation affects the concentration of the investigated contaminants, suggesting that fish physiological conditions should be considered in biomonitoring programs.

During the breeding season, part of the assimilated energy is allocated to gamete production, reducing dependence on energy reserves. Females feed during breeding, indicating that liver lipids stored during maturation are mobilized towards the gonads for reproductive purposes.

Future studies are planned to clarify how lipid dynamics and seasonal and sex-based variations in lipid levels influence the deposition of mercury (Hg) and polycyclic aromatic hydrocarbons (PAHs) in different tissues. These investigations will provide insights into how environmental factors and biological processes affect pollutant storage in aquatic organisms.

Additionally, incorporating vitellogenin analysis could be crucial, as this liver-produced protein, important for ovarian development, can be significantly altered by environmental contaminants. Examining vitellogenin levels across sexes and pollutant exposure will enhance the understanding of contaminants' impact on reproductive health and improve environmental health assessments.

Finally, our results highlight that somatic indexes, like HSI and GSI, are useful indicators of metabolic state and reproductive health, reflecting environmental stress and pollution effects.

Overall, this multidisciplinary approach, including condition indexes, provides a deep understanding of the relationship between reproductive biology and the health of European hake, offering insights into sustainable fishery management and environmental conservation.

Supplementary Materials: The following supporting information can be downloaded at: https://www.mdpi.com/article/10.3390/ani14202938/s1, Table S1: Concentrations of Total Mercury (THg) in muscle tissues of Merluccius merluccius samples from the Adriatic Sea. Sex, biometric data of investigated fish. n, number of fish samples; mean value ± standard deviation; range (min–max) of length, weight, THg; Table S2: Concentrations of Total PAHs in muscle tissues of Merluccius merluccius samples from the Adriatic Sea. Sex, biometric data of investigated fish. n, number of fish samples; mean value ± standard deviation; range (min–max) of length, weight, PAHs; Table S3: Omnibus two-way ANOVA results and post hoc comparisons of Total Hg in relations to the Season and Sex factors along with their interaction; Table S4: Omnibus two-way ANOVA results and post hoc comparisons of Total PAHs in relations to the Season and Sex factors along with their interaction; Table S5: Omnibus two-way ANOVA results and post hoc comparisons of Lipids in relations to the Season and Sex factors along with their interaction; Table S6: Omnibus two-way ANOVA results and post hoc comparisons of GSI index in relations to the Season and Sex factors along with their interaction; Table S7: Omnibus two-way ANOVA results and post hoc comparisons of HSI index in relations to the Season and Sex factors along with their interaction; Table S8: Omnibus two-way ANOVA results and post hoc comparisons of Le Cren CF index in relations to the Season and Sex factors along with their interaction; Table S9: Spring/Female Kendall's tau correlation matrix for the six numerical variables. Bold figures are statistically significant values; Table S10: Spring/Male Kendall's tau correlation matrix for the six numerical variables. Bold figures are statistically significant values; Table S11: Summer/Female Kendall's tau correlation matrix for the six numerical variables. Bold figures are statistically significant values; Table S12: Summer/Male Kendall's tau correlation matrix for the six numerical variables. Bold figures are statistically significant values; Table S13: Autumn/Female Kendall's tau correlation matrix for the six numerical variables. Bold figures are statistically significant values; Table S14: Autumn/Male Kendall's tau correlation matrix for the six numerical variables. Bold figures are statistically significant values; Table S15: Winter/Female Kendall's tau correlation matrix for the six numerical variables. Bold figures are statistically significant values; Table S16: Winter/Male Kendall's tau correlation matrix for the six numerical variables. Bold figures are statistically significant values.

Author Contributions: Conceptualization, M.P., E.F., C.T. and S.C.; methodology, M.P., E.F., C.T. and S.C.; software, S.G.o.G.; validation, C.T., M.M., A.A. and S.I.; formal analysis, S.G.o.G. and F.G.; investigation, M.P., E.F., F.G., A.A., S.I. and S.C.; resources, C.T., M.M. and A.S.; data curation, M.P., S.G.o.G., E.F., C.T., F.G. and S.C.; writing—original draft, M.P. and S.G.o.G.; writing—review and editing, S.G.o.G., E.F., C.T., F.G., M.M., A.A., S.I. and S.C.; supervision, M.P., C.T., M.M. and S.C.; funding acquisition, C.T., M.M. and A.S. All authors have read and agreed to the published version of the manuscript.

Funding: This research was partially supported by the Italian Ministry of Agriculture, Food, and Forestry Policies (MiPAAF) and by the European Commission, in the framework of the "Data Collection Framework—Italian National Program 2017–2019" (GSA17) (J82F17000000007).

Institutional Review Board Statement: The sampling procedures did not include any animal experimentation and animal ethics approval was, therefore, not necessary under the Italian legislation (D.L. 4 March 2014, n. 26, art. 2). The methodology and sampling of this study were based on dead specimens from professional fishing.

Informed Consent Statement: Not applicable.

Data Availability Statement: Upon request due to restrictions, e.g., privacy or ethical.

Acknowledgments: The authors wish to thank Ilaria Costantini from the Institute for Marine Biological Resources and Biotechnology (IRBIM-CNR) for the elaboration of the map of the study area.

Conflicts of Interest: The authors declare no conflicts of interest that could be perceived as prejudicing the impartiality of the research reported.

References

1. Capodiferro, M.; Marco, E.; Grimalt, J.O. Wild fish and seafood species in the western Mediterranean Sea with low safe mercury concentrations. *Environ. Pollut.* **2022**, *314*, 120274. [CrossRef] [PubMed]
2. Colloca, F.; Cardinale, M.; Maynou, F.; Giannoulaki, M.; Scarcella, G.; Jenko, K.; Bellido, J.M.; Fiorentino, F. Rebuilding Mediterranean fisheries: A new paradigm for ecological sustainability. *Fish Fish.* **2013**, *14*, 89–109. [CrossRef]
3. Salomon, M.; Markus, T.; Dross, M. Masterstroke or paper tiger—The reform of the EU's Common Fisheries Policy. *Mar. Pol.* **2014**, *47*, 76–84. [CrossRef]
4. Tsikliras, A.C.; Dinouli, A.; Tsiros, V.Z.; Tsalkou, E. The Mediterranean and Black Sea Fisheries at Risk from Overexploitation. *PLoS ONE* **2015**, *10*, e0121188. [CrossRef] [PubMed]
5. Carpi, P.; Scarcella, G.; Cardinale, M. The Saga of the Management of Fisheries in the Adriatic Sea: History, Flaws, Difficulties, and Successes toward the of the Common Fisheries Policy in the Mediterranean. *Front. Mar. Sci.* **2017**, *4*, 423. [CrossRef]
6. Bahamon, N.; Recasens, L.; Sala-Coromina, J.; Calero, B.; Garcia, J.A.; Rotllant, G.; Maurer, A.; Rojas, A.; Muth, L.; Quevedo, J.; et al. Selectivity-based management for reversing overexploitation of demersal fisheries in North-western Mediterranean Sea. *Mar. Pol.* **2024**, *165*, 106185. [CrossRef]
7. Hala, E.; Bakiu, R. Adriatic Sea Fishery Product Safety and Prospectives in Relation to Climate Change. *Fishes* **2024**, *9*, 160. [CrossRef]
8. Janssen, P.A.H.; Lambert, J.G.D.; Goos, H.J.T. The annual ovarian cycle and the influence of pollution on vitellogenesis in the flounder, *Pleuronectes flesus*. *J. Fish Biol.* **1995**, *47*, 509–523. [CrossRef]
9. Collier, T.K.; Anulacion, B.F.; Arkoosh, M.R.; Dietrich, J.P.; Incardona, J.P.; Johnson, L.L.; Ylitalo, G.M.; Myers, M.S. Effects on fish of polycyclic aromatic hydrocarbons (PAHS) and naphthenic acid exposures. In *Fish Physiology: Organic Chemical Toxicology of Fishes*; Tierney, K.B., Farrell, A.R., Brauner, C.L., Eds.; Academic Press: Cambridge, MA, USA, 2013; Volume 33, pp. 195–255. [CrossRef]
10. Kaddour, A.; Djellouli, F.; Belhoucine, F.; Alioua, A. Heavy metal bioaccumulation and genotoxicity in fish (*Merluccius merluccius*, Linnaeus, 1758) from the Western Algerian Mediterranean coast. *Appl. Ecol. Environ. Res.* **2022**, *20*, 5361–5379. [CrossRef]

11. Van Der Oost, R.; Beyer, J.; Vermeulen, N.P.E. Fish bioaccumulation and biomarkers in environmental risk assessment: A review *Environ. Toxicol. Pharmacol.* **2003**, *13*, 57–149. [CrossRef]
12. Kerdoun, M.A.; Alouk, L.; Rahmani, F.M.; Henni, H.A.; Dali, H.; Kelai, E.; Belkhalfa, H. Mercury in four common fishes sold in Algeria and associated humans risk. *Food Addit. Contam. Part B* **2024**, *17*, 223–229. [CrossRef] [PubMed]
13. Nyeste, K.; Dobrocsi, P.; Czeglédi, I.; Czédli, H.; Harangi, S.; Baranyai, E.; Simon, E.; Nagy, S.A.; Antal, L. Age and diet-specific trace element accumulation patterns in different tissues of chub (*Squalius cephalus*): Juveniles are useful bioindicators of recent pollution. *Ecol. Indic.* **2019**, *101*, 1–10. [CrossRef]
14. Nyeste, K.; Zulkipli, N.; Uzochukwu, I.E.; Somogyi, D.; Nagy, L.; Czeglédi, I.; Harangi, S.; Baranyai, E.; Simon, E.; Nagy, S.A.; et al. Assessment of trace and macroelement accumulation in cyprinid juveniles as bioindicators of aquatic pollution: Effects of diets and habitat preferences. *Sci. Rep.* **2024**, *14*, 11288. [CrossRef]
15. Vagi, M.C.; Petsas, A.S.; Kostopoulou, M.N. Potential Effects of Persistent Organic Contaminants on Marine Biota: A Review on Recent Research. *Water* **2021**, *13*, 2488. [CrossRef]
16. Bolognesi, C.; Hayashi, M. Micronucleus assay in aquatic animals. *Mutagenesis* **2011**, *26*, 205–213. [CrossRef]
17. El-Shehawi, A.M.; Ali, F.K.; Seehy, M.A. Estimation of water pollution by genetic biomarkers in tilapia and catfish species shows species-site interaction. *Afr. J. Biotechnol.* **2007**, *6*, 840–846.
18. Frapiccini, E.; Panfili, M.; Guicciardi, S.; Santojanni, A.; Marini, M.; Truzzi, C.; Annibaldi, A. Effects of biological factors and seasonality on the level of polycyclic aromatic hydrocarbons in red mullet (*Mullus barbatus*). *Environ. Pollut.* **2020**, *258*, 113742. [CrossRef] [PubMed]
19. Amina, K.; Fatma, B.; Amel, A. Integrated use of condition indexes, genotoxic and cytotoxic biomarkers for assessing pollution effects in fish (*Mullus barbatus* L., 1758) on the West coast of Algeria. *South Asian J. Exp. Biol.* **2021**, *11*, 287–299. [CrossRef]
20. Frapiccini, E.; Cocci, P.; Annibaldi, A.; Panfili, M.; Santojanni, A.; Grilli, F.; Marini, M.; Palermo, F.A. Assessment of seasonal relationship between polycyclic aromatic hydrocarbon accumulation and expression patterns of oxidative stress-related genes in muscle tissues of red mullet (*M. barbatus*) from the Northern Adriatic Sea. *Environ. Toxicol. Pharmacol.* **2021**, *88*, 103752. [CrossRef]
21. Pastorelli, A.A.; Baldini, M.; Stacchini, P.; Baldini, G.; Morelli, S.; Sagratella, E.; Zaza, S.; Ciardullo, S. Human exposure to lead, cadmium and mercury through fish and seafood product consumption in Italy: A pilot evaluation. *Food Addit. Contam. Part A* **2012**, *29*, 1913–1921. [CrossRef]
22. Varol, M.; Kaya, G.K.; Sünbül, M.R. Evaluation of health risks from exposure to arsenic and heavy metals through consumption of ten fish species. *Environ. Sci. Pollut. Res. Int.* **2019**, *26*, 33311–33320. [CrossRef] [PubMed]
23. Barone, G.; Storelli, A.; Garofalo, R.; Mallamaci, R.; Storelli, M.M. Residual Levels of Mercury, Cadmium, Lead and Arsenic in some commercially key species from Italian Coasts (Adriatic Sea): Focus on human health. *Toxics* **2022**, *10*, 223. [CrossRef] [PubMed]
24. De Giovanni, A.; Abondio, P.; Frapiccini, E.; Luiselli, D.; Marini, M. Meta-analysis of a new georeferenced database on polycyclic aromatic hydrocarbons in Western and Central Mediterranean seafood. *Appl. Sci.* **2022**, *12*, 2776. [CrossRef]
25. De Giovanni, A.; Iannuzzi, V.; Gallello, G.; Giuliani, C.; Marini, M.; Cervera, M.L.; Luiselli, D. Mercury Intake Estimation in Adult Individuals from Trieste, Italy: Hair Mercury Assessment and Validation of a Newly Developed Food Frequency Questionnaire *Pollutants* **2023**, *3*, 320–336. [CrossRef]
26. Ray, S.; Vashishth, R. From water to plate: Reviewing the bioaccumulation of heavy metals in fish and unravelling human health risks in the food chain. *Emerg. Contam.* **2024**, *10*, 100358. [CrossRef]
27. Faganeli, J.; Horvat, M.; Covelli, S.; Fajon, V.; Logar, M.; Lipej, L.; Cermelj, B. Mercury and methylmercury in the Gulf of Trieste (northern Adriatic Sea). *Sci. Total Environ.* **2003**, *304*, 315–326. [CrossRef]
28. Marini, M.; Frapiccini, E. Persistence of polycyclic aromatic hydrocarbons in sediments in the deeper area of the Northern Adriatic Sea (Mediterranean Sea). *Chemosphere* **2013**, *90*, 1839–1846. [CrossRef]
29. Kotnik, J.; Horvat, M.; Ogrinc, N.; Fajon, V.; Žagar, D.; Cossa, D.; Sprovieri, F.; Pirrone, N. Mercury speciation in the Adriatic Sea *Mar. Pollut. Bull.* **2015**, *96*, 136–148. [CrossRef]
30. Ricci, F.; Capellacci, S.; Casabianca, S.; Grilli, F.; Campanelli, A.; Marini, M.; Penna, A. Variability of hydrographic and biogeochemical properties in the North-western Adriatic coastal waters in relation to river discharge and climate changes *Chemosphere* **2024**, *361*, 142486. [CrossRef]
31. Russo, A.; Artegiani, A. Adriatic sea hydrography. *Sci. Mar.* **1996**, *60*, 33–43.
32. Artegiani, A.; Paschini, E.; Russo, A.; Bregant, D.; Raicich, F.; Pinardi, N. The Adriatic Sea general circulation. Part I: Air-sea interactions and water mass structure. *J. Phys. Oceanogr.* **1997**, *27*, 1492–1514. [CrossRef]
33. Artegiani, A.; Paschini, E.; Russo, A.; Bregant, D.; Raicich, F.; Pinardi, N. The Adriatic Sea general circulation. Part II: Baroclinic circulation structure. *J. Phys. Oceanogr.* **1997**, *27*, 1515–1532. [CrossRef]
34. Rovere, M.; Mercorella, A.; Frapiccini, E.; Funari, V.; Spagnoli, F.; Pellegrini, C.; Bonetti, A.S.; Veneruso, T.; Tassetti, A.N. Dell'Orso, M.; et al. Geochemical and geophysical monitoring of hydrocarbon seepage in the Adriatic Sea. *Sensors* **2020**, *20*, 1504 [CrossRef] [PubMed]
35. Solé, M.; Rodríguez, S.; Papiol, V.; Maynou, F.; Cartes, J.E. Xenobiotic metabolism markers in marine fish with different trophic strategies and their relationship to ecological variables. *Comp. Biochem. Physiol. Part C Toxicol. Pharmacol.* **2009**, *149*, 83–89 [CrossRef] [PubMed]

36. Schirmer, M.; Reinstorf, F.; Leschik, S.; Musolff, A.; Krieg, R.; Strauch, G.; Molson, J.W.; Martienssen, M.; Schirmer, K. Mass fluxes of xenobiotics below cities: Challenges in urban hydrogeology. *Environ. Earth Sci.* **2011**, *64*, 607–617. [CrossRef]
37. Goyer, R.; Clarkson, T. Toxic effects of metals. Casarett & Doull's Toxicology. In *The Basic Science of Poisons*, 5th ed.; Klaassen, C.D., Ed.; McGraw-Hill Health Professions Division: NewYork, NY, USA, 1996.
38. Sharma, R.; Agrawal, M. Biological effects of heavy metals: An overview. *J. Environ. Biol.* **2005**, *26*, 301–313.
39. Amara, R.; Méziane, T.; Gilliers, C.; Hermel, G.; Laffargue, P. Growth and condition indices in juvenile sole *Solea solea* measured to assess the quality of essential fish habitat. *Mar. Ecol. Prog. Ser.* **2007**, *351*, 201–208. [CrossRef]
40. Araújo, F.G.; Morado, C.N.; Parente, T.T.E.; Paumgartten, F.J.; Gomes, I.D. Biomarkers and bioindicators of the environmental condition using a fish species (*Pimelodus maculatus* Lacepède, 1803) in a tropical reservoir in Southeastern Brazil. *Braz. J. Biol.* **2018**, *78*, 351–359. [CrossRef]
41. Kerambrun, E.; Henry, F.; Perrichon, P.; Courcot, L.; Meziane, T.; Spilmont, N.; Amara, R. Growth and condition indices of juvenile turbot, *Scophthalmus maximus*, exposed to contaminated sediments: Effects of metallic and organic compounds. *Aquat. Toxicol.* **2012**, *108*, 130–140. [CrossRef]
42. Pandit, D.N.; Gupta, M.L. Heapto-somatic index, gonado-somatic index and condition factor of Anabas testudineus as bio-monitoring tools of nickel and chromium toxicity. *Int. J. Innov. Eng. Technol.* **2019**, *12*, 25–28.
43. Snyder, S.M.; Pulster, E.L.; Murawski, S.A. Associations between chronic exposure to polycyclic aromatic hydrocarbons and health indices in Gulf of Mexico tilefish (*Lopholatilus chamaeleonticeps*) post Deepwater Horizon. *Environ. Toxicol. Chem.* **2019**, *38*, 2659–2671. [CrossRef]
44. Pulster, E.L.; Gracia, A.; Armenteros, M.; Carr, B.E.; Mrowicki, J.; Murawski, S.A. Chronic PAH exposures and associated declines in fish health indices observed for ten grouper species in the Gulf of Mexico. *Sci. Total Environ.* **2020**, *703*, e135551. [CrossRef]
45. Alves, L.M.; Lemos, M.F.; Cabral, H.; Novais, S.C. Elasmobranchs as bioindicators of pollution in the marine environment. *Mar. Poll. Bull.* **2022**, *176*, 113418. [CrossRef]
46. Kleinkauf, A.; Connor, L.; Swarbreck, D.; Levene, C.; Walker, P.; Johnson, P.J.; Leah, R.T. General condition biomarkers in relation to contaminant burden in European flounder (*Platichthys flesus*). *Ecotox. Environ. Saf.* **2004**, *58*, 335–355. [CrossRef] [PubMed]
47. Morado, C.N.; Araújo, F.G.; Gomes, I.D. The use of biomarkers for assessing effects of pollutant stress on fish species from a tropical river in Southeastern Brazil. *Acta Sci. Biol. Sci.* **2017**, *39*, 431–439. [CrossRef]
48. Łuczyńska, J.; Paszczyk, B.; Łuczyński, M.J. Fish as a bioindicator of heavy metals pollution in aquatic ecosystem of Pluszne Lake, Poland, and risk assessment for consumer's health. *Ecotoxicol. Environ. Saf.* **2018**, *153*, 60–67. [CrossRef]
49. Choudhury, C.; Ray, A.K.; Bhattacharya, S.; Bhattacharya, S. Non lethal concentrations of pesticide impair ovarian function in the freshwater perch, *Anabas testudineus*. *Environ. Biol. Fishes* **1993**, *36*, 319–324. [CrossRef]
50. Sakamoto, K.Q.; Nakai, K.; Aoto, T.; Yokoyama, A.; Ushikoshi, R.; Hirose, H.; Ishizuka, M.; Kazuka, A.; Fujita, S. Cytochrome P450 induction and gonadal status alteration in common carp (*Cyprinus carpio*) associated with the discharge of dioxin contaminated effluent to the Hikiji River, Kanagawa Prefecture, Japan. *Chemosphere* **2003**, *51*, 491–500. [CrossRef]
51. Goede, R.W.; Barton, B.A. Organismic indices and an autopsy-based assessment as indicators of health and condition of fish. *Am. Fish. Soc. Symp.* **1990**, *8*, 93–108.
52. Solé, M.; Antó, M.; Baena, M.; Carrasson, M.; Cartes, J.E.; Maynou, F. Hepatic biomarkers of xenobiotic metabolism in eighteen marine fish from NW Mediterranean shelf and slope waters in relation to some of their biological and ecological variables. *Mar. Env. Res.* **2010**, *70*, 181–188. [CrossRef]
53. Sadekarpawar, S.; Parikh, P. Gonadosomatic and hepatosomatic indices of freshwater fish *Oreochromis mossambicus* in response to a plant nutrient. *World J. Zool.* **2013**, *8*, 110–118.
54. Le Cren, E.D. The Length-Weight Relationship and Seasonal Cycle in Gonad Weight and Condition in the Perch (*Perca fluviatilis*). *J. Anim. Ecol.* **1951**, *20*, 201–219. [CrossRef]
55. Sutton, S.G.; Bult, T.P.; Haedrich, R.L. Relationships among fat weight, body weight, water weight, and condition factors in wild Atlantic salmon parr. *Trans. Am. Fish. Soc.* **2000**, *129*, 527–538. [CrossRef]
56. Barrilli, G.H.C.; Rocha, O.; Negreiros, N.F.; Verani, J.R. Influence of environmental quality of the tributaries of the Monjolinho River on the relative condition factor (Kn) of the local ichthyofauna. *Biota Neotrop.* **2015**, *15*, e20140107. [CrossRef]
57. FAO. *The State of Mediterranean and Black Sea Fisheries—Special Edition*; General Fisheries Commission for the Mediterranean: Rome, Italy, 2023; 52p.
58. FishStat Plus. Universal Software for Fishery Statistical Time Series FAO Fisheries and Aquaculture of the United Nations—Version 4.03.05. 2023.
59. Russo, T.; Bitetto, I.; Carbonara, P.; Carlucci, R.; D'Andrea, L.; Facchini, M.T.; Lembo, G.; Maiorano, P.; Sion, L.; Spedicato, M.T.; et al. A holistic approach to fishery management: Evidence and insights from a central Mediterranean case study (Western Ionian Sea). *Front. Mar. Sci.* **2017**, *4*, 193. [CrossRef]
60. Sion, L.; Zupa, W.; Calculli, C.; Garofalo, G.; Hidalgo, M.; Jadaud, A.; Lefkaditou, E.; Ligas, A.; Peristeraki, P.; Bitetto, I.; et al. Spatial distribution pattern of European hake, *Merluccius merluccius* (Pisces: Merlucciidae), in the Mediterranean Sea. *Sci. Mar.* **2019**, *83*, 21–32. [CrossRef]
61. Maynou, F. Sale price flexibilities of Mediterranean hake and red shrimp. *Mar. Pol.* **2022**, *136*, 104904. [CrossRef]

62. Wiech, M.; Bienfait, A.M.; Silva, M.; Barre, J.; Sele, V.; Bank, M.S.; Bérail, S.; Tessier, E.; Amouroux, D.; Azad, A.M. Organ-specific mercury stable isotopes, speciation and particle measurements reveal methylmercury detoxification processes in Atlantic Bluefin Tuna. *J. Hazard. Mater.* **2024**, *473*, 134699. [CrossRef]
63. Honda, M.; Suzuki, N. Toxicities of Polycyclic Aromatic Hydrocarbons for aquatic animals. *Int. J. Environ. Res. Public Health* **2020**, *17*, 1363. [CrossRef]
64. Bukowska, B.; Mokra, K.; Michałowicz, J. Benzo[a]pyrene—Environmental Occurrence, Human Exposure, and Mechanisms of Toxicity. *Int. J. Mol. Sci.* **2022**, *23*, 6348. [CrossRef]
65. Mai, Y.; Wang, Y.; Geng, T.; Peng, S.; Lai, Z.; Wang, X.; Li, H. A systematic toxicologic study of polycyclic aromatic hydrocarbons on aquatic organisms via food-web bioaccumulation. *Sci. Total Environ.* **2024**, *929*, 172362. [CrossRef] [PubMed]
66. Liao, C.Y.; Fu, J.J.; Shi, J.B.; Zhou, Q.F.; Yuan, C.G.; Jiang, G.B. Methylmercury accumulation, histopathology effects, and cholinesterase activity alterations in medaka (*Oryzias latipes*) following sublethal exposure to methylmercury chloride. *Environ. Toxicol. Pharmacol.* **2006**, *22*, 225–233. [CrossRef] [PubMed]
67. Huang, W.; Gao, L.A.; Shan, X.J.; Lin, L.S.; Dou, S.Z. Toxicity testing of waterborne mercury with red sea bream (*Pagrus major*) embryos and larvae. *Bull. Environ. Contam. Toxicol.* **2011**, *86*, 398–405. [CrossRef] [PubMed]
68. Sfakianakis, D.G.; Renieri, E.; Kentouri, M.; Tsatsakis, A.M. Effect of heavy metals on fish larvae deformities: A review. *Environ. Res.* **2015**, *137*, 246–255. [CrossRef]
69. O'Bryhim, J.R.; Adams, D.H.; Spaet, J.L.Y.; Mills, G.; Lance, S.L. Relationships of mercury concentrations across tissue types, muscle regions and fins for two shark species. *Environ. Pollut.* **2017**, *223*, 323–333. [CrossRef]
70. Zheng, N.A.; Wang, S.; Dong, W.U.; Hua, X.; Li, Y.; Song, X.; Chu, Q.; Hou, S.; Li, Y. The toxicological effects of mercury exposure in marine fish. *Bull. Environ. Contam. Toxicol.* **2019**, *102*, 714–720. [CrossRef]
71. Girolametti, F.; Panfili, M.; Colella, S.; Frapiccini, E.; Annibaldi, A.; Illuminati, S.; Marini, M.; Truzzi, C. Mercury levels in Merluccius merluccius muscle tissue in the central Mediterranean Sea: Seasonal variation and human health risk. *Mar. Poll. Bull.* **2022**, *176*, 113461. [CrossRef]
72. Mascoli, A. Studio della Gonade Maschile di *Merluccius merluccius*: Cambiamenti Macroscopici e Molecolari Durante il Ciclo Riproduttivo. Master's Thesis, Università Politecnica delle Marche, Ancona, Italy, 2018.
73. Illuminati, S.; Truzzi, C.; Annibaldi, A.; Migliarini, B.; Carnevali, O.; Scarponi, G. Cadmium bioaccumulation and metallothionein induction in the liver of the Antarctic teleost *Trematomus bernacchii* during an on-site short-term exposure to the metal via sea water. *Toxicol. Eviron. Chem.* **2010**, *92*, 617–640. [CrossRef]
74. Illuminati, S.; Annibaldi, A.; Bau, S.; Scarchilli, C.; Ciardini, V.; Grigioni, P.; Girolametti, F.; Vagnoni, F.; Scarponi, G.; Truzzi, C. Seasonal evolution of size-segregated particulate mercury in the atmospheric aerosol over Terra Nova Bay, Antarctica. *Molecules* **2020**, *25*, 3971. [CrossRef]
75. Roveta, C.; Pica, D.; Calcinai, B.; Girolametti, F.; Truzzi, C.; Illuminati, S.; Annibaldi, A.; Puce, S. Hg levels in marine Porifera of Montecristo and Giglio Islands (Tuscan archipelago, Italy). *Appl. Sci.* **2020**, *10*, 4342. [CrossRef]
76. Truzzi, C.; Annibaldi, A.; Girolametti, F.; Giovannini, L.; Riolo, P.; Ruschioni, S.; Olivotto, I.; Illuminati, S. A chemically safe way to produce insect biomass for possible application in feed and food production. *Int. J. Environ. Res. Public Health* **2020**, *17*, 2121. [CrossRef] [PubMed]
77. Girolametti, F.; Annibaldi, A.; Illuminati, S.; Damiani, E.; Carloni, P.; Truzzi, C. Essential and potentially toxic elements (PTEs) content in European tea (*Camellia sinensis*) leaves: Risk assessment for consumers. *Molecules* **2023**, *28*, 3802. [CrossRef]
78. Annibaldi, A.; Truzzi, C.; Carnevali, O.; Pignalosa, P.; Api, M.; Scarponi, G.; Illuminati, S. Determination of Hg in farmed and wild Atlantic bluefin tuna (*Thunnus thynnus* L.) muscle. *Molecules* **2019**, *24*, 1273. [CrossRef] [PubMed]
79. Caroselli, E.; Frapiccini, E.; Franzellitti, S.; Palazzo, Q.; Prada, F.; Betti, M.; Goffredo, S.; Marini, M. Accumulation of PAHs in the tissues and algal symbionts of a common Mediterranean coral: Skeletal storage relates to population age structure. *Sci. Total Environ.* **2020**, *743*, 140781. [CrossRef] [PubMed]
80. João Ramalhosa, M.; Paíga, P.; Morais, S.; Delerue-Matos, C.; Prior Pinto Oliveira, M.B. Analysis of polycyclic aromatic hydrocarbons in fish: Evaluation of a quick, easy, cheap, effective, rugged, and safe extraction method. *J. Sep. Sci.* **2009**, *32*, 3529–3538. [CrossRef]
81. ICH Steering Committee. *ICH Harmonized Tripartite Guideline, Validation of Analytical Procedures, Text and Methodology Q2(R1)*; International Conference on Harmonization of Technical Requirements for Registration of Pharmaceuticals for Human Use: London, UK, 1996.
82. Zarantoniello, M.; Randazzo, B.; Nozzi, V.; Truzzi, C.; Giorgini, E.; Cardinaletti, G.; Freddi, L.; Ratti, S.; Girolametti, F.; Osimani, A.; et al. Physiological responses of Siberian sturgeon (*Acipenser baerii*) juveniles fed on full-fat insect-based diet in an aquaponic system. *Sci. Rep.* **2021**, *11*, 1057. [CrossRef] [PubMed]
83. Truzzi, C.; Annibaldi, A.; Illuminati, S.; Antonucci, M.; Api, M.; Scarponi, G.; Lombardo, F.; Pignalosa, P.; Carnevali, O. Characterization of the fatty acid composition in cultivated atlantic bluefin tuna (*Thunnus thynnus* L.) muscle by gas chromatography-mass spectrometry. *Anal. Lett.* **2018**, *51*, 2981–2993. [CrossRef]
84. Truzzi, C.; Illuminati, S.; Antonucci, M.; Scarponi, G.; Annibaldi, A. Heat shock influences the fatty acid composition of the muscle of the Antarctic fish *Trematomus bernacchii*. *Mar. Environ. Res.* **2018**, *139*, 122–128. [CrossRef]
85. Uyanto, S.S. An Extensive Comparisons of 50 Univariate Goodness-of-fit Tests for Normality. *Austrian J. Stat.* **2022**, *51*, 45–97. [CrossRef]

86. Kay, M.; Elkin, L.; Higgins, J.; Wobbrock, J. ARTool: Aligned Rank Transform for Nonparametric Factorial ANOVAs. R Package Version 0.11.1. 2021. Available online: https://github.com/mjskay/ARTool (accessed on 8 October 2024). [CrossRef]
87. Elkin, L.; Kay, M.; Higgins, J.; Wobbrock, J. An Aligned Rank Transform Procedure for Multifactor Contrast Tests. In Proceedings of the UIST '21: The 34th Annual ACM Symposium on User Interface Software and Technology, Virtual Event, 10–14 October 2021. [CrossRef]
88. Quinn, G.P.; Keough, M.J. *Experimental Design and Data Analysis for Biologists*; Cambridge University Press: Cambridge, UK, 2002.
89. McPherson, G. *Statistics in Scientific Investigation*; Springer: New York, NY, USA, 1990.
90. Anderson, T.W.; Finn, J.D. *The New Statistical Analysis of Data*; Springer: New York, NY, USA, 1996.
91. R Core Team. *R: A Language and Environment for Statistical Computing*; R Foundation for Statistical Computing: Vienna, Austria, 2024.
92. Candelma, M.; Marisaldi, L.; Bertotto, D.; Radaelli, G.; Gioacchini, G.; Santojanni, A.; Colella, S.; Carnevali, O. Aspects of Reproductive Biology of the European Hake (*Merluccius merluccius*) in the Northern and Central Adriatic Sea (Gsa 17-Central Mediterranean Sea). *J. Mar. Sci. Eng.* **2021**, *9*, 389. [CrossRef]
93. Mascoli, A.; Candelma, M.; Santojanni, A.; Carnevali, O.; Colella, S. Reproductive biology of male European hake (*Merluccius merluccius*) in Central Mediterranean Sea: An overview from macroscopic to molecular investigation. *Biology* **2023**, *12*, 562. [CrossRef] [PubMed]
94. Freund, R.J.; Wilson, W.J.; Mohr, D.L. *Statistical Methods*, 3rd ed.; Academic Press: Cambridge, MA, USA, 2010.
95. Adams, D.H.; Onorato, G.V. Mercury concentrations in red drum, *Sciaenops ocellatus*, from estuarine and offshore waters of Florida. *Mar. Pollut. Bull.* **2005**, *50*, 291–300. [CrossRef]
96. Bank, M.S.; Chesney, E.; Shine, J.P.; Maage, A.; Senn, D.B. Mercury bioaccumulation and trophic transfer in sympatric snapper species from the Gulf of Mexico. *Ecol. Appl.* **2007**, *17*, 2100–2110. [CrossRef] [PubMed]
97. Tremain, D.M.; Adams, D.H. Mercury in groupers and sea basses from the Gulf of Mexico: Relationships with size, age, and feeding ecology. *Trans. Am. Fish. Soc.* **2012**, *141*, 1274–1286. [CrossRef]
98. Murua, H.; Motos, L.; Lucio, P. Reproductive modality and batch fecundity of the European hake (*Merluccius merluccius* L.) in the Bay of Biscay. *Calif. Coop. Ocean. Fish. Investig. Rep.* **1998**, *39*, 196–203.
99. Korta, M.; Murua, H.; Kurita, Y.; Kjesbu, O.S. How are the oocytes recruited in an indeterminate fish? Applications of stereological techniques along with advanced packing density theory on European hake (*Merluccius merluccius* L.). *Fish. Res.* **2010**, *104*, 56–63. [CrossRef]
100. Domínguez-Petit, R.; Saborido-Rey, F. New bioenergetic perspective of European hake (*Merluccius merluccius* L.) reproductive ecology. *Fish. Res.* **2010**, *104*, 83–88. [CrossRef]
101. Murua, H.; Lucio, P.; Santurtun, M.; Motos, L. Seasonal variation in egg production and batch fecundity of European hake *Merluccius merluccius* (L.) in the Bay of Biscay. *J. Fish. Biol.* **2006**, *69*, 1304–1316. [CrossRef]
102. Kjesbu, O.S.; Murua, H.; Witthames, P.; Saborido-Rey, F. Method development and evaluation of stock reproductive potential of marine fish. *Fish. Res.* **2010**, *104*, 1–7. [CrossRef]
103. Murua, H. The biology and fisheries of European hake, *Merluccius merluccius*, in the north-east Atlantic. *Adv. Mar. Biol.* **2010**, *58*, 97–154. [CrossRef]
104. Pineiro, C.; Sainza, M. Age estimation, growth and maturity of the European hake (*Merluccius merluccius* (Linnaeus, 1758) from Iberian Atlantic waters. *ICES J. Mar. Sci.* **2003**, *60*, 1086–1102. [CrossRef]
105. Recasens, L.; Chiericoni, V.; Belcari, P. Spawning Pattern and Batch Fecundity of the European Hake (*Merluccius merluccius* (Linnaeus, 1758)) in the Western Mediterranean. *Sci. Mar.* **2008**, *72*, 721–732. [CrossRef]
106. Soykan, O.; Ilkyaz, A.T.; Metan, G.; Kinacigal, H.T. Age, Growth and Reproduction of European Hake (*Merluccius merluccius* (Linn., 1758)) in the Central Aegean Sea, Turkey. *J. Mar. Biol. Assoc. UK* **2015**, *95*, 829–837. [CrossRef]
107. Carbonara, P.; Porcu, C.; Donnaloia, M.; Pesci, P.; Sion, L.; Spedicato, M.T.; Zupa, W.; Vitale, F.; Follesa, M.C. The spawning strategy of European hake (*Merluccius merluccius*, L. 1758) across the Western and Central Mediterranean Sea. *Fish. Res.* **2019**, *219*, 105333. [CrossRef]
108. Zorica, B.; Isajlović, I.; Vrgoč, N.; Kec, V.Č.; Medvešek, D.; Vuletin, V.; Radonić, I.; Cvitanić, R.; Lepen Pleić, I.; Šestanović, M. Reproductive Traits of the European Hake, *Merluccius merluccius* (L. 1758), in the Adriatic Sea. *Acta Adriat.* **2021**, *62*, 183–198. [CrossRef]
109. Burger, J.; Fossi, C.; Mc Clellan-Green, P.; Orlando, E.F. Methodologies, bioindicators, and biomarkers for assessing gender-related differences in wildlife exposed to environmental chemicals. *Environ. Res.* **2007**, *104*, 135–152. [CrossRef]
110. Grgec, A.S.; Kljaković-Gašpić, Z.; Orct, T.; Tičina, V.; Sekovanić, A.; Jurasović, J.; Piasek, M. Mercury and selenium in fish from the eastern part of the Adriatic Sea: A risk-benefit assessment in vulnerable population groups. *Chemosphere* **2020**, *261*, 127742. [CrossRef]
111. Storelli, M.M.; Storelli, A.; Giacominelli-Stuffler, R.; Marcotrigiano, G.O. Mercury speciation in the muscle of two commercially important fish, hake (*Merluccius merluccius*) and striped mullet (*Mullus barbatus*) from the Mediterranean Sea: Estimated weekly intake. *Food Chem.* **2005**, *89*, 295–300. [CrossRef]
112. Aksu, A.; Balkis, N.; Taşkin, O.S.; Erşan, M.S. Toxic metal (Pb, Cd, As and Hg) and organochlorine residue levels in hake (*Merluccius merluccius*) from the Marmara Sea, Turkey. *Environ. Monit. Assess.* **2011**, *182*, 509–521. [CrossRef]

113. Hornung, H.; Zismann, L.; Oren, O.H. Mercury in twelve Mediterranean trawl fishes of Israel. *Environ. Int.* **1980**, *3*, 243–248. [CrossRef]
114. Kontas, A. Mercury in the Izmir Bay: An assessment of contamination. *J. Mar. Syst.* **2006**, *61*, 67–78. [CrossRef]
115. Jureša, D.; Blanuša, M. Mercury, arsenic, lead and cadmium in fish and shellfish from the Adriatic Sea. *Food Addit. Contam.* **2003**, *20*, 241–246. [CrossRef] [PubMed]
116. Perugini, M.; Visciano, P.; Manera, M.; Zaccaroni, A.; Olivieri, V.; Amorena, M. Heavy metal (As, Cd, Hg, Pb, Cu, Zn, Se) concentrations in muscle and bone of four commercial fish caught in the central Adriatic Sea, Italy. *Environ. Monit. Assess.* **2014**, *186*, 2205–2213. [CrossRef]
117. Hammerschmidt, C.R.; Sandheinrich, M.B. Maternal diet during oogenesis is the major source of methylmercury in fish embryos *Environ. Sci. Technol.* **2005**, *39*, 3580–3584. [CrossRef]
118. Del Carmen Alvarez, M.; Murphy, C.A.; Rose, K.A.; McCarthy, I.D.; Fuiman, L.A. Maternal body burdens of methylmercury impair survival skills of offspring in Atlantic croaker (*Micropogonias undulatus*). *Aquat. Toxicol.* **2006**, *80*, 329–337. [CrossRef]
119. Sackett, D.K.; Aday, D.D.; Rice, J.A.; Cope, W.G. Maternally transferred mercury in wild largemouth bass, *Micropterus salmoides* *Environ. Poll.* **2013**, *178*, 493–497. [CrossRef]
120. Girolametti, F.; Frapiccini, E.; Annibaldi, A.; Illuminati, S.; Panfili, M.; Marini, M.; Santojanni, A.; Truzzi, C. Total Mercury (THg) Content in Red Mullet (*Mullus barbatus*) from Adriatic Sea (Central Mediterranean Sea): Relation to Biological Parameters, Sampling Area and Human Health Risk Assessment. *Appl. Sci.* **2022**, *12*, 10083. [CrossRef]
121. Belhoucine, F.; Alioua, A.; Bouhadiba, S.; Boutiba, Z. Impact of some biotics and abiotics factors on the accumulation of heavy metals by a biological model *Merluccius merluccius* in the bay of Oran in Algeria. *J. Biodivers. Environ. Sci.* **2014**, *5*, 33–44.
122. Mille, T.; Soulier, L.; Caill-Milly, N.; Cresson, P.; Morandeau, G.; Monperrus, M. Differential micropollutants bioaccumulation in European hake and their parasites *Anisakis* sp. *Environ. Pollut.* **2020**, *265*, 115021. [CrossRef]
123. Mille, T.; Bisch, A.; Caill-Milly, N.; Cresson, P.; Deborde, J.; Gueux, A.; Morandeau, G.; Monperrus, M. Distribution of mercury species in different tissues and trophic levels of commonly consumed fish species from the south Bay of Biscay (France). *Mar. Pollut. Bull.* **2021**, *166*, 112172. [CrossRef]
124. Candelma, M.; Valle, L.D.; Colella, S.; Santojanni, A.; Carnevali, O. Cloning, characterization, and molecular expression of gonadotropin receptors in European hake (*Merluccius merluccius*), a multiple-spawning species. *Fish Physiol. Biochem.* **2018**, *44*, 895–910. [CrossRef] [PubMed]
125. Soares-Gomes, A.; Neves, R.L.; Aucélio, R.; Van Der Ven, P.H.; Pitombo, F.B.; Mendes, C.L.T.; Ziolli, R.L. Changes and variations of polycyclic aromatic hydrocarbon concentrations in fish, barnacles and crabs following an oil spill in a mangrove of Guanabara Bay, Southeast Brazil. *Mar. Pollut. Bull.* **2010**, *60*, 1359–1363. [CrossRef] [PubMed]
126. Marrone, R.; Smaldone, G.; Pepe, T.; Mercogliano, R.; De Felice, A.; Anastasio, A. Polycyclic Aromatic Hydrocarbons (Pahs) in Seafoods Caught in Corigliano Calabro Gulf (CS, Italy). *Ital. J. Food Saf.* **2012**, *1*, 41–46. [CrossRef]
127. Perugini, M.; Visciano, P.; Manera, M.; Turno, G.; Lucisano, A.; Amorena, M. Polycyclic Aromatic Hydrocarbons in Marine Organisms from the Gulf of Naples, Tyrrhenian Sea. *J. Agric. Food Chem.* **2007**, *55*, 2049–2054. [CrossRef]
128. Moraleda-Cibrián, N.; Carrassón, M.; Rosell-Melé, A. Polycyclic aromatic hydrocarbons, polychlorinated biphenyls and organochlorine pesticides in European hake (*Merluccius merluccius*) muscle from the Western Mediterranean Sea. *Mar. Poll. Bull* **2015**, *95*, 513–519. [CrossRef]
129. Bodiguel, X.; Loizeau, V.; Le Guellec, A.M.; Roupsard, F.; Philippon, X.; Mellon-Duval, C. Influence of sex, maturity and reproduction on PCB and pp′ DDE concentrations and repartitions in the European hake (*Merluccius merluccius*, L.) from the Gulf of Lions (NW Mediterranean). *Sci. Total Environ.* **2009**, *408*, 304–311. [CrossRef] [PubMed]
130. Zhao, Z.; Zhang, L.; Cai, Y.; Chen, Y. Distribution of polycyclic aromatic hydrocarbon (PAH) residues in several tissues of edible fishes from the largest freshwater lake in China, Poyang Lake, and associated human health risk assessment. *Ecotoxicol. Environ. Saf.* **2014**, *104*, 323–331. [CrossRef]
131. Frapiccini, E.; Annibaldi, A.; Betti, M.; Polidori, P.; Truzzi, C.; Marini, M. Polycyclic aromatic hydrocarbon (PAH) accumulation in different common sole (*Solea solea*) tissues from the North Adriatic Sea peculiar impacted area. *Mar. Poll. Bull.* **2018**, *137*, 61–68 [CrossRef] [PubMed]
132. Xu, F.L.; Wu, W.J.; Wang, J.J.; Qin, N.; Wang, Y.; He, Q.S.; Tao, S. Residual levels and health risk of polycyclic aromatic hydrocarbons in freshwater fishes from Lake Small Bai-Yang-Dian, Northern China. *Ecol. Model.* **2011**, *222*, 275–286. [CrossRef]
133. Mashroofeh, A.; Bakhtiari, A.R.; Pourkazemi, M. Distribution and composition pattern of polycyclic aromatic hydrocarbons in different tissues of sturgeons collected from Iranian coastline of the Caspian Sea. *Chemosphere* **2015**, *120*, 575–583. [CrossRef]
134. Jafarabadi, A.R.; Bakhtiari, A.R.; Yaghoobi, Z.; Yap, C.K.; Maisano, M.; Cappello, T. Distributions and compositional patterns of polycyclic aromatic hydrocarbons (PAHs) and their derivatives in three edible fishes from Kharg coral Island, Persian Gulf, Iran *Chemosphere* **2019**, *215*, 835–845. [CrossRef]
135. Qin, N.; He, W.; Liu, W.; Kong, X.; Xu, F.; Giesy, J.P. Tissue distribution, bioaccumulation, and carcinogenic risk of polycyclic aromatic hydrocarbons in aquatic organisms from Lake Chaohu, China. *Sci. Total Environ.* **2020**, *749*, 141577. [CrossRef] [PubMed]
136. Stow, C.A.; Jackson, L.J.; Amrhein, J.F. An examination of the PCB: Lipid relationship among individual fish. *Can. J. Fish. Aquat Sci.* **1997**, *54*, 1031–1038. [CrossRef]

137. Devier, M.H.; Augagneur, S.; Budzinski, H.; Le Menach, K.; Mora, P.; Narbonne, J.F.; Garrigues, P. One-year monitoring survey of organic compounds (PAHs, PCBs, TBT), heavy metals and biomarkers in blue mussels from the Arcachon Bay, France. *J. Environ. Monit.* **2005**, *7*, 224–240. [CrossRef]
138. García-Hernández, J.; Cadena-Cárdenas, L.; Betancourt-Lozano, M.; García-De-La-Parra, L.M.; García-Rico, L.; Márquez-Farías, F. Total mercury content found in edible tissues of top predator fish from the Gulf of California, Mexico. *Toxicol. Environ. Chem.* **2007**, *89*, 507–522. [CrossRef]
139. Al-Sulaiti, M.M.; Soubra, L.; Ramadan, G.A.; Ahmed, A.Q.S.; Al-Ghouti, M.A. Total Hg levels distribution in fish and fish products and their relationships with fish types, weights, and protein and lipid contents: A multivariate analysis. *Food Chem.* **2023**, *421*, 136163. [CrossRef]
140. Lloret, J.; Planes, S. Condition, feeding and reproductive potential of white seabream (*Diplodus sargus*) as indicators of habitat quality and the effect of protection in the northwestern Mediterranean. *Mar. Ecol. Prog. Ser.* **2003**, *248*, 197–208. [CrossRef]
141. Lloret, J.; Galzin, R.; Gil de Sola, A.; Souplet, A.; Demestre, M. Habitat related differences in lipid reserves of some exploited fish species in the north-western Mediterranean continental shelf. *J. Fish. Biol.* **2005**, *67*, 51–67. [CrossRef]
142. Lloret, J.; Demestre, M.; Sánchez-Pardo, J. Lipid reserves of red mullet (*Mullus barbatus*) in the north-western Mediterranean. *Sci. Mar.* **2007**, *71*, 269–277. [CrossRef]
143. Pethybridge, H.; Daley, R.; Virtue, P.; Butler, E.C.V.; Cossa, D.; Nichols, P.D. Lipid and mercury profiles of 61 mid-trophic species collected off south-eastern Australia. *Mar. Freshw. Res.* **2010**, *61*, 1092–1108. [CrossRef]
144. Łuczyńska, J.; Łuczyński, M.J.; Nowosad, J.; Kowalska-Góralska, M.; Senze, M. Total mercury and fatty acids in selected fish species on the Polish market: A risk to human health. *Int. J. Environ. Res. Public Health* **2022**, *19*, 10092. [CrossRef] [PubMed]
145. Shulman, G.E.; Love, R.M. *Advances in Marine Ecology. The Biochemical Ecology of Marine Fishes*; Southward, A.J., Tyler, P.A., Young, C.M., Eds.; Academic Press: London, UK, 1999.
146. Ferrer Maza, D.; Lloret, J.; Muñoz, M.; Faliex, E.; Vila, S.; Sasal, P. Parasitism, Condition and Reproduction of the European Hake (*Merluccius merluccius*) in the Northwestern Mediterranean Sea. *J. Mar. Sci.* **2014**, *71*, 1088–1099. [CrossRef]

animals

MDPI

Article

Stock Dynamics of Female Red King Crab in a Small Bay of the Barents Sea in Relation to Environmental Factors

Alexander G. Dvoretsky and Vladimir G. Dvoretsky *

Murmansk Marine Biological Institute of the Russian Academy of Sciences (MMBI RAS), 183038 Murmansk, Russia; ag-dvoretsky@yandex.ru
* Correspondence: v-dvoretsky@yandex.ru

Simple Summary: The current state of knowledge regarding the impact of climate on the number, weight, and size of female red king crabs in the Barents Sea is insufficient for drawing any conclusions. In this study, we examined the relationship between long-term fluctuations in female stock indices and the average weight of an individual crab, as well as temperature conditions. Our analysis demonstrated a robust correlation between these variables at varying time lags (6–10 years), supporting the hypothesis that female maturation occurs more rapidly during periods of elevated temperatures. Our findings indicate that warmer water conditions are conducive to the survival and growth of young crabs. The most significant factors influencing female abundance and biomass are seawater temperatures between June and August. Our findings could prove valuable for fishery managers in anticipating periods of high crab productivity and abundance, as well as for coastal managers in predicting abundance fluctuations of red king crab in the Barents Sea.

Abstract: Stock–recruitment relationships depend on the total abundance of females, their fecundity, and patterns of their maturation. However, the effects of climatic conditions on the abundance, biomass, and mean weight of female red king crabs, *Paralithodes camtschaticus*, from the introduced population (Barents Sea) have not yet been studied. For this reason, we analyzed long-term fluctuations in stock indices and the average weight of an individual crab in a small bay of the Barents Sea and related these parameters to the dynamics of temperature conditions (temperature in January–December, mean yearly temperature, and temperature anomaly) in the sea. The average weight of a crab at age 6–9 had strong negative correlations with water temperature at lags 8 and 9, indicating faster female maturation in warm periods. Positive relationships were registered between temperature and stock indices for 15–19-year-old females at lag 4 and for 10–14-year-old females at lag 10, supporting the idea of higher survival rates of juveniles and their rapid development being a response to a pool of warm waters. Both redundancy and correlation analyses revealed seawater temperatures in June–August being the most important predictors of female abundance and biomass, indicating that favorable temperature conditions in the first 3 months of crab benthic life result in high survivorship rates for red king crabs.

Keywords: red king crab; females; Barents Sea; temperature conditions; climate; abundance

1. Introduction

The red king crab, *Paralithodes camtschaticus* (Tilesius, 1815), is an important species in terms of its commercial value and the ecological role it plays in benthic communities throughout its native areas in the western and eastern Bering Sea, USA, and the Sea of Okhotsk, Russia [1]. The success of red king fisheries in native habitats initiated Soviet scientists to introduce this lithodid crab into the Barents Sea. Although this region is known to have high productivity owing to interactions between cold Arctic and warm Atlantic waters and a wide range of environmental conditions [2–5], there are no native commercially important crab species here. The transoceanic introduction experiment conducted in the 1960s was successful, and the establishment of a new self-sustaining population of red king crab in the Barents Sea was reported in the mid-1990s. For this reason, a Russian–Norwegian joint research fishery for *P. camtschaticus* occurred from 1994 to 2001 [6]. Commercial fisheries were opened in 2002 in Norwegian and in 2004 in Russian waters of the Barents Sea [1]. The red king crab supports a viable fishery with annual stocks (landings) accounting for 199, 192, and 159 thousand metric tons (11,629, 12,529, and 10,420 t) in 2021, 2022, and 2023, respectively [7–9].

The population dynamics of the Barents Sea red king crab have been extensively studied, owing to the species' invasive status and high commercial importance [1,10–17]. Many factors affect the development and distribution of crustaceans [18]. Among these, climatic fluctuations play an important role in the functioning of benthic communities in the Arctic regions [19–23]. Some studies indicate that the biomass of the Barents Sea benthos may fluctuate considerably in response to large-scale climatic forcing [24–26], providing evidence that both abundance and biomass of the whole bottom community and/or its members or groups may be used as indicators of climate change [5]. Recent investigations have documented pronounced climatic shifts in the Arctic seas with a warming trend starting in the early 2000s [27–29]. Climatic variations in the Barents Sea can be detected as a series of temperature anomalies along standard transects of the sea [30,31].

The response of commercially important species to changing climate may vary widely because of exploitation pressure and other direct and indirect effects of environmental variations [32,33]. Recently, we analyzed inter-annual fluctuations in the total number of recruits and commercially sized male red king crabs and found some correlations between these indices and environmental factors, such as water temperature and the North Atlantic oscillation and Arctic oscillation, with time lags [34]. We concluded that the abundance of males is a poor indicator of climate shift because of fishing pressure on this group. In contrast, both the abundance and biomass of juvenile red king crabs have been shown to relate to environmental factors such as water temperature and temperature anomalies, as well as to cod predation. These correlations can vary significantly depending on the age group of juvenile red king crabs [21]. At the same time, the relationships between environmental factors and stock indices of adult female red king crabs in the Barents Sea are far less evident. Females are considered to be the most important functional groups in each red king crab population because their number and individual fecundity directly affect the number of released larvae, the number of juvenile crabs, and, therefore, the commercial stock in the area [35].

In the coastal Barents Sea, the abundance of egg-bearing females in some years may reach 19,000,000 individuals, with mean catch per unit efforts (CPUEs) for these females ranging from 1.9 to 7.3 individuals per pot [16]. Although time-series analysis has demonstrated that the introduction of red king crab and their subsequent population growth have not adversely affected local fish and shellfish stocks [36], negative consequences for coastal benthic communities associated with increased crab abundance have been observed. These consequences include a reduction in species diversity, habitat disturbance, and a simplification of benthic community structure [37–42].

Thus, knowledge about relationships between environmental drivers and stock indices of adult female red king crabs could have important implications, not only for crab fishery management, but also for the prediction of possible changes in benthic communities associated with feeding activity of the crabs and their competition for food with native inhabitants. For this reason, the aim of our paper was to investigate the effect of temperature conditions on the abundance characteristics of female red king crabs in a typical bay of the Barents Sea.

2. Material and Methods

2.1. Red King Crab Stock Indices

Red king crabs were caught by divers at depths of 8–40 m in Dalnezelenetskaya Bay in July–August, 2003–2017. A detailed description of the study area is presented in previous papers [43–45].

Each crab was sexed and weighed and the carapace length (CL, the straight-line distance across the carapace from the posterior margin of the right eye orbit to the medial-posterior margin of the carapace) was measured using calipers. In our study, we considered adult female red king crabs. The crabs were divided into three age groups: 6–9-year-olds (95–125 mm CL), 10–14-year-olds (126–150 mm CL), and 15–19-year-olds (151–170 mm CL), according to published size-at-age data [14,35,46,47].

The area of the bay was divided into several parts with the same types of biological communities (benthic organisms living on hard and soft grounds, kelps, and sand). A standardized transect grid comprising 10 to 25 transects was established to encompass a diverse range of depths (from 3 to 42 m) and various types of benthic communities [12,21]. When duplicate transects were analyzed, the obtained values were averaged to ensure reliable representations of the data. The number of crabs and their biomass on each transect line (observation width: 15 m) were used to estimate the total abundance (the total number of crabs) and the total biomass using a spline approximation method (the approximate representation of a function in a given class from incomplete information using splines) with the Chartmaster Software Tool ver. 1.0 [48]. In addition to the stock indices, the analyses were also applied for averaged individual weights (g) of different aged crabs.

2.2. Environmental Data

It is accepted that climatic conditions of the southern Barents Sea are determined by temperature characteristics of Murmansk coastal waters (stations 1–3 of the standard transect called "Kola Section" 33°30′ E, 69°–78° N, 0–200 m depth). Mean seawater temperature in the Kola Section (averaged month values for each year and the averaged year value for 50–200 m depths, which reflects near-bottom temperatures) in 1984–2017 and data on temperature anomaly in the section were derived from a long-term dataset collected by Polar Research Institute of Marine Fisheries and Oceanography (www.pinro.ru accessed on 20 March 2018). The calculation of anomaly values involved the subtraction of the mean from 1951 to 2017 from the original data and dividing the result by the standard deviation. Initially, we also tested the North Atlantic oscillation indices and salinity data as environmental predictors of the red king crab stock indices, but these factors had no effects on *P. camtschaticus* females and were excluded from further consideration.

2.3. Data Analysis

Firstly, we calculated simple pairwise correlations (Pearson's correlation coefficient) between crab abundance indices and environmental factors. We used both data for the current year and data with time lags of up to 19 years according to the maximum age of a female crab in our study. The normality and homogeneity of variances were tested before performing the statistical analyses. The assumption of homogeneity of variance was verified using Levene's test and the assumption of normality was verified using the Shapiro–Wilk test. The Benjamini–Hochberg procedure was used to correct p-values for multiple testing [49]. In each case, the total number of tests was 14 and a false discovery rate was set at 10%. Since simple correlation overstates the significance of associations in the case of significant autocorrelation in time-series data [50], we tested the data for temporal autocorrelations using the Box–Pierce test and found that the *p*-values were higher than the critical value of 0.05.

Secondly, to analyze the complex factor loadings and for visualization of our results, we used multivariate techniques. The influence of water temperature between 50 and 200 m depth on the biomass of females at different ages was examined using redundancy analysis (RDA). Biomass was used as a metric instead of abundance because this parameter explained a higher proportion of the total variation. RDA is an analytical tool that explicitly models response variables as a function of explanatory variables [51]. This approach can be applied to assess the effects of environmental variables on the stock characteristics of red king crabs. This method was used because a preliminary detrended correspondence analysis showed that the lengths of gradient were all less than 3 [52]. The graphical output of RDA consists of a simple biplot showing the associations between the quantitative explanatory variables and the response variables. This analysis was applied only to the dataset for which the correlation analyses indicated significant results. Prior to RDA, quantitative data were log-transformed.

Correlation testing and redundancy analyses were carried out with the PAST 3.26 and CANOCO 4.5 software packages.

3. Results

3.1. Environmental Conditions

During the period of 1984–2017, temperature conditions varied strongly in the southern Barents Sea. Monthly and mean temperature values as well as temperature anomalies are presented in Figure 1.

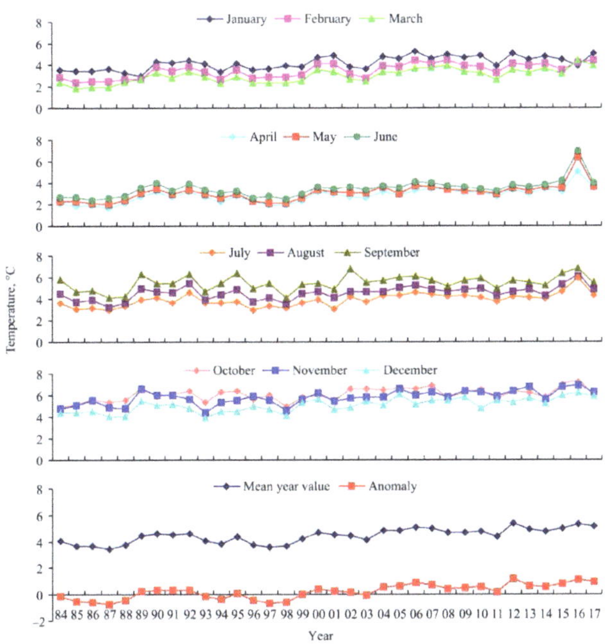

Figure 1. Near-bottom temperature fluctuations in the coastal zone of the Barents Sea (stations 1–3 of the Kola section at 50–200 m depth) in 1984–2017.

March and April were the coldest months, while October and November were the warmest, at a depth of 50–200 m. A warming trend in the Barents Sea began to be registered in 1989 when negative temperature anomalies were reversed and the mean temperature increased by 1 °C in comparison to previous years. At the beginning of the twenty-first century, the advection of the North Atlantic current continued to increase. In the Barents Sea, the warm temperature anomaly peaked (0.86) in 2006, followed by a decrease in the next years (Figure 2).

A second peak (1.14) was registered in 2012. During the subsequent years, water temperature anomalies remained high (0.52–1.1). The highest levels of seawater temperature in the Kola section were found in 2006, 2012, 2016, and 2017, reaching 5.08, 5.36, 5.32, and 5.15 °C, respectively.

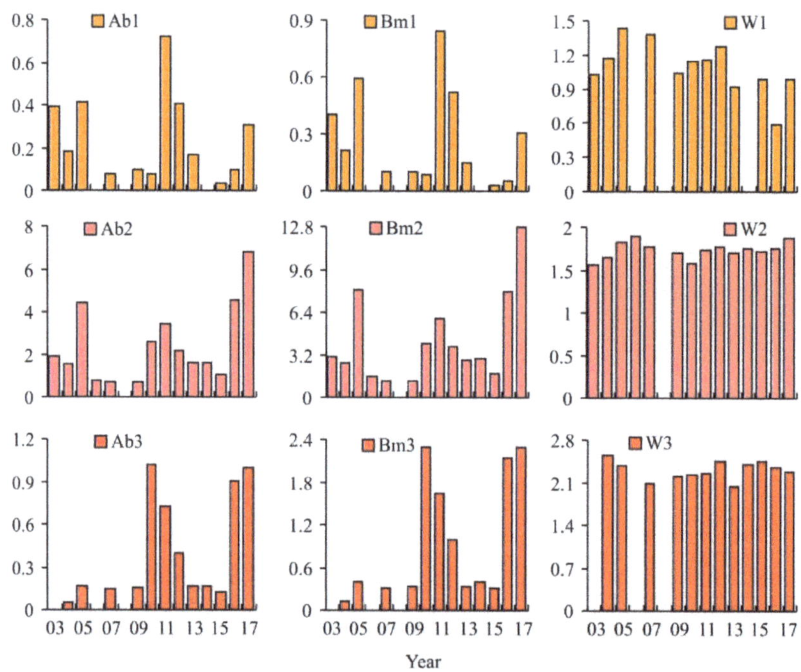

Figure 2. Variations in female red king crab stock indices and mean weight over 2003–2017. Ab—abundance (thousand crabs), Bm—biomass (metric tons), W—weight (kg). 1—aged 6–9 years, 2—aged 10–14 years, 3—aged 15–19 years.

3.2. Red King Crab Stock Indices

During the study period, the 6–9-year-old group had the lowest abundance among adult female red king crabs. The maximum abundance of 725 crabs (biomass 0.84 t) was registered in 2010, while intermediate numbers of 400 crabs (0.4–0.6 t) were found in 2003, 2005, and 2011. The mean weight increased from 1.03 kg in 2003 to a maximum of 1.43 in 2005. A decreasing trend was registered in subsequent years, with a minimum value of 0.59 kg found in 2016.

A total of three peaks were registered in the abundance and biomass of *P. camtschaticus* at ages 10–14. These peaks occurred almost every 6 years across the study period: in 2005 (4415 crabs, 8.05 t), in 2011 (3460 crabs, 5.99 t), and in 2017 (6835 crabs, 12.78 t). The mean weight of a female increased from 1.57 kg in 2003 to 1.90 kg in 2006, then decreased to 1.58 kg in 2010, before increasing again to 1.87 kg in 2017.

The abundance and biomass of 15–19-year-old females were the similar between 2005 and 2009 (150–170 crabs, 0.35–0.41 t). A sharp increase in these indices was observed in 2010 (1030 crabs, 2.29 t). Over the next 3 years, the parameters decreased to the initial levels and increased again to 900–1000 crabs or 2.15–2.18 t in 2016–2017. The mean weight of a crab in this group decreased from 2.56 kg in 2004 to 2.10 in 2007. A smooth increase was observed in the next 4 years. In 2012–2015, mean weights were stable (2.40–2.45 kg) but a lower value (2.04 kg) was observed in 2013.

3.3. Relationships Between Crab Abundance and Environmental Variables

In the case of the 6–9-year-old group, correlation analyses indicated no significant associations between stock indices and environmental variables, while strong negative relationships were registered for the average weight of a crab from this group at time lags 8 and 9 years (Table 1).

Table 1. Correlation coefficients (r) between red king crab abundance, stock and weight characteristics, and environmental conditions in the Barents Sea in 2003–2017.

Parameter	Ab1	Ab2	Ab3	Bm1	Bm2	Bm3	W1	W2	W3
					Lag 4				
T1	0.104	0.333	0.501	0.127	0.312	0.508	−0.211	−0.154	0.261
T2	0.281	0.408	**0.524**	0.298	0.388	**0.532**	−0.055	−0.108	0.359
T3	0.360	0.366	**0.514**	0.375	0.349	**0.522**	0.010	−0.067	0.283
T4	0.242	0.306	**0.546**	0.255	0.293	**0.556**	−0.142	0.043	0.256
T5	0.089	0.203	**0.556**	0.102	0.187	**0.554**	0.000	0.002	0.018
T6	0.168	0.260	**0.643**	0.167	0.245	**0.637**	0.107	0.063	−0.130
T7	−0.151	−0.116	0.435	−0.223	−0.130	0.425	−0.292	−0.152	−0.257
T8	−0.149	−0.014	0.466	−0.217	−0.031	0.447	−0.330	−0.220	−0.445
T9	−0.427	−0.280	0.041	−0.464	−0.278	0.027	−0.463	0.161	−0.490
T10	−0.173	−0.184	0.248	−0.220	−0.186	0.228	−0.261	0.060	−0.602
T11	−0.042	0.348	0.476	−0.147	0.355	0.473	−0.471	0.065	−0.246
T12	0.129	0.012	0.141	0.039	0.013	0.137	0.187	−0.224	−0.261
T	0.024	0.444	**0.739**	−0.029	0.423	**0.743**	−0.373	−0.059	0.024
Ta	0.024	0.444	**0.739**	−0.029	0.423	**0.743**	−0.373	−0.059	0.024
					Lag 8				
T1	−0.380	0.003	0.076	−0.423	0.022	0.094	**−0.717**	0.176	0.057
T2	−0.408	−0.022	0.063	−0.458	−0.010	0.081	**−0.701**	0.022	0.061
T3	−0.336	0.013	0.161	−0.393	0.022	0.178	**−0.666**	−0.007	0.078
T4	−0.194	0.050	0.267	−0.273	0.045	0.282	**−0.649**	−0.192	0.109
T5	−0.070	0.043	0.310	−0.128	0.038	0.322	**−0.593**	−0.173	0.105
T6	−0.120	0.103	0.367	−0.172	0.096	0.375	**−0.638**	−0.183	−0.006
T7	−0.104	0.218	0.387	−0.144	0.212	0.394	**−0.530**	−0.042	−0.035
T8	0.072	0.189	0.312	0.008	0.177	0.310	**−0.569**	−0.249	−0.315
T9	0.190	0.192	0.341	0.160	0.156	0.330	−0.158	−0.592	−0.230
T10	0.298	0.184	0.278	0.272	0.162	0.274	−0.205	−0.375	−0.040
T11	0.058	0.245	0.244	0.012	0.247	0.244	−0.266	−0.182	−0.099
T12	0.009	0.249	0.338	−0.049	0.257	0.336	−0.207	0.026	−0.381
T	−0.253	−0.036	0.208	−0.318	−0.035	0.215	**−0.647**	−0.154	−0.124
Ta	−0.253	−0.036	0.208	−0.318	−0.035	0.215	**−0.647**	−0.154	−0.124
					Lag 9				
T1	−0.416	0.179	0.421	−0.471	0.181	0.415	−0.414	−0.060	−0.174
T2	−0.364	0.259	0.469	−0.432	0.257	0.466	−0.461	−0.090	−0.070
T3	−0.265	0.351	0.529	−0.347	0.347	0.530	0.500	0.081	−0.058
T4	−0.072	0.439	0.558	−0.151	0.430	0.563	−0.485	−0.140	0.097
T5	−0.016	0.274	0.471	−0.099	0.258	0.478	−0.389	−0.248	0.062
T6	−0.004	0.284	0.485	−0.100	0.269	0.492	−0.508	−0.202	0.112
T7	−0.012	0.138	0.154	−0.112	0.141	0.165	**−0.609**	0.014	0.096
T8	−0.052	0.029	0.105	−0.128	0.024	0.121	**−0.615**	−0.163	0.451
T9	0.312	−0.006	−0.028	0.262	−0.021	0.020	−0.354	−0.153	0.432
T10	0.161	0.061	0.038	0.100	0.043	0.059	**−0.560**	−0.127	0.419
T11	−0.073	0.229	0.140	−0.085	0.234	0.159	**−0.688**	0.185	0.356
T12	−0.120	0.224	0.190	−0.136	0.244	0.208	**−0.690**	0.290	0.219
T	−0.182	0.188	0.387	−0.252	0.178	0.393	**−0.645**	−0.205	0.102
Ta	−0.182	0.188	0.387	−0.252	0.178	0.393	**−0.645**	−0.205	0.102
					Lag 10				
T1	-	**0.554**	**0.718**	-	**0.531**	**0.725**	-	−0.026	0.084
T2	-	**0.670**	**0.786**	-	**0.646**	**0.792**	-	−0.029	0.140

Table 1. *Cont.*

Parameter	Ab1	Ab2	Ab3	Bm1	Bm2	Bm3	W1	W2	W3
T3	-	**0.712**	**0.795**	-	**0.689**	**0.800**	-	−0.057	0.147
T4	-	**0.763**	**0.726**	-	**0.743**	**0.734**	-	0.020	0.202
T5	-	**0.731**	**0.706**	-	**0.708**	**0.715**	-	−0.058	0.152
T6	-	**0.729**	**0.711**	-	**0.707**	**0.724**	-	−0.079	0.255
T7	-	**0.514**	0.471	-	**0.508**	**0.493**	-	−0.013	0.415
T8	-	**0.562**	0.512	-	**0.553**	**0.527**	-	0.052	0.327
T9	-	0.390	0.248	-	0.394	0.270	-	0.302	0.487
T10	-	**0.514**	0.348	-	**0.524**	0.364	-	0.317	0.327
T11	-	0.315	0.476	-	0.326	**0.479**	-	0.402	0.015
T12	-	0.157	0.414	-	0.162	0.411	-	0.121	−0.140
T	-	**0.657**	**0.693**	-	**0.643**	**0.707**	-	0.051	0.291
Ta	-	**0.657**	**0.693**	-	**0.643**	**0.707**	-	0.051	0.291

Note: temperature variables: T1–T12—water temperature at stations 1–3 of the Kola section at 50–200 m depth from January to December, respectively, T—averaged year water temperature, Ta—temperature anomaly. Crab indices: Ab—abundance, Bm—biomass, W—weight. Age groups: 1—aged 6–9, 2—aged 10–14, 3—aged 15–19. Bold font indicates significant *p*-values confirmed by the Benjamini–Hochberg procedure. Correlation coefficients for the 0–3, 5–7, and 11–19 lag datasets were insignificant.

In contrast, the abundance and biomass of females at ages 10–14 had strong positive correlations with seawater temperature in the period from January to August and in October, as well as the average annual temperature and temperature anomaly at lag 10. Strong relationships were found between pairs of stock parameters for 15–19-year-old females and environmental variables. Both abundance and biomass showed positive associations with seawater temperature from February to June, mean yearly temperature, and its anomaly at lag 4. Similar patterns were registered for the abundance and biomass of this age group at lag 10 but, in this case, significant correlations were also found for abundance in January and for biomass in January, July, and August (Table 1).

In general, multivariate RDA confirmed the results of simple correlation analyses. For the lag 4 dataset, the largest portion of environmental variables was negatively associated with the first RDA axis (Figure 3a), explaining 65.6% of the total variation in the crab indices. Along Axis 2, the biomass data were positively scaled with the majority of environmental variables (Figure 3a). This axis explained 14.6% of the total variation. Using the data obtained for environmental variables and red king crab indices at lag 8 as input parameters for RDA, we found that the first two axes accounted for 72.8% of the total variance of the crab parameters. The first axis was negatively correlated with almost all variables, while the second axis was positively correlated with the weight and negatively correlated with the biomass data (Figure 3b). The forward selection of environmental factors with Monte Carlo permutation tests reveals mean yearly temperature as the main factor contributing to the observed variability in the weight data, explaining 19% of the total variation.

In the case of the lag 9 dataset, the first two RDA axes explained 79.7% of the total variance in the crab indices and environmental factors. The RDA showed that the most important predictors were seawater temperature in April and the average water temperature, explaining 22% (F = 3.63, $p = 0.033$) and 18% (F = 3.64, $p = 0.021$) of variability in the average crab weight, respectively. These factors were positively scaled mainly with the data for 6–9-year-old crabs (Figure 3c).

RDA applied to the lag 10 dataset showed that environmental variables were negatively associated with the first axis and positively with the second axis, explaining 72.7% and 16.0% of the data variation, respectively. Based on RDA results, seawater temperatures demonstrated strong associations with some crab indices (Figure 3d). Significant factor contributions were found for June (F = 14.20, $p = 0.001$), July (F = 2.71, $p = 0.049$), and August (F = 3.08, $p = 0.017$). These factors explained 52%, 9%, and 9% of the total variation, respectively.

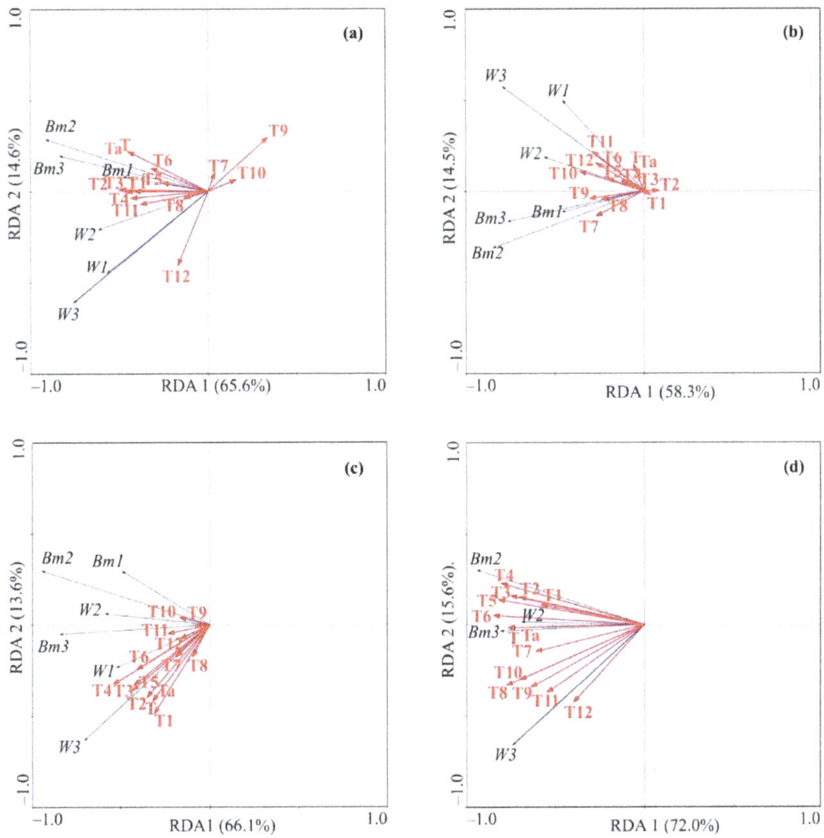

Figure 3. Redundancy analysis ordination plots showing female red king crab characteristics in relation to temperature variables in Dalnezelenetskaya Bay at lag 4 (**a**), lag 8 (**b**), lag 9 (**c**), and lag 10 (**d**). T1–T12—water temperature at stations 1–3 of the Kola section at 50–200 m depth in January–December, T—averaged year water temperature, Ta—temperature anomaly. Crab indices: Bm—biomass, W—weight. Age groups: 1—aged 6–9, 2—aged 10–14, 3—aged 15–19.

4. Discussion

Previous research has indicated that the fluctuations in the abundance of both fish and crabs in Dalnezelenetskaya Bay accurately reflect the broader ecological patterns observed in the Barents Sea [14,21,53]. Consequently, this area can be regarded as a reference site for studying the impacts of environmental factors on specific species, including commercially significant red king crabs and gadid fish. We found that the abundance and biomass of *P. camtschaticus* females varied considerably among age groups. The youngest female crabs were less abundant than older specimens. This pattern seems to reflect the higher mortality rate of 6–9-year-old crabs, owing to their higher vulnerability to predators. In our previous paper, we showed that smaller *P. camtschaticus* juveniles are consumed by the Northeast Arctic cod *Gadus morhua*, a major red king crab predator in the Barents Sea, in substantially higher amounts than larger king crabs [10,21]. Thus, smaller ovigerous females seem to be more vulnerable to predators' attacks including cod and marine mammals. It is important to emphasize that smaller female red king crabs exhibit a tendency to remain within a specific area throughout the year [11]. Therefore, the observed fluctuations are unlikely to be related to changes in migratory patterns. In addition, smaller females are less competitive with their larger conspecifics, both females and males. Competition between different-sized red king crabs is supported by the fact that no aggregations of mixed age classes were found during the study period in the area. This result is in contrast to the patterns registered for *P. camtschaticus* in Alaskan waters, where ovigerous female red king crabs exhibited "podding" behavior, similar to that reported for juveniles [54]. It is important to note that, while larger female red king crabs exhibit seasonal migrations, departing the area in autumn and returning in spring for egg release and spawning [11], there is currently no evidence to support significant variations in the migratory patterns and routes of female red king crabs in the study area [14]. Consequently, the annual influx of newly migrated crabs reflects the general population fluctuations driven by natural mortality and recruitment, both of which are influenced by temperature fluctuations and other temperature-related indices, such as the biomass of predators [21]. It is also not surprising, that the oldest females were less abundant than the females from the 10–14-year age group. Aging as an animal means degradation of all physiological functions, a decrease in mobility, high vulnerability to disease and parasites, and, finally, higher mortality [11,32] [Harman, 2003]. Illegal fishing could also be considered as a factor responsible for the higher mortality of older females because larger and heavier crabs are preferable for fishermen than smaller specimens [14,55]. It is noteworthy that within the studied area, significant fishery activities were not observed throughout the study period. The interannual fluctuations observed were primarily attributed to variations in the local temperature regime. Since 10–14-year-old females predominated in terms of their abundance and biomass, natural fluctuations in their stock indices were clearly expressed. We found that abundant cohorts occur every 6 years. This pattern reflects red king crab recruitment oscillations and recruit–stock relationships in this species and is in good accordance with the patterns reported in the Sea of Okhotsk [56].

The effects of climate shifts on marine communities, in general, and their components, in particular, have been the subject of many studies [57–60]. Responses of benthic assemblages to cooling or warming have been shown to depend on their origin and biogeographical status. For example, in cold years, the abundance of true arctic species increases, while the abundance of boreal and arctic-boreal species decreases [5,24,25]. So, after cooling, some cold-water polychaetes have been shown to shift their distribution to the northern areas of the Barents Sea while, after warming, such species disappeared from the coastal areas where they were abundant in cold and normal years. As a rule, the responses of benthic animals, especially long-lived, to environmental forcing factors are registered with time lags, owing to a cascading effect from the physical stimulus to benthic biological response, which is presumably mediated through constraints on primary production [58,61]. Red king crabs are also affected by climatic conditions [21,34] and, in general, their responses to a changing environment are similar to those observed for other benthic species [62–64].

In Dalnezelenetskaya Bay, a strong positive relationship was observed between temperature conditions and the total number of 15–19-year-old females at lag 4. This result, in our opinion, reflects direct temperature effects on the biomass of benthic animals consumed by *P. camtschaticus*. In the coastal areas of the Barents Sea, echinoderms (sea urchins, sea stars, and brittle stars) play an important role in the diet of adult *P. camtschaticus* because they provide red king crabs with calcium, which is crucial for shell hardening after the crabs molt [37,65,66]. According to the data reported by Frolova et al. [25], the biomasses of echinoderms are directly linked to temperature conditions, but an increase in these parameters as a response to climate forcing is registered at lag 3 for brittle stars and lag 4 for sea stars, which can explain the pattern observed for our lag 4 dataset.

We found negative relationships between the average weight of a female crab at ages 6–9 and environmental conditions at lags 8 and 9. At the same time, we found no negative associations between abundance and water temperatures. This result means that, in warm years, the cohorts of 0–1-year-old crabs are composed of smaller and lighter specimens than in cold and normal years. Growth and molting in red king crabs are temperature-dependent [67], and the maturation of female red king crabs in warmer waters is faster than in colder waters; therefore, females living in more favorable conditions reach sexual maturity earlier than the same-aged individuals from colder environments [68]. In addition, a predator-induced mortality rate in smaller juveniles (age 0–2) is lower than in larger individuals (age 0–5) due to differences between their behavior patterns which are closely related to their vulnerability to predators [21].

The most obvious results were obtained for the abundance and biomass of crabs at ages 15–19 and, especially, 10–14. Strong positive correlations were found for the lag 10 dataset. Oscillations in the climate, coupled with shifts in advection, are likely to affect the survival rates of pelagic larvae and reproductive patterns of red king crab populations [12,69,70]. In cold periods, when the ice cover in the Barents Sea is extended, the period available for the growth of primary producers (microalgae) tends to be shorter [71]. Consequently, cold periods lead to decreasing biomasses of phytoplankton, their predators, and dissolved organic carbon [72]. The pool of food sources for benthic animals, including prey for red king crabs, in such periods is smaller than in warm periods. Furthermore, a complex of biological processes at shallow water sites of the Barents Sea plays a pivotal role in the life cycle of *P. camtschaticus* because adult crabs use the coastal zone for spawning, and juveniles spend their first years of life exclusively at depths not exceeding 100 m [12,35]. Suspended sediments from upland areas are delivered into the nearshore zone of the Barents Sea by river discharge and coastal erosion [73]. These processes are known to have a close association with climate change, which is expressed in warmer years [74]. Thus, an increase in the total terrestrial inputs of particulate organic carbon leads to the increased productivity of benthic marine communities in the Arctic [75–77], including red king crab abundance [34,78]. All these reasons can explain why the abundance of female red king crabs tended to increase after warming and decrease after cooling. Similar patterns have been recorded for stocks of commercial male red king crabs both in their native area (Gulf of Alaska and the eastern Bering Sea, [79]) and in the area of introduction [34]. In addition to abundance, changes in climatic conditions may affect the distribution of female red king crabs. For example, in Bristol Bay, a pool of very cold water led to a shift in the center of abundance from the south-west part to the center of the area [80].

Both simple correlation analyses and RDA applied to the lag 10 dataset indicated that temperature conditions in the summer period are the most important predictors of female abundance (age 10–19) in the coastal zone of the Barents Sea. It is known that the reproductive cycle of *P. camtschaticus* lasts for 11–11.5 months, with a spawning peak in April, a peak of larval release during May–April, and a peak of larval settlement, as well as a transition from a swimming planktonic larva to a crawling benthic juvenile [11]. Since red king crabs have a relatively narrow temperature tolerance [67,81], their survival rates at the first 3 months of life on the seafloor have close associations with temperature and temperature-related conditions such as feeding habits. Other important factors are the availability and characteristics of suitable benthic habitats. During this critical period when they are highly vulnerable, post-settlement red king crabs are also exposed to a new suite of predators and competitors [82,83]. Thus, the temperature regime in coastal waters of the Kola peninsula directly influences mortality, recruitment patterns [21], and, hence, population abundance of red king crabs [70].

Fishing activities exert a significant impact on stock biomass, potentially leading to adverse effects on recruitment when stock biomass levels are critically depleted. This phenomenon is particularly relevant for the male component of a crab population in male-only fisheries [84]. From this perspective, the abundance of females becomes a more suitable metric for examining the relationship between stock dynamics and climatic variations. As the modeling of stock dynamics in the Barents Sea incorporates female abundance of the red king crab, our data hold significant implications for fishery management strategies in this region. Historically, the inability to sustain productive crab fisheries over extended periods has necessitated the adoption of progressively more conservative management approaches [9]. Consequently, this has precipitated dramatic declines in red king crab abundance, both within their native distribution areas and in the Barents Sea [1,12,55,85]. However, in the latter region, timely conservation measures have facilitated the recovery of the stock. Our study has identified time lags of 4 and 10 years in the response of female red king crabs to changes in temperature regimes. These findings provide a crucial foundation for more accurate forecasting of female abundance, thereby informing and refining the management practices for the sustainable exploitation and conservation of the red king crab stock in the Barents Sea.

5. Conclusions

The abundance and biomass of female red king crabs, as well as their mean weight, are closely associated with temperature conditions. Climate forcing leads to a decrease in mortality and provides favorable food conditions for juvenile red king crabs. As a result, a peak in abundance is expected after 10 years of population development. This pattern is also relevant for adult males and, therefore, can be used for the prediction of the total population number, abundance of recruits, and the commercial stock of *P. camtschaticus* in the Barents Sea. Taking into account that fluctuations in abundance are accompanied by fluctuations in distribution, our data may be useful for scientists and managers involved in tracking the range expansion of *P. camtschaticus* in the Barents Sea and evaluating its impact on local benthic assemblages. Although red king crabs are well adapted to the new habitat conditions and their impact on the major fish stocks is considered to be neutral, some authors suggest that there are negative consequences caused by these invaders on benthic fauna at several coastal sites and on the recruitment of non-commercial fish. Indirect effects of crab predation have been shown to include a positive cascade effect on macroalgae due to predation on herbivorous sea urchins, a negative effect on benthic-feeding birds, and the dispersion of crab-associated fish leeches that can increase transmission of trypanosomes to cod. Because the increased abundance and range expansion of red king crabs into new areas can be expected as a response to climate forcing, the managers of alien species should be prepared to undertake appropriate steps and measures to prevent negative scenarios.

Author Contributions: A.G.D.: conceptualization, methodology, investigation, data curation, validation, writing—original draft, and writing—review and editing; V.G.D.: data curation, formal analysis, visualization, software, project administration, supervision, writing—original draft, and writing—review and editing. All authors have read and agreed to the published version of the manuscript.

Funding: This study was funded by the Ministry of Science and Higher Education of the Russian Federation.

Institutional Review Board Statement: This study was approved by the Institutional Review of Murmansk Marine Biological Institute RAS (No. 188-1252/14, approved on 19 December 2023).

Informed Consent Statement: Not applicable.

Data Availability Statement: The data presented in this study are available on request from the corresponding author (the data are not publicly available due to privacy restrictions).

Acknowledgments: The English text of paper and its readability was greatly improved thanks to careful editing and valuable remarks of A. Trebitz. We are also grateful to two anonymous reviewers for their constructive comments.

Conflicts of Interest: The authors declare no conflicts of interest.

References

1. Dvoretsky, A.G.; Dvoretsky, V.G. Red king crab (*Paralithodes camtschaticus*) fisheries in Russian waters: Historical review and present status. *Rev. Fish Biol. Fish.* **2018**, *28*, 331–353. [CrossRef]
2. Loeng, H.; Ozhigin, V.; Ådlandsvik, B. Water fluxes through the Barents Sea. *ICES J. Mar. Sci.* **1997**, *54*, 310–317. [CrossRef]
3. Dvoretsky, V.G.; Dvoretsky, A.G. Local variability of Arctic mesozooplankton biomass and production: A case summer study. *Environ. Res.* **2024**, *241*, 117416. [CrossRef] [PubMed]
4. Wassmann, P.; Reigstad, M.; Haug, T.; Rudels, B.; Carroll, M.L.; Hop, H.; Gabrielsen, G.W.; Falk-Petersen, S.; Denisenko, S.G.; Arashkevich, E.; et al. Food Webs and Carbon Flux in the Barents Sea. *Prog. Oceanogr.* **2006**, *71*, 232–287. [CrossRef]
5. Dvoretsky, A.G.; Dvoretsky, V.G. Filling knowledge gaps in Arctic marine biodiversity: Environment, plankton, and benthos of Franz Josef Land, Barents Sea. *Ocean Coast. Manag.* **2024**, *249*, 106987. [CrossRef]
6. Kuzmin, S.A.; Olsen, S. Barents sea king crab *Paralithodes camtschatica*. The transplantation experiments were successful. *ICES CM. K* **1994**, *12*, 1–12.
7. Dvoretsky, A.G.; Dvoretsky, V.G. Epibionts of an introduced king crab in the Barents Sea: A second five-year study. *Diversity* **2023**, *15*, 29. [CrossRef]
8. Bakanev, S.V.; Stesko, A.V. Red king crab. In *Materials of Total Allowable Catches of Water Biological Resources in Fishing Areas in Inland Seas of the Russian Federation, on the Continental Shelf of the Russian Federation, in the Exclusive Economical Zone of the Russian Federation, in the Azov and Caspian Seas in 2023*; Goryanina, S.V., Ed.; FGBUN VNIRO: Murmansk, Russia, 2023; pp. 5–19. (In Russian)
9. Sokolov, K.M. *Status of the Living Marine Resources in the Barents, White and Kara Seas and the North Atlantic in 2024*; PINRO: Murmansk, Russia, 2024. (In Russian)
10. Dvoretsky, A.G.; Dvoretsky, V.G. Limb autotomy patterns in *Paralithodes camtschaticus* (Tilesius, 1815), an invasive crab, in the coastal Barents Sea. *J. Exp. Mar. Biol. Ecol.* **2009**, *377*, 20–27. [CrossRef]
11. Kuzmin, S.A.; Gudimova, E.N. *Introduction of the Kamchatka (Red King) Crab in the Barents Sea. Peculiarities of Biology, Perspectives of Fishery*; KSC RAS Press: Apatity, Russia, 2002. (In Russian)
12. Dvoretsky, A.G.; Dvoretsky, V.G. Population dynamics of the invasive lithodid crab, *Paralithodes camtschaticus*, in a typical bay of the Barents Sea. *ICES J. Mar. Sci.* **2013**, *70*, 1255–1262. [CrossRef]
13. Sokolov, V.I.; Milyutin, D.M. Distribution, size-sex composition, and reserves of the red king crab (*Paralithodes camtschaticus*) in the upper sublittoral of the Kola Peninsula (the Barents Sea). *Zool. Zhurnal* **2006**, *85*, 158–171. (In Russian with English abstract)
14. Dvoretsky, A.G.; Dvoretsky, V.G. *Ecology of Red King Crab in the Coastal Barents Sea*; SSC RAS Press: Rostov-on-Don, Russia, 2018. (In Russian)
15. Dvoretsky, A.G.; Dvoretsky, V.G. Epibiotic communities of common crab species in the coastal Barents Sea: Biodiversity and infestation patterns. *Diversity* **2022**, *14*, 6. [CrossRef]
16. Stesko, A.V. Distribution and status of the king crab stock in the Russian territorial waters of the Barents Sea. *Probl. Fish.* **2015**, *16*, 175–192, (In Russian with English abstract). [CrossRef]
17. Dvoretsky, A.G.; Dvoretsky, V.G. Ecology and distribution of red king crab larvae in the Barents Sea: A review. *Water* **2022**, *14*, 2328. [CrossRef]
18. Zheng, J.; Kruse, G.H. King Crab Stock Assessments in Alaska. In *King Crabs of the World: Biology and Fisheries Management*; Stevens, B.G., Ed.; CRC Press: Boca Raton, FL, USA, 2014; pp. 539–558.
19. Piepenburg, D. Recent research on Arctic benthos: Common notions need to be revised. *Polar Biol.* **2005**, *28*, 733–755. [CrossRef]
20. Drinkwater, K. The influence of climate variability and change on the ecosystems of the Barents Sea and adjacent waters: Review and synthesis of recent studies from the NESSAS project. *Prog. Oceanogr.* **2011**, *90*, 47–61. [CrossRef]
21. Dvoretsky, A.G.; Dvoretsky, V.G. Effects of environmental factors on the abundance, biomass, and individual weight of juvenile red king crabs in the Barents Sea. *Front. Mar. Sci.* **2020**, *7*, 726. [CrossRef]
22. Oleszczuk, B.; Grzelak, K.; Kędra, M. Community structure and productivity of Arctic benthic fauna across depth gradients during springtime. *Deep Sea Res. I* **2021**, *170*, 103457. [CrossRef]
23. Pavlova, L.V.; Dvoretsky, A.G.; Frolov, A.A.; Zimina, O.L.; Evseeva, O.Y.; Dikaeva, D.R.; Rumyantseva, Z.Y.; Panteleeva, N.N. The impact of sea ice loss on benthic communities of the Makarov Strait (northeastern Barents Sea). *Animals* **2023**, *13*, 2320. [CrossRef]

24. Denisenko, S.G. The Barents Sea zoobenthos under the conditions of changing climate and anthropogenic impact. In *Dynamics of Marine Ecosystems and Current Preservation Problems of Biological Potential of the Seas of Russia*; Tarasov, V.G., Ed.; Dalnauka: Vladivostok, Russia, 2007; pp. 418–511. (In Russian)
25. Frolova, E.A.; Lyubina, O.S.; Dikayeva, D.R.; Ahmetchina, O.Y.; Frolov, A.A. Effect of climatic changes on the zoobenthos of the Barents Sea (on the example of several abundant species). *Dokl. Biol. Sci.* **2007**, *416*, 349–351. [CrossRef]
26. Renaud, P.E.; Ambrose, W.G.; Węsławski, J.M. Benthic communities in the polar night. In *Polar Night Marine Ecology: Life and Light in the Dead of Night*; Berge, J., Johnsen, G., Cohen, J., Eds.; Springer: Cham, Switzerland, 2020; pp. 161–179.
27. Turner, J.; Overland, J. Contrasting climate change in the two polar regions. *Polar Res.* **2009**, *28*, 146–164. [CrossRef]
28. Johannesen, E.; Ingvaldsen, R.B.; Bogstad, B.; Dalpadado, P.; Eriksen, E.; Gjøsæter, H.; Knutsen, T.; Skern-Mauritzen, M.; Stiansen, J.E. Changes in Barents Sea ecosystem state, 1970–2009: Climate fluctuations, human impact, and trophic interactions. *ICES J. Mar. Sci.* **2012**, *69*, 880–889. [CrossRef]
29. Smedsrud, L.H.; Muilwijk, M.; Brakstad, A.; Madonna, E.; Lauvset, S.K.; Spensberger, C.; Born, A.; Eldevik, T.; Drange, H.; Jeansson, E.; et al. Nordic Seas heat loss, Atlantic inflow, and Arctic sea ice cover over the last century. *Rev. Geophys.* **2022**, *60*, e2020RG000725. [CrossRef]
30. Dikaeva, D.R.; Dvoretsky, A.G. Spatial patterns and environmental control of polychaete communities in the southwestern Barents Sea. *Biology* **2024**, *13*, 924. [CrossRef] [PubMed]
31. Dvoretsky, V.G.; Dvoretsky, A.G. Effects of water temperature on zooplankton abundance and biomass in the southwestern Barents Sea: Implications for Arctic monitoring and management. *Ocean Coast. Manag.* **2025**, *261*, 107506. [CrossRef]
32. Russell, B.D.; Connell, S.D.; Mellin, C.; Brook, B.W.; Burnell, O.W.; Fordham, D.A. Predicting the distribution of commercially important invertebrate stocks under future climate. *PLoS ONE* **2012**, *7*, e46554. [CrossRef]
33. Heath, M.R.; Neat, F.C.; Pinnegar, J.K.; Reid, D.G.; Sims, D.W.; Wright, P.J. Review of climate change impacts on marine fish and shellfish around the UK and Ireland. *Aquat. Conserv.* **2012**, *22*, 337–367. [CrossRef]
34. Dvoretsky, A.G.; Dvoretsky, V.G. Inter-annual dynamics of the Barents Sea red king crab (*Paralithodes camtschaticus*) stock indices in relation to environmental factors. *Polar Sci.* **2016**, *10*, 541–552. [CrossRef]
35. Dvoretsky, A.G.; Dvoretsky, V.G. Size at maturity of female red king crab, *Paralithodes camtschaticus*, from the costal zone of Kola Peninsula (southern Barents Sea). *Cah. Biol. Mar.* **2015**, *56*, 49–54.
36. Dvoretsky, A.G.; Dvoretsky, V.G. Commercial fish and shellfish in the Barents Sea: Have introduced crab species affected the population trajectories of commercial fish? *Rev. Fish Biol. Fish.* **2015**, *25*, 297–322. [CrossRef]
37. Pavlova, L.V. Red king crab trophic relations and its influence on bottom biocenoses. In *Biology and Physiology of the Red King Crab from the Coastal Zone of the Barents Sea*; Matishov, G.G., Ed.; Publishing Kola Scientific Centre Russian Academy of Sciences: Apatity, Russia, 2008; pp. 77–104. (In Russian)
38. Britayev, T.A.; Rzhavsky, A.V.; Pavlova, L.V.; Dvoretskij, A.G. Studies on impact of the alien red king crab (*Paralithodes camtschaticus*) on the shallow water benthic communities of the Barents Sea. *J. Appl. Ichthyol.* **2010**, *26* (Suppl. S2), 66–73. [CrossRef]
39. Falk-Petersen, J.; Renaud, P.; Anisimova, N. Establishment and ecosystem effects of the alien invasive red king crab (*Paralithodes camtschaticus*) in the Barents Sea—A review. *ICES J. Mar. Sci.* **2011**, *68*, 479–488. [CrossRef]
40. Oug, E.; Cochrane, S.K.J.; Sundet, J.H.; Norling, K.; Nilsson, H.C. Effects of the invasive red king crab (Paralithodes camtschaticus) on soft-bottom fauna in Varangerfjorden, northern Norway. *Mar. Biodivers.* **2011**, *41*, 467–479. [CrossRef]
41. Oug, E.; Sundet, J.H.; Cochrane, S.K.J. Structural and functional changes of soft-bottom ecosystems in northern fjords invaded by the red king crab (*Paralithodes camtschaticus*). *J. Mar. Syst.* **2018**, *180*, 255–264. [CrossRef]
42. Pavlova, L.V.; Dvoretsky, A.G. Prey selectivity in juvenile red king crabs from the coastal Barents Sea. *Diversity* **2022**, *14*, 568. [CrossRef]
43. Dvoretsky, A.G.; Dvoretsky, V.G. Aquaculture of green sea urchin in the Barents Sea: A brief review of Russian studies. *Rev. Aquac.* **2020**, *12*, 2080–2090. [CrossRef]
44. Evseeva, O.Y.; Ishkulova, T.G.; Dvoretsky, A.G. Environmental drivers of an intertidal bryozoan community in the Barents Sea: A case study. *Animals* **2022**, *12*, 552. [CrossRef]
45. Dvoretsky, A.G.; Dvoretsky, V.G. New echinoderm-crab epibiotic associations from the coastal Barents Sea. *Animals* **2021**, *11*, 917. [CrossRef]
46. Pinchukov, M.A. Estimation of age in females of the red king crab *Paralithodes camtschaticus* (Decapoda, Lithodidae) from the Barents Sea Based on their size composition. *Proc. Kazan Univ. Nat. Sci. Ser.* **2017**, *159*, 480–491. (In Russian)
47. Dvoretsky, A.G.; Dvoretsky, V.G. Size-at-age of juvenile red king crab (*Paralithodes camtschaticus*) in the coastal Barents Sea. *Cah. Biol. Mar.* **2014**, *55*, 43–48.
48. Bizikov, V.A.; Goncharov, S.M.; Polyakov, A.V. The geographical informational system CardMaster. *Rybn. Khoz.* **2007**, *1*, 96–99. (In Russian)
49. McDonald, J.H. *Handbook of Biological Statistics*, 2nd ed.; Sparky House Publishing: Baltimore, MD, USA, 2009.

50. Pyper, B.J.; Peterman, R.M. Comparison of methods to account for autocorrelation analyses of fish data. *Can. J. Fish. Aquat. Sci.* **1998**, *55*, 2127–2140. [CrossRef]
51. Leps, J.; Smilauer, P. *Multivariate Analysis of Ecological Data Using CANOCO*; Cambridge University Press: Cambridge, UK, 2003.
52. Ter Braak, C.J.F.; Smilauer, P. *CANOCO Reference Manual and CanoDraw for Windows User's Guide: Software for Canonical Community Ordination (Version 4.5)*; Microcomputer Power: New York, NY, USA, 2002.
53. Karamushko, O.V.; Zhuravleva, N.G.; Karamushko, L.I.; Kudryavtseva, O.Y.; Raskhozheva, E.V.; Smirnova, E.V. Modern research of the Barents Sea and adjacent waters ichthyofauna. In *Marine Ecosystems and Communities in the Conditions of Current Climate Change*; Matishov, G.G., Ed.; Renome: Saint Petersburg, Russia, 2014; pp. 223–243. (In Russian)
54. Stone, R.P.; O'Clair, C.E.; Shirley, T.C. Aggregating behavior of ovigerous female red king crab, *Paralithodes camtschaticus*, in Auke Bay, Alaska. *Can. J. Fish. Aquat. Sci.* **1993**, *50*, 750–758. [CrossRef]
55. Dvoretsky, A.G.; Dvoretsky, V.G. Renewal of the amateur red king crab fishery in Russian waters of the Barents Sea: Potential benefits and costs. *Mar. Policy* **2022**, *136*, 104916. [CrossRef]
56. Slizkin, A.G.; Safronov, S.G. *Commercial Crabs of the Waters of Kamchatka*; North Pacific: Petropavlovsk-Kamchatsky, Russia, 2000.
57. Bahr, G.; Gulliksen, B. Variation of the epifauna on pier-pilings between 1980 and 1992 near the city of Tromsø, Northern Norway. *Polar Biol.* **2001**, *24*, 282–291.
58. Beuchel, F.; Bahr, G.; Gulliksen, B. Long-term patterns of rocky bottom macrobenthic community structure in an Arctic fjord (Kongsfjorden, Svalbard) in relation to climate variability (1980–2003). *J. Mar. Syst.* **2003**, *63*, 35–48. [CrossRef]
59. Sanchez-Rubio, G.; Perry, H.M.; Biesiot, P.M.; Johnson, D.R.; Lipcius, R.N. Climate-related hydrological regimes and their effects on abundance of juvenile blue crabs (*Callinectes sapidus*) in the northcentral Gulf of Mexico. *Fish. Bull.* **2011**, *396*, 139–146.
60. Noskovich, A.E.; Dvoretsky, A.G. Spatial distribution and growth patterns of a common bivalve mollusk (*Macoma calcarea*) in Svalbard fjords in relation to environmental factors. *Animals* **2024**, *14*, 3352. [CrossRef]
61. Hiddink, J.G.; Burrows, M.T.; García Molinos, J. Temperature tracking by North Sea benthic invertebrates in response to climate change. *Glob. Change Biol.* **2015**, *21*, 117–129. [CrossRef]
62. Tremblay, J.É.; Robert, D.; Varela, D.E.; Lovejoy, C.; Darnis, G.; Nelson, R.J.; Sastri, A.R. Current state and trends in Canadian Arctic marine ecosystems: I. Primary production. *Clim. Change* **2012**, *115*, 161–178. [CrossRef]
63. Renaud, P.E.; Sejr, M.K.; Bluhm, B.A.; Sirenko, B.; Ellingsen, I.H. The future of Arctic benthos: Expansion, invasion, and biodiversity. *Prog. Oceanog.* **2015**, *139*, 244–257. [CrossRef]
64. Roy, V.; Iken, K.; Gosselin, M.; Tremblay, J.É.; Bélanger, S.; Archambault, P. Benthic faunal assimilation pathways and depth-related changes in food-web structure across the Canadian Arctic. *Deep Sea Res I* **2015**, *102*, 55–71. [CrossRef]
65. Anisimova, N.A.; Manushin, I.E. Anisimova, N.A.; Manushin, I.E. A diet of red king crab in the Barents Sea. In *The Red King Crab in the Barents Sea*; Berenboim, B.I., Ed.; PINRO Press: Murmansk, Russia, 2003; pp. 170–189. (In Russian)
66. Pedersen, T.; Fuhrmann, M.M.; Lindstrøm, U.; Nilssen, E.M.; Ivarjord, T.; Ramasco, V.; Jørgensen, L.L.; Sundet, J.H.; Sivertsen, K.; Källgren, E.; et al. Effects of the invasive red king crab on food web structure and ecosystem properties in an Atlantic fjord. *Mar. Ecol. Prog. Ser.* **2018**, *596*, 13–31. [CrossRef]
67. Stevens, B.G. Temperature-dependent growth of juvenile red king crab (*Paralithodes camtschatica*) and its effects on size-at-age and subsequent recruitment in the eastern Bering Sea. *Can. J. Fish. Aquat. Sci.* **1990**, *47*, 1307–1317. [CrossRef]
68. Webb, J. Reproduction Ecology of Commercially Important Lithodid Crabs. In *King Crabs of the World: Biology and Fisheries Management*; Stevens, B.G., Ed.; CRC Press: Boca Raton, FL, USA, 2014; pp. 285–314.
69. Tyler, A.V.; Kruse, G.C. Conceptual modeling of brood strength of red king crabs in the Bristol Bay region of the Bering Sea. In *High Latitude Crabs: Biology, Management and Economics*; University of Alaska Sea Grant Program Report 96-02; University of Alaska: Fairbanks, AK, USA, 1996; pp. 511–544.
70. Zheng, J.; Kruse, G.H. Recruitment Patterns of Alaskan Crabs in Relation to Decadal Shifts in Climate and Physical Oceanography. *ICES J. Mar. Sci.* **2000**, *57*, 438–451. [CrossRef]
71. Barber, D.G.; Lukovich, J.V.; Keogak, J.; Baryluk, S.; Fortier, L.; Henry, G.H.R. The changing climate of the Arctic. *Arctic* **2008**, *61* (Suppl. S1), 7–26. [CrossRef]
72. Koenig, Z.; Fer, I.; Chierici, M.; Fransson, A.; Jones, E.; Kolås, E. Diffusive and advective cross-frontal fluxes of inorganic nutrients and dissolved inorganic carbon in the Barents Sea in autumn. *Prog. Oceanogr.* **2023**, *219*, 103161. [CrossRef]
73. Jakobsen, T.; Ozhigin, V.K. (Eds.) *The Barents Sea: Ecosystem, Resources, Management: Half a Century of Russian-Norwegian Cooperation*; Tapir Academic Press: Trondheim, Norway, 2011.
74. Rawlins, M.A.; Karmalkar, A.V. Regime shifts in Arctic terrestrial hydrology manifested from impacts of climate warming. *Cryosphere* **2024**, *18*, 1033–1052. [CrossRef]
75. Nixon, S.W.; Fulweiler, R.W.; Buckley, B.A.; Granger, S.L.; Nowicki, B.L.; Henry, K.M. The impact of changing climate on phenology, productivity, and benthic–pelagic coupling in Narragansett Bay. *Estuar. Coast. Shelf Sci.* **2009**, *82*, 1–18. [CrossRef]
76. Sakshaug, E.; Johnsen, G.; Kovacs, K. (Eds.) *Ecosystem Barents Sea*; Tapir Academic Press: Trondheim, Norway, 2009.

77. McGovern, M.; Poste, A.E.; Oug, E.; Renaud, P.E.; Trannum, H.C. Riverine impacts on benthic biodiversity and functional traits: A comparison of two sub-Arctic fjords. *Estuar. Coast. Shelf Sci.* **2020**, *240*, 106774. [CrossRef]
78. Fuhrmann, M.M.; Pedersen, T.; Ramasco, V.; Nilssen, E.M. Macrobenthic biomass and production in a heterogenic subarctic fjord after invasion by the red king crab. *J. Sea Res.* **2015**, *106*, 1–13. [CrossRef]
79. Mueter, F.J.; Norcross, B.L.; Royer, T.C. Do cyclic temperatures cause cyclic fisheries? *Can. Spec. Publ. Fish. Aquat. Sci.* **1995**, *121*, 119–129.
80. Loher, T.; Armstrong, D.A. Historical changes in the abundance and distribution of ovigerous red king crabs (*Paralithodes camtschaticus*) in Bristol Bay (Alaska), and potential relationship with bottom temperature. *Fish. Oceanogr.* **2005**, *14*, 292–306. [CrossRef]
81. Stevens, B.G. Biology and Ecology of Juvenile King Crabs. In *King Crabs of the World: Biology and Fisheries Management*; Stevens, B.G., Ed.; CRC Press: Boca Raton, FL, USA, 2014; pp. 261–401.
82. Stevens, B.G. Development and Biology of King Crab Larvae. In *King Crabs of the World: Biology and Fisheries Management*; Stevens, B.G., Ed.; CRC Press: Boca Raton, FL, USA, 2014; pp. 233–260.
83. Long, W.C.; Daly, B.J.; Cummiskey, P.A. Optimizing release strategies for red king crab stock enhancement: Effects of release timing. *Fish. Res.* **2024**, *274*, 106975. [CrossRef]
84. Armstrong, D.A.; Kruse, G.H.; Hines, A.H.; Orensanz, J.M.; McDonald, P.S. A crab for all seasons: The confluence of fisheries and climate as drivers of crab abundance and distribution. In *Biology and Management of Exploited Crab Populations under Climate Change*; Kruse, G.H., Eckert, G.L., Foy, R.J., Lipcius, R.N., Sainte-Marie, B., Stram, D.L., Woodby, D., Eds.; Alaska Sea Grant Publications: Fairbanks, AL, USA, 2010; pp. 1–48.
85. Dvoretsky, A.G.; Dvoretsky, V.G. Distribution of amphipods *Ischyrocerus* on the red king crab, *Paralithodes camtschaticus*: Possible interactions with the host in the Barents Sea. *Estuar. Coast. Shelf Sci.* **2009**, *82*, 390–396. [CrossRef]

MDPI AG
Grosspeteranlage 5
4052 Basel
Switzerland
Tel.: +41 61 683 77 34

Animals Editorial Office
E-mail: animals@mdpi.com
www.mdpi.com/journal/animals

www.ingramcontent.com/pod-product-compliance
Lightning Source LLC
Chambersburg PA
CBHW062033220926
51850CB00010B/2531